James McMahon, Virgil Snyder

Elements of the Differential Calculus

James McMahon, Virgil Snyder

Elements of the Differential Calculus

ISBN/EAN: 9783337813178

Printed in Europe, USA, Canada, Australia, Japan

Cover: Foto ©berggeist007 / pixelio.de

More available books at **www.hansebooks.com**

ELEMENTS

OF THE

DIFFERENTIAL CALCULUS

BY

JAMES McMAHON, A.M. (DUBLIN)
ASSISTANT PROFESSOR OF MATHEMATICS IN CORNELL UNIVERSITY

AND

VIRGIL SNYDER, Ph.D. (GÖTTINGEN)
INSTRUCTOR IN MATHEMATICS IN CORNELL UNIVERSITY

NEW YORK ∴ CINCINNATI ∴ CHICAGO
AMERICAN BOOK COMPANY

PREFACE

THIS book is primarily designed as a text-book for the classes in the Calculus at Cornell University and other institutions in which the object and extent of the work are similar. For the engineering students at Cornell, Differential Calculus is taught during the winter term of the freshman year; the students are then familiar with Analytic Geometry, and many properties of the conics can be supposed known.

When use is made of Cartesian coördinates, they are always assumed to be rectangular.

As an apology for adding still another work to a field in which the literature is already extensive, it may be said that probably no other book has just the scope of this one. Many of the works are too brief, and omit rigorous proofs as being too difficult for the average student, while the more extensive treatises have too much for a student to master in the allotted time.

While the chapter on fundamental principles is a long one, nothing more is introduced than is necessary for subsequent parts of the work; and it is hoped that the matter is so arranged that the student will not find it difficult reading.

In the chapter on expansion of functions unusual stress is laid upon convergence and the calculation of the remainder; and numerous examples are discussed to illustrate the principles.

The chapter on asymptotes is perhaps unusually long, as the subject is so presented that the form of any infinite branch can be readily determined from its approximate equation, and the process is fully illustrated both in this chapter and in a later one on curve tracing.

Quite a full discussion is given to the form of a curve in the vicinity of a singular point, the method of expansion being extensively used.

No list of "higher plane curves" has been prepared, since the subject, as usually given, is properly a part of Analytic Geometry. A chapter on that subject is contained in the Analytic Geometry of this series. The occasional marginal references [A. G.] are to this book.

No specific acknowledgments to other works have been given; for although various works have been consulted, the main inspiration has come from the class room and from extensive consultation with our colleagues.

Many of the examples have been selected from other books, but a large number are new. When original examples have been taken from recent works, acknowledgment of the source is made.

We acknowledge our indebtedness to the other authors of this series for their hearty coöperation; to our colleagues, Dr. J. I. Hutchinson and Dr. G. A. Miller, for the keen interest they have taken in the work and for their assistance in verifying examples and reading proof; to Mr. Peter Field, Fellow in Mathematics, for solving the entire list of exercises; and to Mr. V. T. Wilson, Instructor in Drawing in Sibley College, for drawing the figures. Every figure in the book is new, and drawn to scale, except that in some cases vertical ordinates are proportionately foreshortened to fit the page.

CONTENTS

INTRODUCTION

ARTICLE | PAGE
1. Number 1
2. Operations 1
3. Expressions 3
4. Functions 4
5. Constants and variables 7
6. Continuous variable; continuous function 7

CHAPTER I

FUNDAMENTAL PRINCIPLES

7. Limit of a variable 9
8. Infinitesimals and infinites 10
9. Fundamental theorems concerning infinitesimals, and limits in general 11
10. Comparison of variables 17
11. Comparison of infinitesimals, and of infinites. Orders of magnitude 18
12. Order of magnitude of expressions involving infinitesimals or infinites 20
13. Useful illustrations of infinitesimals of different orders . . 25
14. Continuity of functions 28
15. Comparison of simultaneous infinitesimal increments of two related variables 33
16. Definition of a derivative 37
17. Geometric illustrations of a derivative 38
18. The operation of differentiation . . . 41
19. Increasing and decreasing functions . . . 43
20. Algebraic test of the intervals of increasing and decreasing 45
21. Differentiation of a function of a function . . 46
22. Differentiation of inverse functions 47

vii

CHAPTER II

DIFFERENTIATION OF THE ELEMENTARY FORMS

ARTICLE	PAGE
23. Recapitulation	50
24. Differentiation of the product of a constant and a variable	50
25. Differentiation of a sum	51
26. Differentiation of a product	52
27. Differentiation of a quotient	53
28. Differentiation of a commensurable power of a function	54
29. Elementary transcendental functions	58
30. Differentiation of $\log_a x$ and $\log_a u$	59
31. Differentiation of the simple exponential function	61
32. Differentiation of the general exponential function	61
33. Differentiation of an incommensurable power	62
34. Differentiation of $\sin u$	63
35. Differentiation of $\cos u$	63
36. Differentiation of $\tan u$	64
37. Differentiation of $\cot u$	64
38. Differentiation of $\sec u$	65
39. Differentiation of $\csc u$	65
40. Differentiation of $\operatorname{vers} u$	65

DIFFERENTIATION OF THE INVERSE TRIGONOMETRIC FUNCTIONS

41. Differentiation of $\sin^{-1} u$	66
42. Differentiation of $\cos^{-1} u$	67
43. Differentiation of $\tan^{-1} u$	68
44. Differentiation of $\cot^{-1} u$	68
45. Differentiation of $\sec^{-1} u$	69
46. Differentiation of $\csc^{-1} u$	69
47. Differentiation of $\operatorname{vers}^{-1} u$	70
LIST OF ELEMENTARY FORMS	71

CHAPTER III

SUCCESSIVE DIFFERENTIATION

48. Definition of nth derivative	73
49. Expression for the nth derivative in certain cases	74
50. Leibnitz's theorem concerning the nth derivative of a product	75
51. Successive x-derivatives of y when neither variable is independent	77

CHAPTER IV

Expansion of Functions

ARTICLE	PAGE
52. Introductory	81
53. Convergence and divergence of series	82
54. General test for interval of convergence . . .	83
55. Interval of equivalence. Remainder after n terms .	86
56. Maclaurin's expansion of a function in power-series .	87
57. Development of $f(x)$ in powers of $x-a$.	93
58. Remainder	95
59. Rolle's theorem	95
60. Form of remainder in development of $f(x)$ in powers of $x-a$. Lagrange's form	95
61. Another expression for the remainder. Cauchy's	97
62. Form of remainder in Maclaurin's series .	99
63. Lemma. $\dfrac{x^n}{n!} \doteq 0,\ n \doteq \infty$	100
64. Remainder in the development of a^x, $\sin x$, $\cos x$.	100
65. Taylor's series	101
Proof of the binomial theorem . .	102
Calculation of natural logarithms	105
66. Theorem of mean value. Increment of function in terms of increment of variable	107
Increment of the increment	108
$\lim\limits_{\Delta x \doteq 0} \dfrac{\Delta^2 y}{\Delta x^2} = \dfrac{d^2 y}{dx^2}$	100
67. Illustration: to find the development of a function when that of its derivative is known	109
Series for $\tan^{-1} x$, and calculation of π	110
Series for $\sin^{-1} x$, and calculation of π	112

CHAPTER V

Indeterminate Forms

68. Definition of an indeterminate form	115
69. Indeterminate forms may have determinate values . .	116
70. Evaluation by transformation	118
71. Evaluation by development	119
72. Evaluation by differentiation . .	121
73. Evaluation of the indeterminate form $\dfrac{\infty}{\infty}$ . . .	125

CONTENTS

ARTICLE		PAGE
74.	Evaluation of the form $\infty \cdot 0$	126
75.	Evaluation of the form $\infty - \infty$	126
76.	Evaluation of the form 1^∞	130
77.	Evaluation of the forms 0^0, ∞^0	130

CHAPTER VI

MODE OF VARIATION OF FUNCTIONS OF ONE VARIABLE

78.	Review of increasing and decreasing functions	132
79.	Turning values of a function	132
80.	Critical values of the variable	134
81.	Method of determining whether $\phi'(x)$ changes sign	134
82.	Second method of determining whether $\phi'(x)$ changes sign in passing through zero	137
83.	Conditions for maxima and minima derived from Taylor's theorem	139
84.	Application to rational polynomials	141
85.	Maxima and minima occur alternately	143
86.	Simplifications that do not alter critical values	143
87.	Geometric problems in maxima and minima	144

CHAPTER VII

RATES AND DIFFERENTIALS

88.	Rates. Time as independent variable	151
89.	Abbreviated notation for rates	154
90.	Differentials often substituted for rates	156

CHAPTER VIII

DIFFERENTIATION OF FUNCTIONS OF MORE THAN ONE VARIABLE

91.	Definition of continuity	158
92.	Rate of variation. Partial derivatives	159
93.	Geometric illustration	160
94.	Simultaneous variation of x and y; total rate of variation of z	161
95.	Language of differentials	164
96.	One variable a function of the other	165
97.	Differentiation of implicit functions; relative variation that keeps z constant	166

CONTENTS

ARTICLE		PAGE
98.	Functions of more than two variables	168
99.	One or two relations between the three variables x, y, z.	169
100.	Euler's theorem; relation between a homogeneous function and its partial derivatives	171

CHAPTER IX

Successive Partial Differentiation

101.	Successive differentiation of functions of two variables.	173
102.	Order of differentiation indifferent.	175
103.	Extension of Taylor's theorem to expansion of two variables.	178
104.	Significance of remainder.	180
105.	Form corresponding to Maclaurin's theorem.	180

CHAPTER X

Maxima and Minima of Functions of Two Variables

106.	Definition of maximum and minimum of functions of two variables	183
107.	Determination of maxima and minima.	183
108.	Conditional maxima and minima	191
	Implicit functions	193

CHAPTER XI

Change of the Variable

109.	Interchange of dependent and independent variables	198
110.	Change of the dependent variable.	199
111.	Change of the independent variable	199
112.	Change of two independent variables	201
113.	Change of three independent variables.	204
114.	Application to higher derivatives.	205

APPLICATIONS TO GEOMETRY

CHAPTER XII

TANGENTS AND NORMALS

ARTICLE		PAGE
115.	Geometric meaning of $\dfrac{dy}{dx}$	208
116.	Equation of tangent and normal at a given point	208
117.	Length of tangent, normal, subtangent, subnormal	209

POLAR COÖRDINATES

118.	Meaning of $\rho \dfrac{d\theta}{d\rho}$	213
119.	Relation between $\dfrac{dy}{dx}$ and $\rho \dfrac{d\theta}{d\rho}$	213
120.	Length of tangent, normal, polar subtangent, and polar subnormal	213

CHAPTER XIII

DERIVATIVE OF AN ARC, AREA, VOLUME AND SURFACE OF REVOLUTION

121.	Derivative of an arc	216
122.	Trigonometric meaning of $\dfrac{ds}{dx}$, $\dfrac{ds}{dy}$	217
123.	Derivative of the volume of a solid of revolution	218
124.	Derivative of a surface of revolution	218
125.	Derivative of arc in polar coördinates	219
126.	Derivative of area in polar coördinates	220

CHAPTER XIV

ASYMPTOTES

127.	Hyperbolic and parabolic branches	221
128.	Definition of a rectilinear asymptote	221

DETERMINATION OF ASYMPTOTES

129.	Method of limiting intercepts	221
130.	Method of inspection. Infinite ordinates, asymptotes parallel to axes	224
131.	Infinite ordinates are asymptotes	226
132.	Method of substitution. Oblique asymptotes	227

CONTENTS

ARTICLE		PAGE
133.	Number of asymptotes	229
134.	Method of expansion. Explicit functions	230
135.	Method of expansion. Implicit functions	234
136.	Curvilinear asymptotes	236
137.	Examples of asymptotes of transcendental curves	237
138.	Asymptotes in polar coördinates	239
139.	Determination of asymptotes to polar curves	240

CHAPTER XV

DIRECTION OF BENDING. POINTS OF INFLEXION

140.	Concavity upward and downward	243
141.	Algebraic test for positive and negative bending	244
142.	Analytical proof of the test for the direction of bending	247
143.	Concavity and convexity towards the axis	248
144.	Concavity and convexity; polar coördinates	249

CHAPTER XVI

CONTACT AND CURVATURE

145.	Order of contact	252
146.	Number of conditions implied by contact	253
147.	Contact of odd and of even order	254
148.	Circle of curvature	255
149.	Length of radius of curvature; coördinates of center of curvature	255
150.	Second method	257
151.	Direction of radius of curvature	259
152.	Other forms for R	260
153.	Total curvature of a given arc; average curvature	261
154.	Measure of curvature at a given point	261
155.	Curvature of an arc of a circle	262
156.	Curvature of osculating circle	263
157.	Direct derivation of the expressions for κ and R in polar coördinates	265

EVOLUTES AND INVOLUTES

158.	Definition of an evolute	267
159.	Properties of the evolute	269

CHAPTER XVII
Singular Points

ARTICLE | PAGE
160. Definition of a singular point	275
161. Determination of singular points of algebraic curves	275
162. Multiple points	277
163. Cusps	278
164. Conjugate points	281

CHAPTER XVIII
Curve Tracing

165. Statement of problem	283
166. Trace of curve $x^4 - y^4 + 6xy^2 = 0$	284
167. Form of a curve near the origin	288
168. Another proof	292
169. Oblique branch through origin. Expansion of y in ascending powers of x	292
170. Branches touching either axis	295
171. Two branches oblique; a third touching x-axis	297
172. Approximation to form of infinite branches	299
173. Curve tracing; polar coördinates	303

CHAPTER XIX
Envelopes

174. Family of curves	307
175. Envelope of a family of curves	308
176. Envelope touches every curve of the family	309
177. Envelope of normals of a given curve	310
178. Two parameters, one equation of condition	311
Appendix	315
Answers	327
Index	335

INTRODUCTION

In this general introductory chapter some terms of frequent use in subsequent work will be briefly recalled to mind and illustrated, and their meaning somewhat extended.

1. Number. Mathematics is concerned with the study of numbers and their relations to each other. Numbers are represented by letters, as a, b, c, x, y, z, etc., or by figures, 1, 2, 7, $\sqrt{11}$, -8, $\frac{2}{3}$, etc.

2. Operations. The process of obtaining a number from other numbers by any definite rule is called an *operation*. Addition, subtraction, multiplication, division, raising to integral powers and extracting roots are called *algebraic* operations. All other operations are *transcendental;* thus taking the cosine or the logarithm of a given number is a transcendental operation.

An *inverse* operation is the undoing of what was done by the corresponding direct operation, and it ends where the direct operation began. It may also be defined as an operation the effect of which the direct operation simply annuls.

E.g., if $y = ax + b$, then y is produced by performing a certain operation upon x, viz. multiplying it by a and adding b to the product. The inverse operation consists in expressing x in terms of y, and this is done by subtracting b from y and dividing the difference by a.

Again, if $y = \sin x$, then $x = \sin^{-1} y$, which is variously read, "x is the angle whose sine is y," "x is the inverse sine of y," "x is the anti-sine of y." The relation between the direct and inverse operators may be shown by the identity

$$\sin(\sin^{-1} y) = y,$$

which expresses the truism that y is the sine of any of the angles whose sine is y.

Similarly, $2 = \log_3 9$, which is read the logarithm of 9 to the base 3, hence $9 = \log_3^{-1} 2$, read the anti-logarithm of 2 to the base 3;

in which as before the operating symbol $\log_3$ is transferred from one member to the other by annexing the index of inversion (-1), in conventional analogy with the familiar transference of a multiplier in such equations as $y = mx$, $x = m^{-1} y$. Here m^{-1} has a meaning in itself; but $\log_3^{-1}$ has no meaning unless followed by a number. Thus $(\log_3 2)^{-1}$ and $\log_3^{-1} 2$ have distinct meanings; the former inverts the number $\log_3 2$, while the latter inverts the operator $\log_3$.

Two operations are said to be *successive* when one is performed upon the result of the other; *e.g.*, log sin x means: take the sine of x and then the logarithm of sin x.

Successive operations are called *commutative* when their sequence may be altered without changing the result.

Thus any successive operations of multiplication and division are commutative; *e.g.*, $a \cdot b \cdot c \cdot \dfrac{1}{d} = a \cdot \dfrac{1}{d} \cdot c \cdot b$; but taking the sine and the logarithm are not commutative, for sin log $x \neq$ log sin x.

Another property of certain operations may be first illustrated numerically:

$$3(12 + 6) = 3 \cdot 12 + 3 \cdot 6,$$
$$12^2 \times 6^2 = (12 \times 6)^2.$$

In the first illustration, the result of multiplying $12 + 6$ by 3 is the same whether the number 12 and 6 be first added, and then their sum be multiplied by 3, or 12 and 6 be multiplied separately by 3, and the results added. In the second illustration, the operation of squaring may be performed upon the separate factors, and the results multiplied, or it may be performed after the multiplication.

These facts are expressed by saying that multiplication is distributive as to addition, and that involution is distributive as to multiplication.

The general definition of distributive operations may now be stated as follows :

If one operation consists in combining several elements into one result, a second operation is *distributive* as to the first when the final result is the same, whether the second operation be performed upon the result of the first operation, or upon the several elements of the first operation, and then these results combined by the first operation.

Since $(12 \times 6)+3$ is not equal to $(12+3)(6+3)$, and $\log(216+36)$ is not equal to $\log 216 + \log 36$, hence addition is not distributive as to multiplication, nor taking the logarithm, as to addition.

3. Expressions. Any combination of letters and symbols used to denote a number is an *expression*. It may be called an expression or a number, according as the thought is of the symbol, or of the value which the symbol represents.

An expression is *algebraic* when there are implied no other operations upon the numbers of which it is composed than algebraic operations, and when none of these operations is repeated an infinite number of times. An infinite series involves, in general, only the operations of addition and multiplication, but it is not called an algebraic expression.

E.g., $3x$, $4y+17xyz$, a^4, y^6+x^2, $\sqrt[5]{a^2x^3y^4}$ are algebraic expressions.

Expressions which imply other operations than a finite number of algebraic ones are called *transcendental* expressions.

E.g., $\sin x$, 7^x, $\log(2x+y)$, are transcendental expressions.

An algebraic expression is *rational*, when it does not contain radicals; *surd*, or *irrational*, when it contains radicals or fractional exponents; *entire*, or *integral*, when it has only a numerical denominator. An expression may be integral as to some of its letters, and fractional as to others.

E.g., $\dfrac{7x - 4ay + b}{a - 4c}$ is integral as to x, y, b, but fractional as to a and to c.

An expression is *symmetric* as to any of its letters, when its value remains the same, however these letters be interchanged.

E.g., xyz, $x + y + z$ are symmetric as to x, y, z, or to any two of them; $w + x - y - z$ is symmetric as to w and x, and as to y and z, but not as to x and y, w and y, w and z, nor x and z.

An expression is said to be *transformed* when it is changed in form but not in value; it is *developed* or *expanded* when transformed into a series.

4. Functions. If a number is so related to other numbers that its value depends upon their values, it is a *function* of those numbers; the function is *explicit* when directly expressed in terms of those numbers; *implicit*, when not so expressed.

Thus y is an explicit function of x when the equation of relation between x and y is solved for y.

The numbers upon which the value of the function depends are called the *arguments* of the function.

E.g., in $u = 3xy$, u is an explicit function of the arguments x and y; x is an implicit function of the arguments u, y; y is an implicit function of the arguments u, x.

Again, if the letters x, y, z be related to each other by means of such an equation as
$$z^3 + xy^2 + 4yx + 5x^2 = 0,$$
then each letter is an implicit function of the other two.

An explicit function of one or more numbers is known, given, or determined in terms of those numbers. It is symmetric, algebraic, rational, transcendental, etc., according as the expression which gives it its value is symmetric, algebraic, etc.

E.g., if $y = \dfrac{a + \sqrt{b}}{m + n + c^x}$, then y is irrational in b, symmetric as to m and n, transcendental as to x, algebraic as to m, n, a, b, also as to c when x is commensurable.

If one number (or function) depends on its arguments in the same way as another number depends on its arguments; *i.e.*, if the expressions involved are of the same form, then the first number is said to be the same function of its arguments as the second number is of its arguments.

E.g., if $\qquad ax^2 + bx = c, \quad dy^2 + ey = f,$
then c is the same explicit function of a, b, x as f is of d, e, y; and x is the same implicit function of a, b, c as y is of d, e, f.

A function may be denoted by any convenient letter or symbol, as f, F, ϕ, $\cdots$ with or without indices, and followed by the argument, inclosed in parentheses. During any investigation the same functional symbol will stand for the same operation or series of operations.

E.g., if $f(x) = x^2 - ax$, then $f(y) = y^2 - ay$, $f(xy) = x^2y^2 - axy$.

If $\phi(x, y) \equiv \phi(y, x)$, then (Art. 3) ϕ is a symmetric function of x and y.

If $y = F(x)$, x is often denoted by $F^{-1}(y)$, and is called the inverse F-function of y. This notation is illustrated in connection with "inverse operations" in Art. 2.

EXERCISES

1. If $f(x, y) \equiv ax^2 + bxy + cy^2$, write $f(y, x)$; $f(x, x)$; $f(y, y)$.

2. What relation must exist between the coefficients in exercise 1 to make it a symmetric function?

3. If $\phi(x, y) \equiv 4xy + x^2 + x + 4y - 7$, show that $\phi(1, 2) = \phi(2, 1)$. Does this prove $\phi(x, y)$ to be a symmetric function?

4. Let $\psi(x, y) \equiv Ax + By + C$. Show that $\psi(x, y) = 0$, $\psi(y, -x) = 0$ are the equations of two perpendicular lines.

What curves are represented by

$$\psi(y, 0) = 0? \quad \psi(x^2, y) = 0? \quad \psi\left(\frac{y^2}{A}, \frac{x^2}{B}\right) = 0? \quad \psi\left(\frac{x^2}{A}, -\frac{y^2}{B}\right) = 0?$$

5. If $f(x) = 2x\sqrt{1-x^2}$, show that $f\left(\sin\frac{x}{2}\right) = \sin x$.

Find the value of $f\left(\cos\frac{x}{2}\right)$.

6. What functions satisfy the relations

$\phi(x + y) = \phi(x) \cdot \phi(y)$?	exponential functions.
$f(x) + f(y) = f(xy)$?	logarithms.
$\psi(2x) = 2\psi(x)\sqrt{1 - [\psi(x)]^2}$?	sine.
$\chi(2x) = \dfrac{2\chi(x)}{1 - [\chi(x)]^2}$?	tangent.

7. If $f(x) = x^2 + 3$, and $F(x) = 2 - \sqrt{x}$, find $f[F(x)]$, and $F[f(x)]$.

8. In the last example, find $f[f(x)]$ or $f^2(x)$, $F[F(x)]$ or $F^2(x)$, $f^{-1}(x)$, $F^{-1}(x)$.

9. With the same notation, calculate $f^2 + 2fF - 3F^2$.

10. If $\phi(x) = \dfrac{x-1}{x+1}$, show that $\dfrac{\phi(x) - \phi(y)}{1 + \phi(x) \cdot \phi(y)} = \dfrac{x - y}{1 + xy}$.

11. Given $f(x) = \log\dfrac{1-x}{1+x}$, show that $f(x) + f(y) = f\left(\dfrac{x+y}{1+xy}\right)$.

12. If $f(x) = \sqrt{1-x^2}$, what is $f(\sqrt{1-x^2})$?

Does it follow that if $\phi(x) = y$, $\phi(y) = x$?

Give examples of cases in which this is not true; in which it is true.

13. If $f(xy) = f(x) + f(y)$, prove that $f(1) = 0$.

14. If $f(x + y) = f(x) + f(y)$, show that $f(1) = 0$, and $pf(x) = f(px)$, where p is any positive integer.

15. Using the same function as in the last example, prove that $f(mx) = mf(x)$, where m is any rational fraction.

16. If $y = \log_b(x + \sqrt{1+x^2})$, express x as a function of y.

17. Given $f(x) = a^x$, find $f(a)$, $f(1)$, $f(0)$; show that in this function $[f(x)]^2 = f(2x)$.

18. Given $xy - 2x + y = n$; show that y is not a function of x when $n = 2$.

19. If $y = \phi(x) = \dfrac{2x-1}{3x-2}$, show that $x = \phi(y)$, and that $x = \phi^2(x)$.

20. If $y = f(x) = \dfrac{1+x}{1-x}$ and $z = f(y)$, find z as a function of x.

5. Constants and variables. Usually, during an investigation, some of the numbers that enter into it preserve their values unchanged; while all the other numbers take a series of different values.

A *constant* number is one that always remains the same throughout the investigation.

A *variable* number is one that changes its value, so that at different stages it requires different numerals to express it.

The word number will usually be omitted, and the words constant and variable will be used alone, in problems where it is necessary to distinguish between them.

If y be expressed in terms of x by the relation $y = \phi(x)$, then, if a numerical value be given to x, the corresponding value of y may be computed; and if another value be given to x, a new value can be found for y, and so on. In this equation, both x and y are variables, but if the value of x at any instant be given, the resulting value of y is known. In such a case, x is called the *independent* variable, and y the *dependent* variable. The argument is the independent variable, the function is the dependent one.

6. Continuous variable; continuous function. When the variable, in passing from one value to another, passes through every intermediate value in order, then it is *continuous*.

A function $f(x)$ of a continuous variable x is called a *continuous function* in the interval from $x = a$ to $x = b$, if it has the following properties:

It remains real and finite when x takes any real value in the assigned interval.

For each value of x, the function has either a single value or any number of determinate values.

As x changes from m to n (two arbitrary numbers within the interval), the function $f(x)$, if single-valued, changes from $f(m)$ to $f(n)$ by passing through every intermediate value, in order, at least once; and, if $f(x)$ is multiple-valued, each value of $f(x)$ changes from a particular value of $f(m)$ to a corresponding particular value of $f(n)$, in such a way as to pass through every intermediate value, in order, at least once.

If, at a value $x = h$, any one of these conditions fail, the function is said to have *a discontinuity* at $x = h$.

The *increment* taken by a variable, in passing from one value to another, is the difference obtained by subtracting the first value from the second. An increment of x will be expressed by the symbol Δx.

It is implied in the definition of a continuous function that for any small increment of the variable, the increment of the function is also small, and that to the variable an increment can always be given, so small that the corresponding increment of the function shall be smaller than any number that may be assigned, no matter how small.

E.g., if $y = f(x)$ is a continuous function of x in the vicinity of the value $x = x_1$, an increment Δx can be given to x, such that, if Δy be a corresponding increment of y, then

$$\Delta y = f(x_1 + \Delta x) - f(x_1)$$

can be made smaller than h, where h is any number previously assigned.

CHAPTER I

FUNDAMENTAL PRINCIPLES

This chapter treats of the fundamental ideas of a limit and of an infinitesimal, and uses them to lead up to the notion of a derivative, with which the Calculus is so largely concerned.

7. Limit of a variable. If a variable take successive values that approach nearer and nearer to a given constant, so that the difference between the variable and the constant may become smaller than any number that can be assigned, then the constant is called the *limit* of the variable.

This definition applies whether the variable be always greater or always less, or sometimes greater and sometimes less than the constant.

E.g., the circumference of a circle is the limit of the perimeter of an inscribed polygon, and also the limit of the perimeter of a circumscribed polygon when the length of the sides is made less than any assigned number. Similarly the area of the circle is the common limit of the areas of the inscribed and circumscribed polygons.

The slope of a tangent to a curve is the limit of the slope of a secant, when two points of intersection of the secant with the curve approach coincidence.

Thus far the illustrations apply to either the first or second case, in which the variable is either always less or always greater than its limit. An illustration of the third case, wherein the variable may be sometimes greater and sometimes

9

less than the constant, is furnished by a decreasing geometric progression with a negative common ratio.

For instance, consider the series $1, -\frac{1}{2}, +\frac{1}{4}, -\frac{1}{8}, \cdots$, in which the number of terms is infinite. The sum of n terms of this series approaches $\frac{2}{3}$ as a limit, when n is taken larger and larger. The first term is 1; the sum of the first two terms is $\frac{1}{2}$; the sum of the first three is $\frac{3}{4}$; of the first four, $\frac{5}{8}$; and so on; and these successive sums are alternately greater and less than $\frac{2}{3}$, but any one of them is nearer $\frac{2}{3}$ than any sum preceding. By taking terms enough, a sum can be reached that will differ from $\frac{2}{3}$ by less than any number that may be assigned; for the sum of n terms of this geometric series is

$$s_n = \tfrac{2}{3}[1 - (-\tfrac{1}{2})^n],$$

hence
$$s_n - \tfrac{2}{3} = -\tfrac{2}{3}(-\tfrac{1}{2})^n,$$

which can evidently be made less than any assigned number by sufficiently increasing n.

8. Infinitesimals and infinites. A variable that has zero for its limit is called an *infinitesimal*. In other words, an infinitesimal is a variable that becomes smaller than any number that can be assigned.

The reciprocal of an infinitesimal is then a variable that becomes larger than any number that can be assigned, and is called an *infinite* variable.

E.g., the number $(\tfrac{1}{2})^n$ which presents itself in the last illustration is an infinitesimal when n is taken larger and larger; and its reciprocal 2^n is an infinite variable.

From these definitions of the words "limit" and "infinitesimal" the following useful corollaries are immediate inferences.

Cor. 1. The difference between a variable and its limit is an infinitesimal variable.

Cor. 2. Conversely, if the difference between a constant and a variable be an infinitesimal, then the constant is the limit of the variable.

For convenience, the symbol $\doteq$ will be placed between a variable and a constant to indicate that the variable approaches the constant as a limit; thus the symbolic form $x \doteq a$ is to be read "the variable x approaches the constant a as a limit."

The corollaries just mentioned may accordingly be symbolically stated thus:

1. If $x \doteq a$, then $x = a + \alpha$, wherein $\alpha \doteq 0$;
2. If $x = a + \alpha$, and $\alpha \doteq 0$, then $x \doteq a$.

It will appear that the chief use of Cor. 1 is to convert given "limit relations" into the form of ordinary equations, so that they may be at once combined or transformed by the laws governing the equality of numbers; and then Cor. 2 will serve to express the final result in the original form of a limit-relation.

In all cases, whether a variable actually becomes equal to its limit or not, the important property is that their difference is an infinitesimal. An infinitesimal is not necessarily in all stages of its history a small number. Its essence lies in its power of decreasing numerically, having zero for its limit, and not in the smallness of any of the constant values it may pass through. It is frequently defined as an "infinitely small quantity," but this expression should be interpreted in the above sense. Thus a constant number, however small it may be, is not an infinitesimal.

9. Fundamental theorems concerning infinitesimals, and limits in general. This article will be devoted to a rigorous treatment of the theory of limits so far as necessary to furnish a logical basis for the process of differentiation to which the chapter leads up. Theorems 1–3, which are special theorems relating to infinitesimal variables, are deduced immediately from the definition of an infinitesimal; and are then used in conjunction with the corollaries of Art. 8, to establish the general theorems 4–9 relating to the limits of any variables.

THEOREM 1. The product of an infinitesimal α by any finite constant, k, is an infinitesimal; *i.e.*, if

$$\alpha \doteq 0,$$

then $\quad k\alpha \doteq 0.$

For, let c be any assigned number; then, by hypothesis, α can become less than $\dfrac{c}{k}$; hence $k\alpha$ can become less than c, the arbitrary, assigned number, and is, therefore, infinitesimal.

THEOREM 2. The algebraic sum of any finite number (n) of infinitesimals is an infinitesimal; *i.e.*, if

$$\alpha \doteq 0,\ \beta \doteq 0,\ \cdots,$$

then $\quad \alpha + \beta + \cdots \doteq 0.$

For the sum of the n variables does not numerically exceed n times the largest of them, but this product is an infinitesimal by theorem 1; hence the sum of the n variables is an infinitesimal.

NOTE. The sum of an infinite number of infinitesimals may be infinitesimal, finite, or infinite, according to circumstances.

E.g., let a be a finite constant, and let n be a variable that becomes infinite; then $\dfrac{a}{n^2},\ \dfrac{a}{n},\ \dfrac{a}{n^{\frac{1}{2}}}$, are all infinitesimal variables; but

$$\dfrac{a}{n^2} + \dfrac{a}{n^2} + \cdots \text{ to } n \text{ terms} = \dfrac{a}{n}, \text{ which is infinitesimal,}$$

while $\quad \dfrac{a}{n} + \dfrac{a}{n} + \cdots \text{ to } n \text{ terms} = a,$ which is finite,

and $\quad \dfrac{a}{n^{\frac{1}{2}}} + \dfrac{a}{n^{\frac{1}{2}}} + \cdots \text{ to } n \text{ terms} = an^{\frac{1}{2}},$ which is infinite.

THEOREM 3. The product of two infinitesimal variables is an infinitesimal; *i.e.*, if
$$\alpha \doteq 0, \quad \beta \doteq 0,$$
then $\quad\alpha\beta \doteq 0.$

For, let c be any assigned number < 1; then α, β, can each become less than c; hence $\alpha\beta$ can become less than c^2, which is less than c, since $c < 1$; thus $\alpha\beta$ can become less than any assigned number, and is, therefore, infinitesimal.

NOTE. From theorems 1–3, it follows that, if
$$\alpha \doteq 0, \quad \beta \doteq 0, \quad \gamma \doteq 0,$$
then $\quad a\alpha + b\beta + c\gamma + d\beta\gamma + e\gamma\alpha + f\alpha\beta + g\alpha\beta\gamma \doteq 0$
when a, b, c, d, e, f, g, are finite constants.

THEOREM 4. If two variables (x, y) be always equal, and if one of them (x) approach a limit (a), then the other approaches the same limit; *i.e.*, if
$$x = y, \text{ and } x \doteq a,$$
then $\quad y \doteq a.$

For, by Art. 8, Cor. 1,
$$x = a + \alpha,$$
in which $\quad\alpha \doteq 0;$
hence $\quad y = a + \alpha;$
and, therefore, $\quad y \doteq a,$
by Art. 8, Cor. 2.

THEOREM 5. The limit of the sum of a constant (c) and a variable (x) is equal to the constant plus the limit of the variable; *i.e.*,
$$\lim(c + x) = c + \lim x.$$

For, let $\quad x \doteq a:$

then $\quad x = a + \alpha,\quad$ [Art. 8, Cor. 1.
in which $\quad \alpha \doteq 0$;
therefore $\quad c + x = c + a + \alpha$,
hence $\quad c + x \doteq c + a,\quad$ [Art. 8, Cor. 2.
i.e., $\quad \lim(c + x) = c + a = c + \lim x$.

THEOREM 6. The limit of the product of a constant and a variable is equal to the constant multiplied by the limit of the variable; i.e.,
$$\lim cx = c \lim x.$$

For, using the notation of theorem 5, and multiplying by c,
$$cx = ca + c\alpha;$$
therefore $\quad cx \doteq ca,\quad$ [Art. 8, Cor. 2.
i.e., $\quad \lim cx = ca = c \lim x$.

THEOREM 7. If the sum of a finite number of variables $(x, y, \ldots)$ be variable, then the limit of their sum is equal to the sum of their limits; i.e.,
$$\lim(x + y + \cdots) = \lim x + \lim y + \cdots.$$

For, let $\quad x \doteq a,\ y \doteq b,\ \ldots$;
then $\quad x = a + \alpha,\ y = b + \beta,\ \ldots,\quad$ [Art. 8, Cor. 1.
wherein $\quad \alpha \doteq 0,\ \beta \doteq 0,\ \ldots$;
hence $\quad x + y + \cdots = (a + b + \cdots) + (\alpha + \beta + \cdots)$;
but $\quad \alpha + \beta + \cdots \doteq 0$,
by theorem 2; hence, by Art. 8, Cor. 2,
$$\lim(x + y + \cdots) = a + b + \cdots = \lim x + \lim y + \cdots.$$

NOTE. The limit of the sum of an infinite number of variables may not be equal to the sum of their limits. (Cf. Th. 2.)

E.g., when $n \doteq \infty$, the sum of the limits of $\frac{1}{2} + \frac{1}{n^2},\ \frac{1}{2^2} + \frac{2}{n^2},\ \frac{1}{2^3} + \frac{3}{n^2},\ \ldots,\ \frac{1}{2^n} + \frac{n}{n^2}$, is 1; but the limit of their sum is $1\frac{1}{2}$.

Cor. If the sum of a finite number of variables $(x, y, z, \ldots)$ be constant, then this constant (c) is equal to the sum of their limits; *i.e.*, if
$$x + y + z + \cdots = c,$$
then $\quad \lim x + \lim y + \lim z + \cdots = c.$

For, by transposition,
$$y + z + \cdots = c - x;$$
hence, by theorems 4, 7, and 5,
$$\lim y + \lim z + \cdots = \lim (c - x) = c - \lim x;$$
therefore $\quad \lim x + \lim y + \lim z + \cdots = c.$

Theorem 8. If the product of a finite number of variables $(x, y, z, \ldots)$ be variable, then the limit of their product is equal to the product of their limits; *i.e.*,
$$\lim xyz \cdots = \lim x \lim y \lim z \cdots.$$

For, using the previous notation, and taking the case of three variables,
$$xyz = (a + \alpha)(b + \beta)(c + \gamma)$$
$$= abc + ab\gamma + bc\alpha + ca\beta + b\alpha\gamma + c\alpha\beta + a\beta\gamma + \alpha\beta\gamma;$$
hence, by theorems 1, 2, 3, and Art. 8, Cor. 2,
$$\lim xyz = abc = \lim x \lim y \lim z.$$

Cor. If the product of a finite number of variables (x, y, z) be constant, then this constant is equal to the product of their limits; *i.e.*, if $xyz = c$, then $\lim x \lim y \lim z = c$.

For, let w be any other variable, then
$$xyzw = cw;$$
hence, by theorems 4, 6, 8,
$$\lim x \lim y \lim z \lim w = c \lim w;$$
therefore $\quad \lim x \lim y \lim z = c.$

THEOREM 9. If the quotient of two variables (x, y) be variable, then the limit of their quotient is equal to the quotient of their limits, provided these limits are finite;

i.e., $$\lim \frac{x}{y} = \frac{\lim x}{\lim y}.$$

For, since $x = \frac{x}{y} \cdot y$, hence by theorems 4, 8,

$$\lim x = \lim \frac{x}{y} \cdot \lim y,$$

therefore $$\lim \frac{x}{y} = \frac{\lim x}{\lim y}.$$

COR. 1. If the quotient of two variables (x, y) be constant, then this constant (c) is equal to the quotient of their limits;

i.e., if $$\frac{x}{y} = c, \text{ then } \frac{\lim x}{\lim y} = c.$$

For, since $x = cy$, hence by theorems 4, 6,

$$\lim x = c \lim y,$$

therefore $$\frac{\lim x}{\lim y} = c.$$

COR. 2. The limit of the quotient of a constant (c) and a variable (x) is equal to the constant divided by the limit of the variable;

i.e., $$\lim \frac{c}{x} = \frac{c}{\lim x}.$$

For, let $\frac{c}{x} = y$, then $c = xy$, and by theorem 8, Cor.,

$$c = \lim x \lim y;$$

therefore $\lim y = \frac{c}{\lim x}$; that is, $\lim \frac{c}{x} = \frac{c}{\lim x}$.

10. Comparison of variables. Some of the principles just established will now be used in comparing variables with each other. The relative importance of two variables that are approaching limits is measured by the limit of their ratio.

DEFINITION. One variable (α) is said to be infinitesimal, infinite, or finite, in comparison with another variable (x) when the limit of their ratio ($\alpha : x$) is zero, infinite, or finite.

In the first two cases, the phrase "infinitesimal or infinite in comparison with" is sometimes replaced by the less precise phrase "infinitely smaller or infinitely larger than." In the third case, the variable will be said to be of the same order of magnitude.

The following theorem and corollary are useful in comparing two variables:

THEOREM 10. The limit of the quotient of any two variables (x, y) is not altered by adding to them any two numbers (α, β), which are respectively infinitesimal in comparison with these variables (x, y);

i.e.,
$$\lim \frac{x+\alpha}{y+\beta} = \lim \frac{x}{y},$$

provided
$$\frac{\alpha}{x} \doteq 0, \quad \frac{\beta}{y} \doteq 0.$$

For, since
$$\frac{x+\alpha}{y+\beta} = \frac{x}{y} \cdot \frac{1+\frac{\alpha}{x}}{1+\frac{\beta}{y}};$$

hence, by theorems 4, 8,

$$\lim \frac{x+\alpha}{y+\beta} = \lim \frac{x}{y} \cdot \lim \frac{1+\frac{\alpha}{x}}{1+\frac{\beta}{y}};$$

but, by theorems 9, 5, and hypothesis,

$$\lim \frac{1+\frac{\alpha}{x}}{1+\frac{\beta}{y}} = 1;$$

therefore, $\quad \lim \dfrac{x+\alpha}{y+\beta} = \lim \dfrac{x}{y}.$

Cor. If the difference (δ) between two variables (x, y) be infinitesimal as to either, the limit of their ratio is 1, and conversely;

i.e., if $\quad \dfrac{x-y}{y} \doteq 0,$ then $\dfrac{x}{y} \doteq 1.$

For, since $\quad x - y = \delta,$ then $x = y + \delta,$

and $\quad \lim \dfrac{x}{y} = \lim \dfrac{y+\delta}{y} = \lim \left(1 + \dfrac{\delta}{y}\right) = 1.$

Conversely, if $\quad \dfrac{x}{y} \doteq 1,$ then $\dfrac{x-y}{y} \doteq 0.$

For, by Art. 8, Cor. 1,

$$\frac{x}{y} - 1 \doteq 0; \; i.e., \; \frac{x-y}{y} \doteq 0.$$

11. Comparison of infinitesimals, and of infinites. Orders of magnitude. It has already been stated that any two variables are said to be of the same order of magnitude when the limit of their ratio is a finite number; that is to say, is neither infinite nor zero. In less precise language, two variables are of the same order of magnitude when one variable is neither infinitely larger nor infinitely smaller than the other. For instance, $k\beta$ is of the same order as β when k is any finite number; thus a finite multiplier or divisor does not affect the order of magnitude of any variable, whether infinitesimal, finite, or infinite.

In a problem involving infinitesimals, any one of them, α, may be chosen as a standard of comparison as to magnitude; then α is called the principal infinitesimal of the *first order* of smallness, and its reciprocal α^{-1} is the principal infinite of the first order of largeness; α^2 is called the principal infinitesimal of the *second order* of smallness, and its reciprocal α^{-2} is the principal infinite of the second order of largeness. In general, α^n is called the principal infinitesimal of order n, when n is either a positive integer or a positive fraction; but when n is any negative number, α^n is the principal infinite of the corresponding positive order $(-n)$. An infinitesimal or infinite of order zero is a finite number.

Besides the principal infinitesimal (α^n) of the nth order, there are many other infinitesimals of the same order of smallness, for instance, any infinitesimal of the form $k\alpha^n$, in which k is any finite multiplier.

To test for the order (n) of any given infinitesimal (β) with reference to the standard infinitesimal (α) on which it depends, it is necessary to select an exponent n, such that

$$\lim_{\alpha \doteq 0} \frac{\beta}{\alpha^n} = k, \text{ some finite constant.}$$

E.g., to find the order of the variable $3x^4 - 4x^8$, with reference to x as the base infinitesimal.

Comparing with x^2, x^3, x^4, in succession:

$$\lim_{x \doteq 0} \frac{3x^4 - 4x^8}{x^2} = \lim_{x \doteq 0} (3x^2 - 4x) = 0, \text{ not finite};$$

$$\lim_{x \doteq 0} \frac{3x^4 - 4x^8}{x^3} = \lim_{x \doteq 0} (3x - 4) = -4, \text{ finite};$$

$$\lim_{x \doteq 0} \frac{3x^4 - 4x^8}{x^4} = \lim_{x \doteq 0} \left(3 - \frac{4}{x}\right) = \infty, \text{ not finite};$$

hence $3x^4 - 4x^8$ is an infinitesimal of the same order of smallness as x^3; that is, of the third order.

The order of largeness of an infinite variable can be tested in a similar way. For instance, if x be taken as the base infinite, let it be required to find the order of the variable $3x^4 - 4x^3$. Comparing with x^3 and x^4:

$$\lim_{x \doteq \infty} \frac{3x^4 - 4x^3}{x^3} = \lim_{x \doteq \infty} (3x - 4) = \infty;$$

$$\lim_{x \doteq \infty} \frac{3x^4 - 4x^3}{x^4} = \lim_{x \doteq \infty} \left(3 - \frac{4}{x}\right) = 3;$$

hence $3x^4 - 4x^3$ is an infinite of the same order of largeness as x^4, that is, of the fourth order.

The process of finding the limit of the ratio of two infinitesimals is facilitated by the following principle, based on theorem 10 of Art. 10: The limit of the quotient of two infinitesimals is not altered by adding to them (or subtracting from them) any two infinitesimals of higher order, respectively.

E.g., $$\lim_{x \doteq 0} \frac{3x^2 + x^4}{4x^2 - 2x^{\frac{5}{2}}} = \lim_{x \doteq 0} \frac{3x^2}{4x^2} = \frac{3}{4}.$$

This principle is sometimes called "the fundamental theorem of the Differential Calculus," owing to its use in the "fundamental problem" stated in Art. 15.

12. Order of magnitude of expressions involving infinitesimals or infinites.

THEOREM 11. The product of two infinitesimals is another infinitesimal whose order is the sum of their orders.

For, let β, γ be infinitesimals of orders m, n, with reference to the base infinitesimal α; then, by definition,

$$\lim_{\alpha \doteq 0} \frac{\beta}{\alpha^m} = b, \quad \lim_{\alpha \doteq 0} \frac{\gamma}{\alpha^n} = c, \text{ where } b, c \text{ are finite};$$

hence, multiplying and using theorem 8,

$$\lim_{a \doteq 0} \frac{\beta\gamma}{\alpha^{m+n}} = bc, \text{ a finite number,}$$

therefore $\beta\gamma$ is an infinitesimal of order $m+n$.

COR. 1. The product of two infinites is another infinite whose order is the sum of their orders.

For, let β, γ be infinites of orders m, n; then, since the principal infinites of these orders are α^{-m}, α^{-n},

$$\lim_{a \doteq 0} \frac{\beta}{\alpha^{-m}} = b, \quad \lim_{a \doteq 0} \frac{\gamma}{\alpha^{-n}} = c,$$

therefore $\lim_{a \doteq 0} \frac{\beta\gamma}{\alpha^{-(m+n)}} = bc$, a finite number;

hence $\beta\gamma$ is of the same order as $\alpha^{-(m+n)}$, that is, an infinite of order $m+n$.

COR. 2. The product of an infinitesimal of order m by an infinite of order n is an infinitesimal of order $m-n$ when $m>n$; but it is an infinite of order $n-m$ when $n>m$.

For, since $\lim_{a \doteq 0} \frac{\beta}{\alpha^m} = b, \quad \lim_{a \doteq 0} \frac{\gamma}{\alpha^{-n}} = c;$

hence $\lim_{a \doteq 0} \frac{\beta\gamma}{\alpha^{m-n}} = bc = \lim_{a \doteq 0} \frac{\beta\gamma}{\alpha^{-(n-m)}},$

therefore when $m>n$, $\beta\gamma$ is an infinitesimal of order $m-n$, and when $n>m$, $\beta\gamma$ is an infinite of order $n-m$.

THEOREM 12. The quotient of an infinitesimal of order m by an infinitesimal of order n is an infinitesimal of order $m-n$ when $m>n$; but it is an infinite of order $n-m$ when $n>m$.

For, since $\lim_{a \doteq 0} \frac{\beta}{\alpha^m} = b, \quad \lim_{a \doteq 0} \frac{\gamma}{\alpha^n} = c,$

therefore, dividing and using theorem 9,

$$\lim_{a \doteq 0} \frac{\frac{\beta}{\gamma}}{\alpha^{m-n}} = \frac{b}{c} = \lim_{a \doteq 0} \frac{\frac{\beta}{\gamma}}{\alpha^{-(n-m)}};$$

hence $\frac{\beta}{\gamma}$ is an infinitesimal of order $m-n$, when $m > n$, and it is an infinite of order $n-m$, when $n > m$.

Cor. 1. The quotient of an infinite of order m by an infinite of order n is an infinite of order $m-n$, when $m > n$; but it is an infinitesimal of order $n-m$ when $n > m$.

Cor. 2. The ratio of two infinitesimals is finite, infinitesimal, or infinite according as the antecedent is of the same order, a higher order, or a lower order, than the consequent.

Cor. 3. The ratio of two infinites is finite, infinitesimal, or infinite according as the antecedent is of the same order, a lower order, or a higher order, than the consequent.

Theorem 13. The order of an infinitesimal is not altered by adding or subtracting another infinitesimal of higher order.

For, let β, γ be two infinitesimals of order m, n, in which $m < n$, then

$$\lim_{a \doteq 0} \frac{\beta}{\alpha^m} = b, \quad \lim_{a \doteq 0} \frac{\gamma}{\alpha^n} = c,$$

hence $$\lim \frac{\beta + \gamma}{\alpha^m} = \lim \frac{\beta}{\alpha^m} + \lim \frac{\gamma}{\alpha^m},$$

but $\frac{\gamma}{\alpha^m}$ is an infinitesimal of order $n-m$, by theor. 12;

thus $$\lim \frac{\gamma}{\alpha^m} = 0,$$

and $$\lim \frac{\beta + \gamma}{\alpha^m} = \lim \frac{\beta}{\alpha^m} = b,$$

therefore $\beta + \gamma$ is an infinitesimal of the same order as β.

NOTE. The order of an infinitesimal is not altered by adding, but may be altered by subtracting, another infinitesimal of the same order and sign.

For instance, let $\beta = 3\,\alpha^2 + 4\,\alpha^3$, of second order,

$\gamma = 3\,\alpha^2 - 2\,\alpha^3$, of second order,

then $\beta + \gamma = 6\,\alpha^2 + 2\,\alpha^3$, of second order,

but $\beta - \gamma = 6\,\alpha^3$, of third order.

COR. The sum of a finite number of infinitesimals of the same sign is an infinitesimal of an order equal to the lowest order among the infinitesimals summed.

THEOREM 14. The order of an infinite is not altered by adding or subtracting another infinite of lower order.

NOTE. The order of an infinite is not altered by adding, but may be altered by subtracting, another infinite of the same order and sign. (Proof and illustration as above.)

COR. The sum of a finite number of infinites of the same sign is an infinite of an order equal to the highest order among the infinites summed.

THEOREM 15. The limit of the finite sum of any number of infinitesimals is not altered by replacing any infinitesimal by another that bears to it a ratio whose limit is unity.

For, let $\alpha_1 + \alpha_2 + \cdots + \alpha_n,$

be the sum of n infinitesimals of such a nature that as n increases, each term decreases so that the limit of the sum is finite.

Let there be n other infinitesimals,

$$\beta_1, \beta_2, \cdots, \beta_n,$$

so related to the first set that

$$\lim \frac{\beta_1}{\alpha_1} = 1, \; \lim \frac{\beta_2}{\alpha_2} = 1, \; \cdots \lim \frac{\beta_n}{\alpha_n} = 1,$$

then

$$\frac{\beta_1}{\alpha_1} = 1 + \epsilon_1, \; \frac{\beta_2}{\alpha_2} = 1 + \epsilon_2, \; \cdots \frac{\beta_n}{\alpha_n} = 1 + \epsilon_n, \; [\epsilon_1 \doteq 0, \; \epsilon_2 \doteq 0 \cdots$$

and $\beta_1 = \alpha_1 + \epsilon_1 \alpha_1, \; \beta_2 = \alpha_2 + \epsilon_2 \alpha_2, \; \cdots \beta_n = \alpha_n + \epsilon_n \alpha_n,$
hence

$$\beta_1 + \beta_2 + \cdots + \beta_n = (\alpha_1 + \alpha_2 + \cdots + \alpha_n) + (\epsilon_1 \alpha_1 + \epsilon_2 \alpha_2 + \cdots + \epsilon_n \alpha_n).$$

Next let η be an infinitesimal that is numerically equal to the largest of the ϵ's,

then $\epsilon_1 \alpha_1 + \epsilon_2 \alpha_2 + \cdots + \epsilon_n \alpha_n \, |<| \, \eta (\alpha_1 + \alpha_2 + \cdots + \alpha_n),$*
hence

$$(\beta_1 + \beta_2 + \cdots + \beta_n) - (\alpha_1 + \alpha_2 + \cdots + \alpha_n) \, |<| \, \eta (\alpha_1 + \alpha_2 + \cdots + \alpha_n).$$

Taking limits and remembering that, by hypothesis,

$\lim (\alpha_1 + \alpha_2 + \cdots + \alpha_n)$ is finite, and $\lim \eta = 0,$

it follows that

$$\lim (\beta_1 + \beta_2 + \cdots + \beta_n) = \lim (\alpha_1 + \alpha_2 + \cdots + \alpha_n).$$

NOTE. This theorem may sometimes be conveniently stated as follows: the limit of the finite sum of infinitesimals is not altered if these infinitesimals be replaced by others which differ from them respectively by infinitesimals of higher order.†

---

* The symbol $|<|$ stands for "is numerically less than." (See Art. 54.)
† This is called the "fundamental theorem of the Integral Calculus."

13. Useful illustrations of infinitesimals of different orders.

THEOREM 1. $\lim\limits_{\theta \doteq 0} \dfrac{\sin \theta}{\theta} = 1; \quad \lim\limits_{\theta \doteq 0} \dfrac{\tan \theta}{\theta} = 1.$

With O as a center and $OA = r$ as radius, describe the circular arc AB. Let the tangent at A meet OB produced in D; draw BC perpendicular to OA, cutting OA in C. Let the angle $AOB = \theta$ in radian measure,

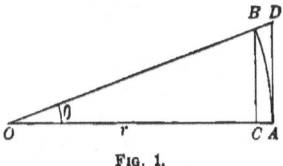

Fig. 1.

then
$$\text{arc } AB = r\theta,$$
$$CB < \text{arc } AB < AD, \quad \text{by geometry,}$$
i.e.,
$$r \sin \theta < r\theta < r \tan \theta,$$
$$\sin \theta < \theta < \tan \theta.$$

By dividing each member of these inequalities by $\sin \theta$,
$$1 < \dfrac{\theta}{\sin \theta} < \sec \theta,$$
but when $\theta \doteq 0$, $\sec \theta \doteq 1$,

hence $\lim\limits_{\theta \doteq 0} \dfrac{\theta}{\sin \theta} = 1,$ and $\lim\limits_{\theta \doteq 0} \dfrac{\sin \theta}{\theta} = 1.$

Similarly, by dividing the inequalities by $\tan \theta$,
$$\cos \theta < \dfrac{\theta}{\tan \theta} < 1,$$

hence $\lim\limits_{\theta \doteq 0} \dfrac{\theta}{\tan \theta} = 1,$ and $\lim\limits_{\theta \doteq 0} \dfrac{\tan \theta}{\theta} = 1.$

Cor. 1. The numbers θ, $\sin \theta$, $\tan \theta$ are infinitesimals of the same order.

Cor. 2. The expressions $\sin \theta - \theta$, $\tan \theta - \theta$ are infinitesimal as to θ.

THEOREM 2. If one angle θ, of a right triangle, be an infinitesimal of the first order, then the hypothenuse r and the adjacent side x are either both finite, or they are infinitesimals of the same order; and the opposite side y is an infinitesimal of order one higher than that of r and x.

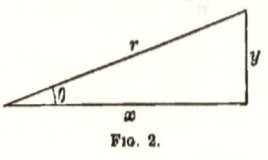

Fig. 2.

For $\dfrac{r}{x} = \cos\theta$, which approaches the value 1 as $\theta \doteq 0$; hence x, r are infinitesimals of the same order; which may be the order zero.

Also $$y = r\sin\theta,$$
and $\sin\theta$ is of order 1; therefore y is of order one higher than r.

Cor. In the same case, if θ be of the first order, and if r and x be of the order n, then the difference between r and x is an infinitesimal of order $n + 2$.

For $$r^2 - x^2 = y^2 = r^2 \sin^2\theta, \quad r - x = \frac{r^2 \sin^2\theta}{r + x};$$
but the orders of r^2, $\sin^2\theta$, $r + x$, are respectively $2n$, 2, n;
$\therefore$ $r - x$ is of order
$$2n + 2 - n = n + 2.$$

THEOREM 3. The difference between the length of an infinitesimal arc of a circle and its chord is of at least the third order when the arc is the first order.

For, let CD be the arc, and CB, DB, tangents at its extremities; then
$$\text{chord } CD < \text{arc } CD < DB + BC.$$

Let the angle $BOD = \theta$ be taken as the principal infinitesimal; then, since arc $CD = 2r\theta$, and r is finite, hence arc CD is of order 1.

13.] FUNDAMENTAL PRINCIPLES 27

Again, since AD is of order 1 (Th. 1, Cor. 1), and angle $ADB = \theta$ is of order 1, hence DB is of order 1, and $DB - DA$ is of order 3 (Th. 2, Cor.); ∴ arc $CD -$ chord CD is of order, at least, three.

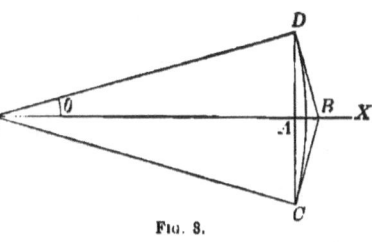

Fig. 3.

Theorem 4. The difference between the length of any infinitesimal arc (of finite curvature), and its chord, is an infinitesimal of, at least, the third order.

Note. The curvature is said to be finite when the limiting ratio of the length of a small chord to the angle between the tangents at its extremities is finite, and not zero.

Thus, in the present case, the chord PQ and the angle TSP are, by hypothesis, infinitesimals of the same order.*

Let the angle TSP be the principal infinitesimal; then, since

$$TSP = SQR + RPS,$$

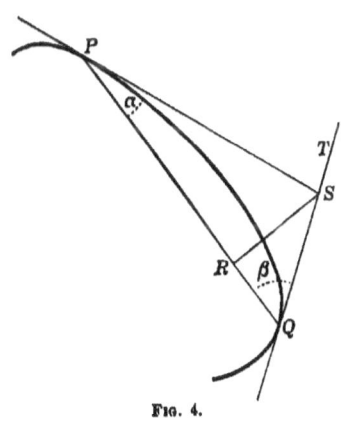

Fig. 4.

it follows that the greater of the latter two angles, say RQS, is of the first order, while the other may be of the first or a higher order. Also, the greater of the two segments RQ, PR, say the latter, is of the first order, while RQ may be of the first or higher order.

---

* If TSP were of higher order than PQ, the curvature would be zero; if of lower order, the curvature would be infinite; the former is the case at an inflection, the latter at a cusp.

Again, by theorem 2, QR, QS are of the same order, and PR, PS are of the same order.

Now arc QP − chord $QP < QS + SP − QP$, [geom.

i.e., $< (QS − QR) + (SP − RP);$

· but since $QS − QR = QS(1 − \cos \beta) = 2QS \sin^2 \frac{\beta}{2}$,

and, similarly, $SP − RP = 2SP \sin^2 \frac{\alpha}{2}$,

and, since each of these products is, at least, of the third order, hence arc QP − chord QP is of, at least, the third order.

EXERCISES

1. Let ABC be a triangle having a right angle at C; draw CD perpendicular to AB, DE perpendicular to CB, EF perpendicular to DB, FG perpendicular to EB; let the angle BAC be an infinitesimal of the first order, AB remaining finite. Prove that:

CD, CB are of order 1;

DB, DE are of order 2;

EB, EF, $(CB − CD)$ are of order 3;

FB, FG, $(DB − DE)$ are of order 4.

2. Of what order is the area of the triangle ABC? BCD? CDE?

3. A straight line, of constant length, slides between two rectangular straight lines, CAA', $CB'B$; let AB, $A'B'$ be two positions of the line. Show that, in the limit, when the two positions coincide,

$$\frac{AA'}{BB'} = \frac{CB}{CA}.$$

14. Continuity of functions. From the foregoing theorems on limits, and the definition of a continuous function, the following theorems relating to continuity are easily derived, and applied to the ordinary classes of functions.

THEOREM 1. If a variable approach a constant, as a limit, according to any given law, then any function of the variable approaches the same function of the constant as a limit if the function be continuous for values of the variable in the vicinity of the constant.

Let $f(x)$ be a continuous function of x, for values of x near a; then when $x \doteq a$

$$f(x) \doteq f(a),$$

regard being had to correspondence of multiple values, if any.

For, let $\qquad x = a + h,$
where $h \doteq 0$; then $\quad f(a+h) - f(a)$
can be made less than any assigned number from the definition of a continuous function (Art. 6); hence

$$f(a+h) = f(x) \doteq f(a).$$

Ex. Prove $\lim f(x) = f(\lim x)$, *i.e.*, the operators f, $\lim$, commutative.

COR. Conversely, any function, $f(x)$, is continuous in the vicinity of $x = a$, if, when $x \doteq a$, $f(x)$ remains real and $\doteq f(a)$, a finite constant.

THEOREM 2. If $y = f(x)$ be a continuous function of x in the vicinity of $x = a$, then the inverse function

$$x = f^{-1}(y)$$

is a continuous function of y in the vicinity of the value

$$y = f(a) = b.$$

For $y = f(x)$ can be represented by a curve which is continuous at (a, b), and

$$x = f^{-1}(y)$$

is represented by the same curve in the vicinity of (a, b).*

COR. If $\qquad f(x) \doteq f(a),$
then one value of x approaches the limit a.

---

* A rigorous algebraic proof of the continuity of an inverse function will be found in the appendix.

Theorem 3. If two functions be continuous at $x = a$, then their sum, difference, and product are continuous functions at $x = a$, and also their quotient, provided the denominator does not vanish at $x = a$. This follows from Th. 8, 9, Art. 9.

Cor. 1. The product of any finite number of functions, each of which is continuous at $x = a$, is continuous at $x = a$.

Cor. 2. If $\phi(x)$ be continuous, and $\phi(a) \neq 0$, $\dfrac{1}{\phi(x)}$ is continuous at $x = a$.

Theorem 4. The algebraic function x^n, in which n is any commensurable number, is continuous for all values of x not infinite.

(1) Let n be a positive integer; then theorem 3 applies.

(2) Let n be the reciprocal of a positive integer; and let

$$y = x^{\frac{1}{q}},$$

then $x = y^q$;

hence, by (1), x is a continuous function of y, and by theorem 2, y is a continuous function of x.

(3) Let n be a positive fraction, $\dfrac{p}{q}$; then $x^{\frac{1}{q}}$ is continuous by (2), and $(x^{\frac{1}{q}})^p$ is continuous by (1).

(4) Let n be any negative commensurable number; then corollary 2 of theorem 3 applies.

Cor. A rational integral function is finite and continuous for all finite values of the variable. (Theorems 3, 4.)

Lemma. When x approaches zero as a limit, then the exponential function a^x approaches unity as a limit:

i.e., if $x \doteq 0$, then $a^x \doteq 1$.

14.] FUNDAMENTAL PRINCIPLES 31

For, let h be any assigned positive number, and let $x = \dfrac{1}{n}$, in which n can become as large as desired; then it is evidently possible to choose n so large that $(1 + h)^n$ shall exceed the number a,

i.e., $$(1 + h)^n > a,$$

and $$1 + h > a^{\frac{1}{n}},$$

hence $$a^x - 1 < h.$$

Thus the exponent x has been chosen so small that $a^x - 1$ is less than the assigned number,

i.e., $\quad a^x - 1 \doteq 0,$ when $x \doteq 0,$

and $\quad a^x \doteq 1,$ when $x \doteq 0.$

THEOREM 5. The exponential function a^x is a continuous function of x, when x is not infinite, provided a is positive,

i.e., $\quad a^{x+h} - a^x \doteq 0,$ when $h \doteq 0.$

For $\quad a^{x+h} - a^x = a^x(a^h - 1),$

but $\quad a^h - 1 \doteq 0,$ when $h \doteq 0,$ by lemma,

hence $\quad a^{x+h} - a^x \doteq 0,$ when $h \doteq 0,$

and a^x is a continuous function of x.

THEOREM 6. The function $\log_a x$ is continuous when x lies between zero and positive infinity where a is positive.

For, let $\quad y = \log_a x,$

then $\quad x = a^y;$

hence, by theorem 5, x is a continuous function of y, when y lies between $-\infty$ and $+\infty$, that is x between 0 and $+\infty$. Therefore, by theorem 2, y is a continuous function of x, when x lies between 0 and $+\infty$.

Cor. 1. If u, v be two continuous variables, then $u^v \doteq a^b$ when $u \doteq a$, where a is positive, and $v \doteq b$.

For $\qquad \log u \doteq \log a,$
and, since $\qquad v \doteq b,$
hence $\qquad v \log u \doteq b \log a,$
that is, $\qquad \log u^v \doteq \log a^b,$
therefore $\quad u^v \doteq a^b$, when $u \doteq a, v \doteq b.$ [Th. 2, Cor.

Cor. 2. If u, v be continuous functions of x, u^v is a continuous function of x. (Th. 1, Cor.)

Cor. 3. If x be a continuous variable, x^n is a continuous function of x, when n is either commensurable or incommensurable. This corollary is a generalization of theorem 5.

THEOREM 7. The functions $\sin x$, $\cos x$ are continuous for all finite values of x;

i.e., $\qquad \sin(x+h) - \sin x \doteq 0$, when $h \doteq 0$,
for $\qquad \sin(x+h) - \sin x = 2\cos(x + \tfrac{1}{2}h)\sin\tfrac{1}{2}h,$
but $\sin\tfrac{1}{2}h \doteq 0$ when $h \doteq 0$, and $\cos(x + \tfrac{1}{2}h)$ is not infinite,
hence $\qquad \sin(x+h) - \sin x \doteq 0$ when $h \doteq 0,$
that is, $\sin x$ is continuous.

Similarly for $\cos x$.

EXERCISES

1. Prove that $\tan x$, $\sec x$ are continuous functions of x for all values except $x = \tfrac{1}{2}(2n+1)\pi$, n being any integer.

2. Prove that $\cot x$, $\csc x$ are continuous functions of x for all values except $x = n\pi$, n being any integer.

3. Find the bounds of continuity of each inverse trigonometric function. Draw the graph, and show the continuity of each of the multiple values.

4. Show that $2^{\frac{1}{x}}$ is not continuous at $x = 0$. Let x successively approach zero from positive and negative values.

15. Comparison of simultaneous infinitesimal increments of two related variables. The last few articles were concerned with the principles to be used in comparing any two infinitesimals. In the illustrations given, the law by which each variable approached zero was assigned, or else the two variables were connected by a fixed relation; and the object was to find the limit of their ratio. The value of this limit gave the relative importance of the infinitesimals.

In the present article the particular infinitesimals compared are not the principal variables (x, y) themselves, but simultaneous increments (h, k) of these variables, as they start out from given values (x_1, y_1) and vary in an assigned manner; as in the familiar instance of the abscissa and ordinate of a given curve.

The variables x, y are then to be replaced by their equivalents $x_1 + h$, $y_1 + k$; in which the increments h, k are themselves variables, and can, if desired, be both made to approach zero as a limit; for since y is supposed to be a continuous function of x, its increment can be made as small as desired by taking the increment of x sufficiently small.

The determination of the limit of the ratio of k to h, as h approaches zero, subject to an assigned relation between x and y, is the fundamental problem of the Differential Calculus.

E.g., let the relation be
$$y = x^2;$$
let x_1, y_1 be simultaneous values of the variables x, y; and when x changes to the value $x_1 + h$, let y change to the value $y_1 + k$; then
$$y_1 = x_1^2,$$
$$y_1 + k = (x_1 + h)^2 = x_1^2 + 2x_1h + h^2;$$
hence
$$k = 2x_1h + h^2.$$

This is a relation connecting the increments h, k.

Here it is to be observed that the relation between the infinitesimals h, k is not directly given, but has first to be derived from the known relation between x and y.

Let it next be required to compare these simultaneous increments by finding the limit of their ratio when they approach the limit zero.

By division,

$$\frac{k}{h} = 2x_1 + h;$$

hence, by Art. 9, theorem 5,

$$\lim_{h \doteq 0} \frac{k}{h} = 2x_1.$$

This result may be expressed in familiar language by saying that when x increases through the value x_1, then y increases $2x_1$ times as much as x; and thus when x continues to increase uniformly, y increases more and more rapidly. For instance, when x passes through the value 4, and y through the value 16, the limit of the ratio of their increments is 8, and hence y is changing 8 times as fast as x; but when x is passing through 5, and y through 25, the limit of the ratio of their increments is 10, and y is changing 10 times as fast as x.

The following table will numerically illustrate the fact that the ratio of the infinitesimal increments h, k approaches nearer and nearer to some definite limit when h and k both approach the limit zero.

Let x_1, the initial value of x, be 4; then y_1, the initial value of y, is 16. Let h, the increment of x, be 1; then k, the corresponding increment of y, is found from

$$16 + k = (4 + 1)^2;$$

thus $k = 9$, and $\dfrac{k}{h} = 9$. Next let h be successively diminished to the values $.8, .6, .4, \cdots$; then the corresponding values of k and of $\dfrac{k}{h}$ are as shown in the table:

$x = 4 + h$	$y = 16 + k$	k	$\dfrac{k}{h}$
$4 + 1$	25	9	9
$4 + .8$	23.04	7.04	8.8
$4 + .6$	21.16	5.16	8.6
$4 + .4$	19.36	3.36	8.4
$4 + .2$	17.64	1.64	8.2
$4 + .1$	16.81	.81	8.1
$4 + .01$	16.0801	.0801	8.01
$\cdots$	$\cdots$	$\cdots$	$\cdots$
$4 + h$	$16 + 8h + h^2$	$8h + h^2$	$8 + h$

Thus the ratio of corresponding increments takes the successive values $8.8, 8.6, 8.4, 8.2, 8.1, 8.01, \cdots$, and can be brought as near to 8 as desired by taking h small enough.

As another example let the relation between x and y be
$$y^2 = x^3,$$
then
$$y_1^2 = x_1^3,$$
$$(y_1 + k)^2 = (x_1 + h)^3,$$
hence, by expansion and subtraction,
$$2 y_1 k + k^2 = 3 x_1^2 h + 3 x_1 h^2 + h^3,$$
$$k(2 y_1 + k) = h(3 x_1^2 + 3 x_1 h + h^2),$$
$$\frac{k}{h} = \frac{3 x_1^2 + 3 x_1 h + h^2}{2 y_1 + k}.$$

Therefore $\displaystyle\lim \frac{k}{h} = \lim \frac{3 x_1^2 + 3 x_1 h + h^2}{2 y_1 + k}$, as $h \doteq 0, k \doteq 0$,

and, by Art. 10, theorem 10,
$$\lim \frac{k}{h} = \frac{3 x_1^2}{2 y_1}.$$

The "initial values" of x, y have been written with subscripts to show that only the increments (h, k) vary during the algebraic process, and also to emphasize the fact that the limit of the ratio of the simultaneous increments depends on the particular values through which the variables are passing, when they are supposed to take these increments. With this understanding the subscripts will hereafter be omitted. Moreover, the increments h, k will, for greater distinctness, be denoted by the symbols Δx, Δy, read "increment of x," "increment of y." The symbol Δ is derived from the initial letter of the word *difference*, as the increment of a variable, in passing from one value to another, is obtained by subtracting the first value from the second.

Ex. 1. If $x^2 + y^2 = a^2$, find $\lim \dfrac{\Delta y}{\Delta x}$. Let the initial values of the variables be denoted by x, y, and let the variables take the respective increments Δx, Δy, so that their new values $x + \Delta x$, $y + \Delta y$ shall still satisfy the given relation, then

$$(x + \Delta x)^2 + (y + \Delta y)^2 = a^2.$$

By expansion, and subtraction,

$$2x \cdot \Delta x + (\Delta x)^2 + 2y \cdot \Delta y + (\Delta y)^2 = 0,$$

hence
$$\Delta x(2x + \Delta x) = -\Delta y(2y + \Delta y),$$

and
$$\frac{\Delta y}{\Delta x} = -\frac{2x + \Delta x}{2y + \Delta y}.$$

Therefore
$$\lim_{\Delta x \doteq 0} \frac{\Delta y}{\Delta x} = -\lim_{\Delta x \doteq 0} \frac{2x + \Delta x}{2y + \Delta y} = -\frac{x}{y}.$$

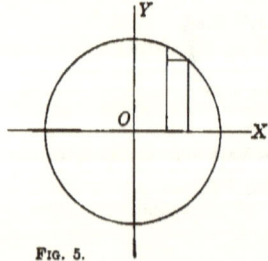

Fig. 5.

The negative sign indicates that when Δx, and the ratio $x:y$, are positive, Δy is negative, that is, an increase in x produces a decrease in y. This may be illustrated geometrically by drawing the circle whose equation is $x^2 + y^2 = a^2$ (Fig. 5).

Ex. 2. If $x^2 + y = y^2 - 2x$, prove
$$\lim_{\Delta x \doteq 0} \frac{\Delta y}{\Delta x} = \frac{2x + 2}{2y - 1}.$$

FUNDAMENTAL PRINCIPLES

Similarly when the relation between x and y is given in the explicit functional form

$$y = \phi(x),$$

then $y + \Delta y = \phi(x + \Delta x),$

and $\Delta y = \phi(x + \Delta x) - \phi(x) = \Delta \phi(x),$

hence $\lim \dfrac{\Delta y}{\Delta x} = \lim \dfrac{\phi(x + \Delta x) - \phi(x)}{\Delta x}.$

When the form of ϕ is given, the limit of this ratio can be evaluated, and expressed as a function of x; and this function is then called the *derivative* of the function $\phi(x)$ with regard to the independent variable x.

The formal definition of the derivative of a function with regard to its variable is given in the next article.

16. Definition of a derivative.

If to a variable a small increment be given, and if the corresponding increment of a continuous function of the variable be determined, then the limit of the ratio of the increment of the function to the increment of the variable, when the latter increment approaches the limit zero, is called the derivative of the function as to the variable.

Let $\phi(x)$ be a finite and continuous function of x, and Δx a small increment given to x, then the derivative of $\phi(x)$ as to x is

$$\lim_{\Delta x \doteq 0} \left\{ \frac{\phi(x + \Delta x) - \phi(x)}{\Delta x} \right\} \equiv \lim_{\Delta x \doteq 0} \frac{\Delta \phi(x)}{\Delta x}.$$

It is important to distinguish between $\lim \dfrac{\Delta \phi(x)}{\Delta x}$ and $\dfrac{\lim \Delta \phi(x)}{\lim \Delta x}$, that is, between the limit of the ratio of two infinitesimals, and the ratio of their limits. The latter is indeterminate of the form $\dfrac{0}{0}$ and may have any value; but the former has usually a determinate value, as illustrated in the examples of the last article.

EXERCISES

1. Find the derivative of $x^2 - 2x$ as to x.
2. Find the derivative of $3x^2 - 4x + 3$ as to x.
3. Find the derivative of $\dfrac{1}{4x}$ as to x.
4. Find the derivative of $x^4 - 2 + \dfrac{3}{x^2}$ as to x.

17. Geometrical illustrations of a derivative.
Some conception of the meaning and use of a derivative will be afforded by one or two geometrical illustrations.

Let $y = \phi(x)$ be a function of x that remains finite and continuous for all values of x between certain assigned constants a and b; and let the variables x, y be taken as the rectangular coördinates of a moving point; then the relation between x and y is represented graphically, within the assigned bounds of continuity by the curve whose equation is

$$y = \phi(x).$$

Let (x_1, y_1), (x_2, y_2) be the coördinates of two points P_1, P_2 on this curve; then it is evident that the ratio

$$\frac{y_2 - y_1}{x_2 - x_1} = \tan \alpha,$$

wherein α is the inclination angle of the secant line P_1P_2, to the x-axis. Let P_2 be moved nearer and nearer to coincidence with P_1, so that $x_2 \doteq x_1$, $y_2 \doteq y_1$; then the secant line P_1P_2 approaches nearer and nearer to coincidence with the tangent line drawn at the point P_1, and the inclination angle

(α) of the secant approaches as a limit the inclination angle (ϕ) of the tangent line.
Hence, by theorem 7, and Ex. 1, Art. 14,

$\tan \alpha \doteq \tan \phi$.

Thus $\dfrac{y_2 - y_1}{x_2 - x_1} \doteq \tan \phi$,

when $x_2 \doteq x_1$, $y_2 \doteq y_1$.

This can also be seen from the similar triangles

KSP_1 and P_1MP_2.

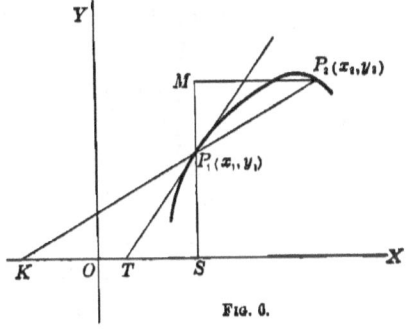

Fig. 0.

The proportion

$$\dfrac{P_1M}{MP_2} = \dfrac{SP_1}{KS}$$

is true, whatever be the position of P_2. When P_2 approaches to coincidence with P_1, $P_1M \doteq 0$, $MP_2 \doteq 0$, but their ratio approaches $\dfrac{SP_1}{TS}$, which is $\tan \phi$.

It may be observed that if x_2 be put directly equal to x_1, and y_2 to y_1, the ratio on the left would, in general, assume the indeterminate form $\dfrac{0}{0}$, as in other cases of finding the limit of the ratio of two infinitesimals; but it has just been shown that the ratio of the infinitesimals $y_2 - y_1$, $x_2 - x_1$ has, nevertheless, a determinate limit measured by $\tan \phi$.

They are thus infinitesimals of the same order except when ϕ is 0 or $\dfrac{\pi}{2}$.

If the differences $x_2 - x_1$, $y_2 - y_1$ be denoted by Δx, Δy, then $\qquad x_2 = x_1 + \Delta x, \quad y_2 = y_1 + \Delta y$;

but, since
$$y = \phi(x),$$
$$y_1 = \phi(x_1), \quad y_2 = \phi(x_2);$$
hence the ratio of the simultaneous increments may be written in the various forms
$$\frac{\Delta y}{\Delta x} = \frac{y_2 - y_1}{x_2 - x_1} = \frac{\phi(x_2) - \phi(x_1)}{x_2 - x_1} = \frac{\phi(x_1 + \Delta x) - \phi(x_1)}{\Delta x}.$$
In the last form, x is regarded as the independent variable, and Δx its independent increment; and the numerator is the increment of the function $\phi(x)$, caused by the change of x from the value x_1 to the value $x_1 + \Delta x$. The limit of this ratio, as $\Delta x \doteq 0$, is the value of the derivative of the function $\phi(x)$, when x has the value x_1. Here x_1 stands for any assigned value of x. Thus the derivative of any continuous function $\phi(x)$ is another function of x which measures the slope of the tangent to the curve $y = \phi(x)$, drawn at the point whose abscissa is x.

Ex. Find the slope of the tangent line to the curve $y = \dfrac{2}{x^2}$ at the point (1, 2).

Here
$$\tan \phi = \lim_{\Delta x \doteq 0} \frac{\frac{2}{(x + \Delta x)^2} - \frac{2}{x^2}}{\Delta x}$$
$$= \lim_{\Delta x \doteq 0} \frac{-2(2x + \Delta x)}{x^2(x + \Delta x)^2} = -\frac{4}{x^3}.$$

Hence $\tan \phi = -4$, when $x = 1$; and the equation of the tangent line at the point (1, 2) is $\quad y - 2 = -4(x - 1).$ [Cf. A.G., Art. 53.

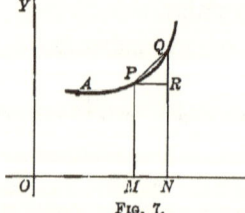

Fig. 7.

As another illustration, let the coördinates of P be (x, y), and those of $Q, (x + \Delta x, y + \Delta y)$; then $MN = PR = \Delta x$, and $PS = RQ = \Delta y$. Let the area $OAPM$ be denoted by z, then z is evidently some function of the abscissa x; also let area $OAQN, = z + \Delta z$, then area $MNQP = \Delta z$, is the increment taken by the function z, when x takes the increment Δx; but $MNPQ$ lies between the rectangles MR, MQ,

hence $\quad y\Delta x < \Delta z < (y + \Delta y)\Delta x,$

and $\quad y < \dfrac{\Delta z}{\Delta x} < y + \Delta y.$

Therefore, when $\Delta x, \Delta y, \Delta z$ all $\doteq 0$,

$$\lim \frac{\Delta z}{\Delta x} = y.$$

Thus if the ordinate and the area be each expressed as functions of the abscissa, the derivative of the area function with regard to the abscissa is equal to the ordinate function.

Ex. If the area included between a curve, the axis of y, and the ordinate whose abscissa is x, be given by the equation

$$z = x^3,$$

find the equation of the curve.

Here $\quad y = \lim \dfrac{\Delta z}{\Delta x} = \lim\limits_{\Delta x \doteq 0} \dfrac{(x + \Delta x)^3 - x^3}{\Delta x}$

$\quad = \lim\limits_{\Delta x \doteq 0} [3x^2 + 3x\Delta x + (\Delta x)^2] = 3x^2.$

18. The operation of differentiation. It has been seen in a number of examples that when, on a given function $\phi(x)$, the operation indicated by

$$\lim_{\Delta x \doteq 0} \frac{\phi(x + \Delta x) - \phi(x)}{\Delta x}$$

is performed, the result of the operation is another function of x. This function may have properties similar to those of $\phi(x)$, or it may be of an entirely different class.

The above indicated operation is for brevity denoted by

the symbol $\dfrac{d\phi(x)}{dx}$, and the resulting derivative function by $\phi'(x)$; thus

$$\dfrac{d\phi(x)}{dx} \equiv \lim_{\Delta x \doteq 0} \dfrac{\Delta \phi(x)}{\Delta x} \equiv \lim_{\Delta x \doteq 0} \dfrac{\phi(x+\Delta x)-\phi(x)}{\Delta x} \equiv \phi'(x).$$

The process of performing this indicated operation is called the *differentiation* of $\phi(x)$ with regard to x. The symbol * $\dfrac{d}{dx}$, when spoken of separately, is called the *differentiating operator*, and expresses that any function written after the d is to be differentiated with regard to x, just as the symbol cos prefixed to $\phi(x)$ indicates that the latter is to have a certain operation performed upon it; namely, that of finding its cosine.

The process of differentiating $\phi(x)$ consists of the following steps:

1. Obtain $\phi(x+\Delta x)$ by changing x into $x+\Delta x$ in $\phi(x)$.
2. Find $\Delta \phi(x)$ by subtracting $\phi(x)$ from $\phi(x+\Delta x)$.
3. Divide this difference $\Delta \phi(x)$ by Δx.
4. Find the limit of the quotient $\dfrac{\Delta \phi(x)}{\Delta x}$ when $\Delta x \doteq 0$.

This series of steps should be memorized. In words, these four steps can be expressed as follows:

1. Give a small increment to the variable.
2. Compute the resulting increment of the function.
3. Divide the increment of the function by the increment of the variable.
4. Obtain the limit of this quotient as the increment of the variable approaches zero.

---

* This symbol is sometimes replaced by the single letter D.

EXERCISES

Find the derivatives of the following functions:

1. $5y^3 - 2y + 6$ as to y;
2. $7t^2 - 4t - 11t^8$ as to t;
3. $8u^3 - 4u + 10$ as to $2u$;
4. $2x^2 - 5x + 6$ as to $x - 3$.

This process will be applied in the next chapter to all the classes of functions whose continuity within certain intervals has been established in Art. 14; and it will be found that for each of them a derivative function exists; that is, that $\lim \dfrac{\Delta\phi(x)}{\Delta x}$ has a determinate and unique value, and that the curve $y = \phi(x)$ has a definite tangent within the range of continuity of the function.

A few curious functions have been devised, which are continuous and yet possess no definite derivative; but they do not present themselves in any of the ordinary uses of the Calculus. Again, there are a few functions for which $\lim \dfrac{\Delta\phi(x)}{\Delta x}$ has a certain value when $\Delta x \doteq 0$ from the positive side, and a different value when $\Delta x \doteq 0$ from the negative side; the derivative is then said to be *non-unique*.

Functions that possess a unique derivative within an assigned interval are said to be *differentiable* in that interval.

Ex. Show that a function is not differentiable at a discontinuity (Art. 6).

19. Increasing and decreasing functions. A good example of the use of the derivative is its application to finding the intervals of increasing or decreasing for a given function.

A function is called an *increasing* function if it increases as the variable increases and decreases as the variable decreases. A function is called a *decreasing* function if it decreases as the variable increases, and increases as the variable decreases.

E.g., the function $x^2 + 4$ decreases as x increases from $-\infty$ to 0, but it increases as x increases from 0 to $+\infty$. Thus $x^2 + 4$ is a decreasing function while x is negative, and an increasing function while x is positive. This is well shown by the locus of the equation $y = x^2 + 4$ (Fig. 8).

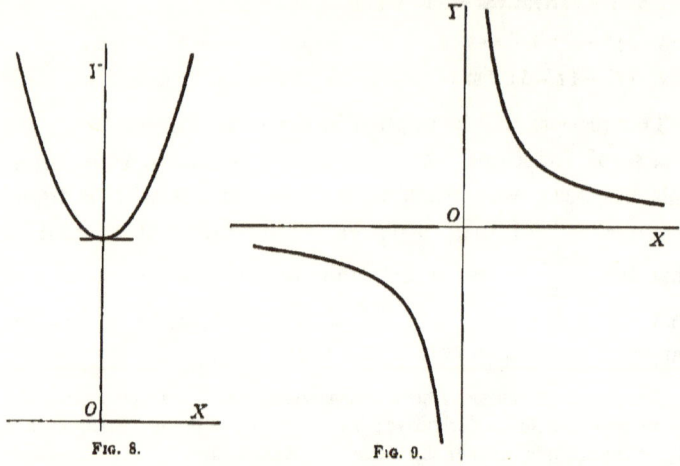

Fig. 8.   Fig. 9.

Again, the form of the curve $y = \dfrac{1}{x}$ shows that $\dfrac{1}{x}$ is a decreasing function, as x passes from $-\infty$ to 0, and also a decreasing function, as x passes from 0 to $+\infty$. When x passes through 0, the function changes discontinuously from the value $-\infty$ to the value $+\infty$ (Fig. 9).

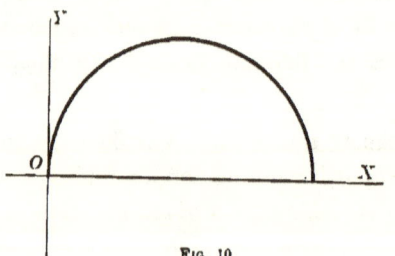

Fig. 10.

Most functions are increasing functions for some values of the variable, and decreasing functions for others.

E.g., $\sqrt[4]{2rx - x^2}$ is an increasing function from $x = 0$ to $x = r$, and a decreasing function from $x = r$ to $x = 2r$ (Fig. 10).

A function is said to be an increasing function in the neighborhood of a given value of x if it increases as x increases through a small interval including this value; similarly for a decreasing function.

20. Algebraic test of the intervals of increasing and decreasing. Let $y = \phi(x)$ be a function of x, and let it be real, continuous and differentiable for all values of x from a to b; then by definition y is increasing or decreasing at a point $x = x_1$, according as

$$\phi(x_1 + \Delta x) - \phi(x_1)$$

is positive or negative, where Δx is a small positive number.

The sign of this expression is not changed if it be divided by Δx, no matter how small Δx may be; hence $\phi(x)$ is an increasing or a decreasing function at the value x_1, according as

$$\frac{dy}{dx} = \lim_{\Delta x \doteq 0} \left\{ \frac{\phi(x_1 + \Delta x) - \phi(x_1)}{\Delta x} \right\} = \phi'(x_1)$$

is positive or negative.

Thus the intervals in which $\phi(x)$ is an increasing function are the same as the intervals in which $\phi'(x)$ is positive.

E.g., to find the intervals in which the function

$$\phi(x) = 2x^3 - 9x^2 + 12x - 6$$

is increasing or decreasing. The derivative is

$$\phi'(x) = 6x^2 - 18x + 12 = 6(x-1)(x-2);$$

hence, as x passes from $-\infty$ to 1, the derived function $\phi'(x)$, is positive and $\phi(x)$ increases from $\phi(-\infty)$ to $\phi(1)$; *i.e.*, from $\phi = -\infty$ to $\phi = -1$; as x passes from 1 to 2, $\phi'(x)$ is negative, and $\phi(x)$ decreases from $\phi(1)$ to $\phi(2)$; *i.e.*, from -1 to -2; and as x passes from 2 to $+\infty$, $\phi'(x)$ is positive, and $\phi(x)$ increases from $\phi(2)$ to $\phi(\infty)$; *i.e.*, from -2 to $+\infty$. The locus of the equation $y = \phi(x)$ is shown in figure 11. At points where $\phi'(x) = 0$, the function $\phi(x)$ is neither increasing nor decreasing. At such points the tangent is parallel to the axis of x. Thus in this illustration, at $x = 1$, $x = 2$, the tangent is parallel to the x-axis.

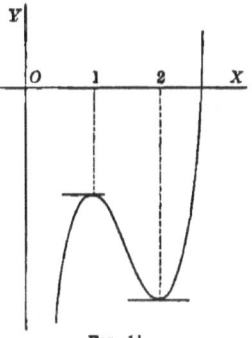

Fig. 11.

EXERCISES

1. Find the intervals of increasing and decreasing for the function
$$\phi(x) \equiv x^3 + 2x^2 + x - 4.$$
Here $\phi'(x) = 3x^2 + 4x + 1 = (3x + 1)(x + 1).$
The function increases from $x = -\infty$ to $x = -1$; decreases from $x = -1$ to $x = -\tfrac{1}{3}$; increases from $x = -\tfrac{1}{3}$ to $x = \infty$.

2. Find the intervals of increasing and decreasing for the function
$$y = x^3 - 2x^2 + x - 4,$$
and show where the curve is parallel to the x-axis.

3. At how many points can the slope of the tangent to the curve
$$y = 2x^3 - 3x^2 + 1$$
be 1? -1? Find the points.

4. Compute the angle at which the following curves intersect:
$$y = 3x^2 - 1, \quad y = 2x^2 + 3. \quad [\text{Cf. A.G., p. 164.}]$$

21. Differentiation of a function of a function. Suppose that y, instead of being given directly as a function of x, is expressed as a function of another variable u, which is itself expressed as a function of x; and let it be required to find the derivative of y with regard to the independent variable x.

Let $y = f(u)$, in which u is a function of x. Suppose that x passes through an assigned value x_1; and let u pass through a corresponding value u_1; and y, in consequence, through a value y_1. When x changes to the value $x_1 + \Delta x$, let u and y, under the given relations, change continuously to the values $u_1 + \Delta u$, $y_1 + \Delta y$; then

$$\frac{\Delta y}{\Delta x} = \frac{\Delta y}{\Delta u} \cdot \frac{\Delta u}{\Delta x} = \frac{f(u + \Delta u) - f(u)}{\Delta u} \cdot \frac{\Delta u}{\Delta x};$$

hence, equating limits,

$$\frac{dy}{dx} = \frac{dy}{du} \cdot \frac{du}{dx} = \frac{df(u)}{du} \cdot \frac{du}{dx};$$

in which the combination of values $(x = x_1, u = u_1, y = y_1)$ is to be substituted.

The derivative of a function of u with regard to x is equal to the product of the derivative of the function with regard to u, and the derivative of u with regard to x; each derivative being estimated at the same combination of corresponding values of the three variables.

The given functions may be multiple-valued, such as $y = \sqrt{a^2 - u^2}$, $u = \sin^{-1} x$. Then when any assigned value x_1 is given to x, the functions u and $\dfrac{du}{dx}$ take multiple values; let one of the branches of u be specified; and let u_1 be the value of u on this branch, corresponding to $x = x_1$. When the value u_1 is given to u, the functions y and $\dfrac{dy}{du}$ take multiple values; let the value of y on a specified branch be y_1. Then, by the theorem, one of the values of $\dfrac{du}{dx}$ taken at $x = x_1$, multiplied by one of the values of $\dfrac{dy}{du}$ taken at $(x = x_1, u = u_1)$, will give one of the values of $\dfrac{dy}{dx}$ taken at $(x = x_1, y = y_1)$, and these are the respective unique values of the three derivatives taken at the specified combination $(x = x_1, u = u_1, y = y_1)$. This combination is represented geometrically in three dimensions by one of the points of intersection of the plane $x = x_1$ with the intersection-curve of the two surfaces that represent the given functions.

Ex. 1. Given $y = 3u^2 - 1,\ u = 3x^2 + 4$; find $\dfrac{dy}{dx}$.

$$\frac{dy}{du} = 6u,\quad \frac{du}{dx} = 6x;$$

$$\frac{dy}{dx} = \frac{dy}{du} \cdot \frac{du}{dx} = 36 ux = 36 x (3 x^2 + 4).$$

Ex. 2. Given $y = 3u^2 - 4u + 5,\ u = 2x^3 - 5$; find $\dfrac{dy}{dx}$.

22. Differentiation of inverse functions. Relation between $\dfrac{dy}{dx}$ and $\dfrac{dx}{dy}$. When $y = f(x)$ is a continuous and differentiable function of x, the symbol $\dfrac{dy}{dx}$ stands for the numerical measure of the limit of the ratio of an increment of y to an assigned increment of x. Next, let y be taken as the independent variable; then the inverse function $x = f^{-1}(y)$ is a continuous function of y; and if a small increment be given to y, it is required to find the limit of the ratio of the resulting increment of x to the assigned increment of y.

Let x, y have the initial values x_1, y_1, and let the variables change, subject to the given relation, so as to assume the values

$$x_1 + \Delta x, \ y_1 + \Delta y;$$

then, since $\qquad \dfrac{\Delta y}{\Delta x} \cdot \dfrac{\Delta x}{\Delta y} = 1,$

hence, by the theory of limits (Art. 7, Th. 9, Cor. 2),

$$\frac{dx}{dy} = \frac{1}{\dfrac{dy}{dx}}$$

in which the two corresponding values, $x = x_1$, $y = y_1$, are understood to be substituted.

Thus if $y = f(x)$ be a differentiable function of x, the inverse function $x = f^{-1}(y)$ is a differentiable function of y, and the derivative of x with regard to y is the reciprocal of the derivative of y with regard to x, each derivative being estimated at the same pair of corresponding values of x and y.

NOTE. Either variable may be a multiple-valued function of the other, as in the familiar relation, $x^2 + y^2 = a^2$.

When any value x_1 is given to x, the functions y and $\dfrac{dy}{dx}$ take multiple values; and, when the corresponding value y_1 is given to y, the functions x and $\dfrac{dx}{dy}$ take multiple values.

One pair of values of $\dfrac{dy}{dx}$ and of $\dfrac{dx}{dy}$ will be reciprocal, and these will be their respective values for the combination

$$(x = x_1,\ y = y_1).$$

In geometrical language, they will belong to the same point (x_1, y_1) of the representative curve.

Ex. From Ex. 1, p. 36, find the values of $\dfrac{dy}{dx}, \dfrac{dx}{dy}$ at the four points $(\pm 3, \pm 4)$ on the circle $x^2 + y^2 = 25$; and write down the equations of the four tangents.

MISCELLANEOUS EXERCISES

1. Find $\lim \dfrac{3(x-1)(x+6)}{2(x^2+3x+2)}$ as $x \doteq \pm \infty;\ \pm 1;\ \pm 2;\ 0$.

2. If $n \doteq \infty$, show whether the theorems of limits apply to:

$$\dfrac{a}{n} + \dfrac{a}{n} + \cdots \text{(to } n \text{ terms)} = a;$$

$$a^{\frac{1}{n}} \times a^{\frac{1}{n}} \times \cdots \text{(to } n \text{ factors)} = a;$$

$$a^{\frac{1}{n^2}} \times a^{\frac{1}{n^2}} \times \cdots \text{(to } n \text{ factors)} = a^{\frac{1}{n}};$$

$$a^{\frac{1}{n^2}} \times a^{\frac{2}{n^2}} \times a^{\frac{3}{n^2}} \times \cdots \times a^{\frac{n}{n^2}} = a^{\frac{n+1}{2n}}.$$

3. Draw graphs of $a^x, \log x, \log (x^2 - x), \tan x$. Show discontinuities.

4. What kinds of discontinuity have $a^{\frac{1}{x}}, \sin \dfrac{1}{x}$, at $x = 0$?

5. What locus has its area proportional to the square of the abscissa?

6. Show that the perimeter of an inscribed regular n-gon equals

$$2nr \sin \dfrac{\pi}{n} = 2\pi r \left[\dfrac{\sin \dfrac{\pi}{n}}{\dfrac{\pi}{n}} \right] \doteq 2\pi r, \text{ as } n \doteq \infty.$$

7. Prove that the derivative of a constant is zero.

CHAPTER II

DIFFERENTIATION OF THE ELEMENTARY FORMS

23. In recent articles, the meaning of the symbol $\frac{dy}{dx}$ was explained and illustrated; and a method of expressing its value, as a function of x, was exemplified, in cases in which y was a simple algebraic function of x, by direct use of the definition. This method is not always the most convenient one in the differentiation of more complicated functions.

The present chapter will be devoted to the establishment of some general rules of differentiation which will, in many cases, save the trouble of going back to the definition.

The next five articles treat of the differentiation of algebraic functions and of algebraic combinations of other differentiable functions.

24. Differentiation of the product of a constant and a variable.

Let $$y = cx;$$
then $$y + \Delta y = c(x + \Delta x),$$
$$\Delta y = c(x + \Delta x) - cx = c\Delta x,$$
$$\frac{\Delta y}{\Delta x} = c;$$
therefore $$\frac{dy}{dx} = c. \qquad [\text{Art. 9, Th. 9.}$$

Cor. 1. If $y = cu$, where u is a function of x, then, by Art. 21,

$$\frac{d(cu)}{dx} = c\frac{du}{dx}. \tag{1}$$

The derivative of the product of a constant and a variable is equal to the constant multiplied by the derivative of the variable.

Cor. 2. The operator $\frac{d}{dx}$ and the constant multiplier c are commutative operators.

Is this true of the operators Δ and c?

25. Differentiation of a sum.

Let $\quad y = f(x) + \phi(x) + \psi(x),$

then $\quad y + \Delta y = f(x + \Delta x) + \phi(x + \Delta x) + \psi(x + \Delta x),$

$$\frac{\Delta y}{\Delta x} = \frac{f(x + \Delta x) - f(x)}{\Delta x} + \frac{\phi(x + \Delta x) - \phi(x)}{\Delta x}$$

$$+ \frac{\psi(x + \Delta x) - \psi(x)}{\Delta x},$$

therefore, by equating the limits of both members,

$$\frac{dy}{dx} = f'(x) + \phi'(x) + \psi'(x). \qquad [\text{Art. 9, Th. 7.}$$

Cor. 1. If $y = u + v + w$, in which u, v, w, are functions of x, then

$$\frac{d}{dx}(u + v + w) = \frac{du}{dx} + \frac{dv}{dx} + \frac{dw}{dx}. \tag{2}$$

The derivative of the sum of a finite number of functions is equal to the sum of their derivatives.

Cor. 2. The operator $\frac{d}{dx}$ is distributive as to addition.

Is this true of the operator Δ?

If the number of functions be infinite, theorem 7 of Art. 9 may not apply, that is, the limit of the sum may not be equal to the sum of the limits; and hence the derivative of the sum may not be equal to the sum of the derivatives. Thus the derivative of an infinite series cannot always be found by differentiating it term by term. (See note, p. 14, and footnote to Art. 156.)

26. Differentiation of a product.

Let $y = f(x)\phi(x)$,

then $\dfrac{\Delta y}{\Delta x} = \dfrac{f(x+\Delta x)\phi(x+\Delta x) - f(x)\phi(x)}{\Delta x}$.

By subtracting and adding $f(x)\phi(x+\Delta x)$ in the numerator, this result may be re-arranged thus:

$$\frac{\Delta y}{\Delta x} = \phi(x+\Delta x)\frac{f(x+\Delta x) - f(x)}{\Delta x} + f(x)\frac{\phi(x+\Delta x) - \phi(x)}{\Delta x}.$$

Equating limits, as $\Delta x \doteq 0$, using Art. 9, theorems 7, 8, and noting that the first factor $\phi(x+\Delta x) \doteq \phi(x)$ since $\phi(x)$ is by hypothesis continuous (Art. 14), it follows that

$$\frac{dy}{dx} = \phi(x)f'(x) + f(x)\phi'(x).$$

COR. 1. By writing $u = \phi(x)$, $v = f(x)$, this result can be more concisely written,

$$\frac{d(uv)}{dx} = u\frac{dv}{dx} + v\frac{du}{dx}. \tag{3}$$

The derivative of the product of two functions is equal to the sum of the products of the first factor by the derivative of the second, and the second factor by the derivative of the first.

This rule for differentiating a product of two functions may be stated thus: Differentiate the product, regarding the first factor as constant, then regarding the second factor as constant, and add the two results.

Since $\frac{d}{dx}(uv) \neq u\frac{d}{dx}v$, the operator $\frac{d}{dx}$ is not commutative with a variable multiplier.

COR. 2. To find the derivative of the product of three functions
$$y = \phi(x)\,\theta(x)\,\psi(x).$$
Let $\qquad f(x) = \theta(x)\,\psi(x),$

then $\qquad y = \phi(x)f(x),$

hence $\qquad \dfrac{dy}{dx} = f(x)\,\phi'(x) + \phi(x)f'(x),$

but $\qquad f'(x) = \theta(x)\,\psi'(x) + \psi(x)\,\theta'(x);$

hence, substituting the values for $f(x), f'(x),$

$$\frac{dy}{dx} = \psi(x)\,\phi(x)\,\theta'(x) + \theta(x)\,\phi(x)\,\psi'(x) + \theta(x)\,\psi(x)\,\phi'(x);$$

and so on, for any finite number of factors.

This result can also be written in the form

$$\frac{d(uvw)}{dx} = uv\frac{dw}{dx} + vw\frac{du}{dx} + wu\frac{dv}{dx}. \qquad (4)$$

The derivative of the product of any finite number of factors is equal to the sum of the products obtained by multiplying the derivative of each factor by all the other factors.

Ex. Show that the operators Δ and $\dfrac{d}{dx}$ are not distributive as to multiplication.

27. Differentiation of a quotient.

Let $\qquad y = \dfrac{f(x)}{\phi(x)},$

then $\qquad y + \Delta y = \dfrac{f(x + \Delta x)}{\phi(x + \Delta x)},$

$$\frac{\Delta y}{\Delta x} = \frac{\dfrac{f(x+\Delta x)}{\phi(x+\Delta x)} - \dfrac{f(x)}{\phi(x)}}{\Delta x}$$

$$= \frac{\phi(x)f(x+\Delta x) - f(x)\phi(x+\Delta x)}{\Delta x\, \phi(x)\, \phi(x+\Delta x)}.$$

By subtracting and adding $\phi(x)f(x)$ in the numerator, this expression may be written

$$\frac{\Delta y}{\Delta x} = \frac{\phi(x)\left\{\dfrac{f(x+\Delta x)-f(x)}{\Delta x}\right\} - f(x)\left\{\dfrac{\phi(x+\Delta x)-\phi(x)}{\Delta x}\right\}}{\phi(x)\,\phi(x+\Delta x)}.$$

Hence, by equating limits,

$$\frac{dy}{dx} = \frac{\phi(x)f'(x) - f(x)\phi'(x)}{[\phi(x)]^2}. \qquad \text{[Art. 9, Ths. 8, 9.}$$

Another form of this result is

$$\frac{d}{dx}\left(\frac{u}{v}\right) = \frac{v\dfrac{du}{dx} - u\dfrac{dv}{dx}}{v^2}. \tag{5}$$

The derivative of the quotient of two functions is equal to the denominator multiplied by the derivative of the numerator minus the numerator multiplied by the derivative of the denominator, divided by the square of the denominator.

28. Differentiation of a commensurable power of a function.

Let $y = u^n$, in which u is a function of x; then there are three cases to consider.

1. n a positive integer.
2. n a negative integer.
3. n a commensurable fraction.

1. n a positive integer.

This is a particular case of (4), the factors $u, v, w, \cdots$ all being equal. Thus

$$\frac{dy}{dx} = nu^{n-1}\frac{du}{dx}.$$

2. n a negative integer.

Let $n = -m$, in which m is a positive integer; then

$$y = u^n = u^{-m} = \frac{1}{u^m},$$

and $\quad \dfrac{dy}{dx} = \dfrac{-mu^{m-1}}{u^{2m}} \cdot \dfrac{du}{dx} \quad$ by (5), and Case (1)

$$= -mu^{-m-1}\frac{du}{dx};$$

hence $\quad \dfrac{dy}{dx} = nu^{n-1}\dfrac{du}{dx}.$

3. n a commensurable fraction.

Let $n = \dfrac{p}{q}$, where p, q are both integers, which may be either positive or negative; then

$$y = u^n = u^{\frac{p}{q}};$$

hence $\quad y^q = u^p,$

and * $\quad \dfrac{d}{dx}(y^q) = \dfrac{d}{dx}(u^p);$

i.e., $\quad qy^{q-1}\dfrac{dy}{dx} = pu^{p-1}\dfrac{du}{dx}.$

Solving for the required derivative,

$$\frac{dy}{dx} = \frac{p}{q} u^{\frac{p}{q}-1} \frac{du}{dx};$$

hence $\quad \dfrac{du^n}{dx} = nu^{n-1}\dfrac{du}{dx}. \qquad (6)$

The derivative of any commensurable power of a function is equal to the exponent of the power multiplied by the power with its exponent diminished by unity, multiplied by the derivative of the function.

---

* If two functions be identical, their derivatives are identical.

These theorems will be found sufficient for the differentiation of any algebraic function; as such functions are made up of the operations of addition, subtraction, multiplication, division, and involution, in which the exponent is an integer or commensurable fraction.

The following examples will serve to illustrate the theorems, and will show the combined application of the general forms (1) to (6).

ILLUSTRATIVE EXAMPLES

1. $y = \dfrac{3x^2 - 2}{x + 1}$; find $\dfrac{dy}{dx}$.

$$\dfrac{dy}{dx} = \dfrac{(x+1)\dfrac{d}{dx}(3x^2 - 2) - (3x^2 - 2)\dfrac{d}{dx}(x+1)}{(x+1)^2}. \qquad \text{by (5)}$$

$$\dfrac{d}{dx}(3x^2 - 2) = \dfrac{d}{dx}(3x^2) - \dfrac{d}{dx}(2) \qquad \text{by (2)}$$

$$= 6x. \qquad \text{by (1), (6), Ex. 7, p. 49.}$$

$$\dfrac{d}{dx}(x+1) = \dfrac{dx}{dx} = 1. \qquad \text{by (2)}$$

Substituting these results in the expression for $\dfrac{dy}{dx}$,

$$\dfrac{dy}{dx} = \dfrac{(x+1)6x - (3x^2 - 2)}{(x+1)^2} = \dfrac{3x^2 + 6x + 2}{(x+1)^2}.$$

2. $u = (3s^2 + 2)\sqrt{1 + 5s^2}$; find $\dfrac{du}{ds}$.

$$\dfrac{du}{ds} = (3s^2 + 2)\dfrac{d}{ds}\sqrt{1 + 5s^2} + \sqrt{1 + 5s^2} \cdot \dfrac{d}{ds}(3s^2 + 2). \qquad \text{by (3)}$$

$$\dfrac{d}{ds}\sqrt{1 + 5s^2} = \dfrac{d}{ds}(1 + 5s^2)^{\frac{1}{2}}$$

$$= \dfrac{1}{2}(1 + 5s^2)^{-\frac{1}{2}}\dfrac{d}{ds}(1 + 5s^2) \qquad \text{by (6)}$$

$$= \dfrac{5s}{\sqrt{1 + 5s^2}}.$$

$$\dfrac{d}{ds}(3s^2 + 2) = 6s. \qquad \text{by (6)}$$

28.]   DIFFERENTIATION OF ELEMENTARY FORMS   57

Substituting these values in the expression for $\dfrac{du}{ds}$,

$$\frac{du}{ds} = \frac{5s(3s^2+2)}{\sqrt{1+5s^2}} + 6s\sqrt{1+5s^2} = \frac{45s^3 + 16s}{\sqrt{1+5s^2}}.$$

3. $y = \dfrac{\sqrt{1+x^2} + \sqrt{1-x^2}}{\sqrt{1+x^2} - \sqrt{1-x^2}};$ find $\dfrac{dy}{dx}$.

First, as a quotient, by (5),
$$\frac{dy}{dx} = \frac{(\sqrt{1+x^2} - \sqrt{1-x^2})\dfrac{d}{dx}(\sqrt{1+x^2} + \sqrt{1-x^2})}{(\sqrt{1+x^2} - \sqrt{1-x^2})^2}$$

$$\frac{-(\sqrt{1+x^2} + \sqrt{1-x^2})\dfrac{d}{dx}(\sqrt{1+x^2} - \sqrt{1-x^2})}{(\sqrt{1+x^2} - \sqrt{1-x^2})^2}$$

$\dfrac{d}{dx}(\sqrt{1+x^2} + \sqrt{1-x^2}) = \dfrac{d}{dx}\sqrt{1+x^2} + \dfrac{d}{dx}\sqrt{1-x^2},$ by (2)

$\dfrac{d}{dx}\sqrt{1+x^2} = \dfrac{d}{dx}(1+x^2)^{\frac{1}{2}} = \dfrac{1}{2}(1+x^2)^{-\frac{1}{2}}\dfrac{d}{dx}(1+x^2).$ by (6)

$\dfrac{d}{dx}(1+x^2) = 2x.$ by (2) and (6)

Similarly for the other terms. Combining the results,

$$\frac{dy}{dx} = \frac{-2}{x^3}\left(1 + \frac{1}{\sqrt{1-x^4}}\right).$$

Ex. 3 may also be worked by first rationalizing denominator.

EXERCISES

Find the x-derivatives of the functions in 1–10.

1. $y = c\sqrt{x}$.

2. $y = \dfrac{x}{\sqrt{a^2 - x^2}}.$

3. $y = \dfrac{1}{(a+x)^m} \cdot \dfrac{1}{(b+x)^n}.$

4. $y = \dfrac{\sqrt{a+x}}{\sqrt{a}+\sqrt{x}}.$

5. $y = \sqrt{\dfrac{1+x}{1-x}}.$

6. $y = \left\{\dfrac{x}{1+\sqrt{1-x^2}}\right\}^n.$

7. $y = (2\,a^{\frac{1}{2}} + x^{\frac{1}{2}})\sqrt{a^{\frac{1}{2}} + x^{\frac{1}{2}}}.$

8. $y = (x-a)(x-b)(x-c)^2.$

9. $y = \sqrt{\dfrac{1-x^2}{(1+x^2)^3}}$. 10. $y = \dfrac{x^n+1}{x^n-1}$.

11. Given, $(a+x)^5 = a^5 + 5\,a^4x + 10\,a^3x^2 + 10\,a^2x^3 + 5\,ax^4 + x^5$; find $(a+x)^4$ by differentiation.

12. Show that the slope of the tangent to the curve $y = x^3$ is never negative. Show where the slope increases or decreases.

13. Given $b^2x^2 + a^2y^2 = a^2b^2$, find $\dfrac{dy}{dx}$: (1) by differentiating as to x; (2) by differentiating as to y; (3) by solving for y and differentiating as to x.

14. Show that (1) in Ex. 13 is a special case of (3).

29. Elementary transcendental functions. Functions that involve operations other than addition, subtraction, multiplication, involution (with integer exponent), and evolution (with integer index) are *transcendental* functions [Art. 4].

The most elementary transcendental functions are:

Simple exponential functions, consisting of a constant number raised to a power whose exponent is variable, as 4^x, a^{x^2};

general exponential functions, involving a variable raised to a power whose exponent is variable, as $x^{\sin x}$;

the *logarithmic* * functions, as $\log_a x$, $\log_b u$;

the *incommensurable powers* of a variable, as $x^{\sqrt{2}}$, u^π;

the *trigonometric* functions, as $\sin u$, $\cos u$;

the *inverse trigonometric* functions, as $\sin^{-1} u$, $\tan^{-1} x$.

There are still other transcendental functions, but they will not be considered in this book.

The next four articles treat of the logarithmic, the two exponential functions, and the incommensurable power.

---

* The more general logarithmic function $\log_v u$ is not classified separately, as it can be reduced to the quotient $\dfrac{\log_a u}{\log_a v}$.

30. Differentiation of $\log_a x$ and $\log_a u$.

Let $\qquad y = \log_a x,$

then $\qquad y + \Delta y = \log_a(x + \Delta x)$

$$\frac{\Delta y}{\Delta x} = \frac{\log_a(x + \Delta x) - \log_a x}{\Delta x},$$

$$= \frac{1}{\Delta x}\log_a\left(\frac{x + \Delta x}{x}\right).$$

For convenience writing h for Δx, and re-arranging,

$$\frac{\Delta y}{\Delta x} = \frac{1}{x} \cdot \frac{x}{h}\log_a\left(1 + \frac{h}{x}\right)$$

$$= \frac{1}{x}\log_a\left(1 + \frac{h}{x}\right)^{\frac{x}{h}},$$

$$\frac{dy}{dx} = \frac{1}{x}\lim_{h \doteq 0}\left[\log_a\left(1 + \frac{h}{x}\right)^{\frac{x}{h}}\right].$$

To evaluate the expression $\left(1 + \frac{h}{x}\right)^{\frac{x}{h}}$ when $h \doteq 0$, expand it by the binomial theorem, supposing $\frac{x}{h}$ to be a large positive integer m.

The expansion may be written

$$\left(1 + \frac{1}{m}\right)^m = 1 + m \cdot \frac{1}{m} + \frac{m(m-1)}{1 \cdot 2} \cdot \frac{1}{m^2} + \frac{m(m-1)(m-2)}{1 \cdot 2 \cdot 3} \cdot \frac{1}{m^3} + \cdots,$$

which can be put in the form

$$\left(1 + \frac{1}{m}\right)^m = 1 + 1 + \frac{1}{1}\frac{\left(1 - \frac{1}{m}\right)}{2} + \frac{1}{1}\frac{\left(1 - \frac{1}{m}\right)\left(1 - \frac{2}{m}\right)}{2\quad 3} + \cdots.$$

Now as m becomes very large, the terms $\frac{1}{m}, \frac{2}{m}, \cdots$ become very small, and when $m \doteq \infty$ the series becomes

$$\lim_{m \doteq \infty}\left(1 + \frac{1}{m}\right)^m = 1 + 1 + \frac{1}{2!} + \frac{1}{3!} + \frac{1}{4!} + \cdots.$$

The numerical value of the sum of this series can be readily calculated to any desired approximation. This sum is an important constant, which is denoted by the letter e, and is equal to 2.7182814 ..., thus

$$\lim_{m \doteq \infty} \left(1 + \frac{1}{m}\right)^m = e = 2.7182814 \ldots.\text{*}$$

The number e is known as the *natural* or Naperian *base;* and logarithms to this base are called natural or Naperian logarithms. Natural logarithms will be written without a subscript, as $\log x$; in other bases a subscript, as in $\log_a x$, will generally be used to designate the base; but the *common* logarithm, $\log_{10} x$, is often written Log x. The logarithm of e to any base a is called the *modulus* of the system whose base is a.

If the value, $\lim_{h \doteq 0}\left(1 + \frac{h}{x}\right)^{\frac{x}{h}} = e$, be substituted in the expression for $\frac{dy}{dx}$, there results [Th. 6, p. 31; Ex. p. 29.

$$\frac{dy}{dx} = \frac{1}{x} \cdot \log_a e.$$

More generally, by Art. 21,

$$\frac{d}{dx}\log_a u = \frac{1}{u} \cdot \log_a e \cdot \frac{du}{dx}. \tag{7}$$

In the particular case in which $a = e$,

$$\frac{d}{dx}\log u = \frac{1}{u}\frac{du}{dx}. \tag{8}$$

The derivative of the logarithm of a function is the product of the derivative of the function and the modulus of the system of logarithms, divided by the function.

---

* This method of obtaining e is rather too brief to be rigorous; it assumes that $\frac{x}{\Delta x}$ is a positive integer, but that is equivalent to restricting Δx to approach zero in a particular way. A rigorous and general proof will be found in the appendix.

31. Differentiation of the simple exponential function.

Let $y = a^u$;

then $\log y = u \log a.$

Differentiating both members of this identity as to x,

$$\frac{1}{y}\frac{dy}{dx} = \log a \cdot \frac{du}{dx}, \text{ by form (8)},$$

$$\frac{dy}{dx} = \log a \cdot y \cdot \frac{du}{dx};$$

therefore $\quad \dfrac{d}{dx} a^u = \log a \cdot a^u \cdot \dfrac{du}{dx},\quad(9)$

and $\quad \dfrac{d}{dx} e^u = e^u \cdot \dfrac{du}{dx}.\quad(10)$

The derivative of an exponential function with a constant base is equal to the product of the function, the natural logarithm of the base, and the derivative of the exponent.

32. Differentiation of the general exponential function.

Let $y = u^v$,

in which u, v are both functions of x.

Take the logarithm of both sides, and differentiate; then

$$\log y = v \log u,$$

$$\frac{1}{y}\frac{dy}{dx} = \frac{dv}{dx}\log u + \frac{v}{u}\frac{du}{dx}, \text{ by forms}(3),(8),$$

$$\frac{dy}{dx} = y\left[\log u \cdot \frac{dv}{dx} + \frac{v}{u}\frac{du}{dx}\right];$$

therefore $\quad \dfrac{d}{dx} u^v = u^v\left[\log u \cdot \dfrac{dv}{dx} + \dfrac{v}{u}\dfrac{du}{dx}\right].\quad(11)$

The derivative of an exponential function in which the base is also a variable is obtained by first differentiating, regarding the base as constant, and, again, regarding the exponent as constant, and adding the results.

33. Differentiation of an incommensurable power.

Let
$$y = u^n,$$
in which n is an incommensurable constant; then

$$\log y = n \log u,$$

$$\frac{1}{y}\frac{dy}{dx} = \frac{n}{u} \cdot \frac{du}{dx},$$

$$\frac{dy}{dx} = n \cdot \frac{y}{u} \cdot \frac{du}{dx},$$

$$\frac{d}{dx}u^n = nu^{n-1}\frac{du}{dx}.$$

This result is of the same form as (6), so that, in the theorem of Art. 27, the qualifying word "commensurable" can now be omitted.

Ex. $y = (4x^2 - 7)^{2+\sqrt{x^2-5}}$, find $\frac{dy}{dx}$.

$\log y = (2 + \sqrt{x^2 - 5}) \log (4x^2 - 7).$

$\frac{1}{y}\frac{dy}{dx} = \frac{x}{\sqrt{x^2-5}} \log(4x^2 - 7) + (2 + \sqrt{x^2-5})\frac{8x}{4x^2-7}.$ [Art. 32.

$\frac{dy}{dx} = (4x^2 - 7)^{2+\sqrt{x^2-5}} \cdot x \left[\frac{\log(4x^2-7)}{\sqrt{x^2-5}} + \frac{8(2+\sqrt{x^2-5})}{4x^2-7}\right].$

The following exercises relate to the differentiation of combinations of algebraic, logarithmic, and exponential functions.

EXERCISES

Find the x-derivatives of the following functions:

1. $y = \log(4x^2 - 7x + 2).$
2. $y = e^{4x+5}.$
3. $y = e^{\frac{1}{1+x}}.$
4. $y = x^n \log x.$
5. $y = \sqrt{x} - \log(\sqrt{x} + 1).$
6. $y = \dfrac{e^x - e^{-x}}{e^x + e^{-x}}.$

32–35.] *DIFFERENTIATION OF ELEMENTARY FORMS* 63

7. $y = \dfrac{e^x}{1+e^x}$.

8. $y = e^x(1 - x^3)$.

9. $y = \log(\log x)$.

10. $y = e^{x^x}$.

11. $y = a^{x^2}$.

12. $y = \log_a(3x^2 - \sqrt{2+x})$.

13. $y = x^x$.

14. $y = \dfrac{(x-1)^{\frac{5}{2}}}{(x-2)^{\frac{3}{4}}(x-3)^{\frac{2}{3}}}$.

In 14, take the logarithm of both members before differentiating.

Articles 34–40 will treat of the differentiation of the Trigonometric Functions within the range of continuity.

34. Differentiation of sin u.

Let $\quad y = \sin u$,

then $\quad \dfrac{\Delta y}{\Delta x} = \dfrac{\sin(u + \Delta u) - \sin u}{\Delta u} \cdot \dfrac{\Delta u}{\Delta x}$

$= \dfrac{2 \cos\frac{1}{2}(2u + \Delta u) \sin\frac{1}{2}\Delta u}{\Delta u} \cdot \dfrac{\Delta u}{\Delta x}$

$= \cos(u + \tfrac{1}{2}\Delta u) \cdot \dfrac{\sin\frac{1}{2}\Delta u}{\frac{1}{2}\Delta u} \cdot \dfrac{\Delta u}{\Delta x};$

but, when $\Delta u \doteq 0$, $\cos(u + \tfrac{1}{2}\Delta u) \doteq \cos u$, by Art. 11, and $\dfrac{\sin\frac{1}{2}\Delta u}{\frac{1}{2}\Delta u} \doteq 1$, by Art. 13; hence, passing to the limit

$$\frac{d}{dx}\sin u = \cos u \frac{du}{dx}. \qquad (12)$$

The derivative of the sine of a function is equal to the product of the cosine of the function and the derivative of the function.

35. Differentiation of cos u.

Let $\quad y = \cos u = \sin\left(\dfrac{\pi}{2} - u\right)$,

then $\dfrac{dy}{dx} = \dfrac{d}{dx}\sin\left(\dfrac{\pi}{2} - u\right) = \cos\left(\dfrac{\pi}{2} - u\right)\dfrac{d}{du}\left(\dfrac{\pi}{2} - u\right)\dfrac{du}{dx}$,

$$\frac{d}{dx}\cos u = -\sin u \frac{du}{dx}. \qquad (13)$$

The derivative of the cosine of a function is equal to minus the product of the sine of the function and the derivative of the function.

36. Differentiation of tan u.

Let
$$y = \tan u = \frac{\sin u}{\cos u},$$

then
$$\frac{dy}{dx} = \frac{\cos u \cdot \dfrac{d}{dx}\sin u - \sin u \cdot \dfrac{d}{dx}\cos u}{\cos^2 u} \qquad \text{by (5)}$$

$$= \frac{\cos^2 u \cdot \dfrac{du}{dx} + \sin^2 u \cdot \dfrac{du}{dx}}{\cos^2 u} = \frac{\dfrac{du}{dx}}{\cos^2 u}, \quad (12), (13)$$

that is,
$$\frac{d}{dx}\tan u = \sec^2 u \frac{du}{dx}. \tag{14}$$

The derivative of the tangent of a function is equal to the product of the square of the secant of the function and the derivative of the function.

37. Differentiation of cot u.

Let
$$y = \cot u = \frac{1}{\tan u},$$

then
$$\frac{dy}{dx} = \frac{-1}{\tan^2 u} \cdot \frac{d}{dx}\tan u = -\frac{\sec^2 u}{\tan^2 u}\frac{du}{dx}, \quad (5), (14)$$

$$\frac{d}{dx}\cot u = -\csc^2 u \frac{du}{dx}. \tag{15}$$

The derivative of the cotangent of a function is equal to minus the product of the square of the cosecant of the function and the derivative of the function.

38. Differentiation of sec u.

Let
$$y = \sec u = \frac{1}{\cos u},$$

then
$$\frac{dy}{dx} = \frac{-1}{\cos^2 u} \cdot \frac{d}{dx}\cos u = -\frac{\sin u}{\cos^2 u}\frac{du}{dx},$$

$$\frac{d}{dx}\sec u = \tan u \sec u \frac{du}{dx}. \qquad (16)$$

The derivative of the secant of a function is equal to the product of the secant of the function, the tangent of the function, and the derivative of the function.

39. Differentiation of csc u.

Let
$$y = \csc u = \frac{1}{\sin u},$$

then
$$\frac{dy}{dx} = \frac{-1}{\sin^2 u} \cdot \frac{d}{dx}\sin u = -\frac{\csc u}{\sin^2 u}\frac{du}{dx}.$$

$$\frac{d}{dx}\csc u = -\csc u \cot u \frac{du}{dx}. \qquad (17)$$

The derivative of the cosecant of a function is equal to minus the product of the cosecant of the function, the cotangent of the function, and the derivative of the function.

40. Differentiation of vers u.

Let
$$y = \operatorname{vers} u = 1 - \cos u,$$

then
$$\frac{dy}{dx} = -\frac{d}{dx}\cos u,$$

$$\frac{d}{dx}\operatorname{vers} u = \sin u \frac{du}{dx}. \qquad (18)$$

The derivative of the versed-sine of a function is equal to the product of the sine of the function and the derivative of the function.

The following exercises relate to the differentiation of combinations of algebraic, logarithmic, exponential, and trigonometric functions.

EXERCISES

Find the x-derivatives of the following functions:

1. $\sin 5x^2$.
2. $\sin^2 7x$.
3. $\tfrac{1}{3}\tan^3 x - \tan x$.
4. $2\sin x \cos x$.
5. $\tan a^{\frac{1}{x}}$.
6. $\log \tan (\tfrac{1}{2}x + \tfrac{1}{4}\pi)$.
7. $\log \cot x$.
8. $\sin nx \sin^n x$.
9. $\tan x - x$.
10. $\sin(u+b)\cos(u-b)$.
11. $x^{\sin x}$.
12. $\sin(\sin u)$.

DIFFERENTIATION OF THE INVERSE TRIGONOMETRIC FUNCTIONS

41. Differentiation of $\sin^{-1} u$.

Let $$y = \sin^{-1} u;$$
then $$\sin y = u;$$
and, by differentiating both members of this identity,

$$\cos y \frac{dy}{dx} = \frac{du}{dx};$$

hence $$\frac{dy}{dx} = \frac{1}{\cos y}\frac{du}{dx} = \frac{1}{\pm\sqrt{1-\sin^2 y}}\frac{du}{dx};$$

i.e., $$\frac{d}{dx}\sin^{-1} u = \pm\frac{1}{\sqrt{1-u^2}}\frac{du}{dx}.$$

The ambiguity of sign accords with the fact that $\sin^{-1} u$ is a many-valued function of u, since, for any value of u between -1 and 1, there is a series of angles whose sine is u; and, when u receives an increase, some of these angles increase and some decrease; hence, for some of them, $\dfrac{d\sin^{-1} u}{du}$ is positive, and for some negative. It will be seen that, when $\sin^{-1} u$ lies in the first or fourth quarter, it increases with u, and, when in the second or third quarter, it decreases with u. Hence, if it be agreed that $\sin^{-1} u$ shall mean the angle between $-\frac{1}{2}\pi$ and $+\frac{1}{2}\pi$, whose sine is u, then

$$\frac{d}{du}\sin^{-1} u = +\frac{1}{\sqrt{1-u^2}}, \quad \frac{d}{dx}\sin^{-1} u = +\frac{1}{\sqrt{1-u^2}}\frac{du}{dx}. \quad (19)$$

Thus the ambiguity in the derivative is removed by specifying that $\sin^{-1} u$ is to mean the numerically smallest angle whose sine is u.

It is well to note the distinction between an ambiguous derivative and a non-unique derivative. In the present case, the ambiguity disappears when any particular branch of the many-valued function is specified, and thus each branch has a unique derivative.

The derivative of the anti-sine of a function is equal to the derivative of the function divided by the square root of unity minus the square of the function.

42. Differentiation of $\cos^{-1} u$.

It may be proved, by the method used in Art. 41, that

$$\frac{d}{dx}\cos^{-1} u = \mp \frac{1}{\sqrt{1-u^2}}\frac{du}{dx}.$$

To discriminate between the two values of this derivative, observe that, when $\cos^{-1} u$ lies in the first or second quarter, it decreases with u, and when in the third or fourth quarter, it increases with u. Hence, if it be agreed that $\cos^{-1} u$ shall mean the angle between 0 and π, whose cosine is u, then

$$\frac{d}{du}\cos^{-1} u = \frac{-1}{\sqrt{1-u^2}}, \quad \frac{d}{dx}\cos^{-1} u = \frac{-1}{\sqrt{1-u^2}}\frac{du}{dx}. \quad (20)$$

Here the ambiguity in the derivative is removed by specifying that $\cos^{-1} u$ is to mean the smallest positive angle whose cosine is u.

The derivative of the anti-cosine of a function is equal to minus the derivative of the function divided by the square root of unity minus the square of the function.

43. Differentiation of $\tan^{-1} u$.

Let $\qquad y = \tan^{-1} u$;

then $\qquad \tan y = u$,

$$\sec^2 y \frac{dy}{dx} = \frac{du}{dx},$$

$$\frac{dy}{dx} = \frac{1}{\sec^2 y}\frac{du}{dx} = \frac{1}{1 + \tan^2 y}\frac{du}{dx};$$

therefore $\qquad \dfrac{d}{dx}\tan^{-1} u = \dfrac{1}{1+u^2}\dfrac{du}{dx}.$ \hfill (21)

The absence of ambiguity accords with the fact that, on each of its branches corresponding to the same value of u, $\tan^{-1} u$ is an increasing function of u. Unless otherwise stated, $\tan^{-1} u$ is specified to mean the numerically smallest angle whose tangent is u.

The derivative of the anti-tangent of a function is equal to the derivative of the function divided by unity plus the square of the function.

44. Differentiation of $\cot^{-1} u$.

It may be proved, by the method used in Art. 43, that

$$\frac{d}{dx}\cot^{-1} u = \frac{-1}{1+u^2}\frac{du}{dx}. \qquad (22)$$

On each of the branches corresponding to the same value of u, $\cot^{-1} u$ is a decreasing function of u. Unless otherwise stated, $\cot^{-1} u$ is specified to mean the numerically smallest angle whose cotangent is u.

The derivative of the anti-cotangent of a function is equal to minus the derivative of the function divided by unity plus the square of the function.

45. Differentiation of $\sec^{-1} u$.

Let
$$y = \sec^{-1} u,$$
then
$$\sec y = u,$$
$$\sec y \tan y \frac{dy}{dx} = \frac{du}{dx},$$
$$\frac{dy}{dx} = \frac{1}{\sec y \tan y} \frac{du}{dx} = \frac{1}{\sec y \sqrt{\sec^2 y - 1}} \frac{du}{dx},$$
$$\frac{d}{dx} \sec^{-1} u = \frac{1}{u\sqrt{u^2-1}} \frac{du}{dx}. \qquad (23)$$

If it be agreed that $\sec^{-1} u$ shall stand for the numerically smallest angle whose secant is u,—that is to say, if when u is positive $\sec^{-1} u$ shall be taken between 0 and $\frac{1}{2}\pi$, and when u is negative $\sec^{-1} u$ shall be taken between $-\frac{1}{2}\pi$ and $-\pi$,—then it will be seen on comparing the directions of algebraic increase of u and $\sec^{-1} u$ that the positive sign should be given to the radical in (23).

The derivative of the anti-secant of a function is equal to the derivative of the function divided by the product of the function and the square root of the square of the function less unity.

46. Differentiation of $\csc^{-1} u$.

It may be proved, by the method of Art. 45, that
$$\frac{d}{dx} \csc^{-1} u = \frac{-1}{u\sqrt{u^2-1}} \frac{du}{dx}. \qquad (24)$$

Ex. Show that the algebraic sign is correct if it be agreed that $\csc^{-1} u$ shall mean the numerically smallest angle whose cosecant is u.

The derivative of the anti-cosecant of a function is equal to minus the derivative of the function divided by the product of the function and the square root of the square of the function less unity.

47. Differentiation of vers⁻¹ u.

Let $\qquad y = \text{vers}^{-1} u;$

then $\qquad \text{vers } y = u,$

$$\sin y \frac{dy}{dx} = \frac{du}{dx},$$

$$\frac{dy}{dx} = \frac{1}{\sin y} \frac{du}{dx} = \frac{1}{\sqrt{1-(1-\text{vers } y)^2}} \frac{du}{dx},$$

$$\frac{d}{dx} \text{vers}^{-1} u = \frac{1}{\sqrt{2u-u^2}} \frac{du}{dx}. \qquad (25)$$

Ex. Show that the sign of the radical is to be taken positive if vers⁻¹u be specified to mean the smallest positive angle whose versed-sine is u.

The derivative of the anti-versed-sine of a function is equal to the derivative of the function divided by the square root of twice the function minus the square of the function.

EXERCISES

Differentiate the following expressions:

1. $x \sin^{-1} x$.
2. $\tan x \tan^{-1} x$.
3. $\sin^{-1} \dfrac{x+1}{\sqrt{2}}$.
4. $\tan^{-1} \dfrac{2x}{1+x^2}$.
5. $\tan^{-1} e^x$.
6. $\cos^{-1}(\log x)$.
7. $\log(\cos^{-1} x)$.
8. $\sin^{-1} 2x^2$.
9. $\text{vers}^{-1} \dfrac{x}{a}$.
10. $\cot^{-1}(x^2 - 5)$.
11. $\sec^{-1} \dfrac{1}{\sqrt{1-x^2}}$.
12. $\csc^{-1} 3\sqrt{x}$.
13. $\sin \log x$.
14. $\log \sin x$.
15. $\sqrt{\sin x^2}$.
16. $e^{\cos \frac{1}{x}}$.
17. $e^{\tan^{-1} x}$.
18. $\sin(\cos x)$.

The results of this chapter are for convenience summarized on pages 71, 72; they will suffice to differentiate any combination of algebraic, logarithmic, exponential, trigonometric, and inverse trigonometric functions.

DIFFERENTIATION OF ELEMENTARY FORMS

$$\frac{d(cu)}{dx} = c\frac{du}{dx}. \tag{1}$$

$$\frac{d}{dx}(u+v+w) = \frac{du}{dx} + \frac{dv}{dx} + \frac{dw}{dx}. \tag{2}$$

$$\frac{d(uv)}{dx} = u\frac{dv}{dx} + v\frac{du}{dx}. \tag{3}$$

$$\frac{d}{dx}(uvw) = uv\frac{dw}{dx} + uw\frac{dv}{dx} + vw\frac{du}{dx}. \tag{4}$$

$$\frac{d}{dx}\frac{u}{v} = \frac{v\frac{du}{dx} - u\frac{dv}{dx}}{v^2}. \tag{5}$$

$$\frac{d}{dx}u^n = nu^{n-1}\frac{du}{dx}. \tag{6}$$

$$\frac{d}{dx}\log_a u = \frac{\log_a e}{u}\frac{du}{dx}. \tag{7}$$

$$\frac{d}{dx}\log_e u = \frac{1}{u}\frac{du}{dx}. \tag{8}$$

$$\frac{d}{dx}a^u = \log_e a \cdot a^u \cdot \frac{du}{dx}. \tag{9}$$

$$\frac{d}{dx}e^u = e^u \frac{du}{dx}. \tag{10}$$

$$\frac{d}{dx}u^v = u^v \cdot \log_e u \cdot \frac{dv}{dx} + vu^{v-1}\frac{du}{dx}. \tag{11}$$

$$\frac{d}{dx}\sin u = \cos u \frac{du}{dx}. \tag{12}$$

$$\frac{d}{dx}\cos u = -\sin u \frac{du}{dx}. \tag{13}$$

$$\frac{d}{dx}\tan u = \sec^2 u \frac{du}{dx}. \tag{14}$$

$$\frac{d}{dx}\cot u = -\csc^2 u \frac{du}{dx}. \tag{15}$$

$$\frac{d}{dx}\sec u = \sec u \tan u \frac{du}{dx}. \tag{16}$$

$$\frac{d}{dx}\csc u = -\csc u \cot u \frac{du}{dx}. \tag{17}$$

$$\frac{d}{dx}\text{vers } u = \sin u \frac{du}{dx}. \tag{18}$$

$$\frac{d}{dx}\sin^{-1} u = \frac{1}{\sqrt{1-u^2}}\frac{du}{dx}. \tag{19}$$

$$\frac{d}{dx}\cos^{-1} u = \frac{-1}{\sqrt{1-u^2}}\frac{du}{dx}. \tag{20}$$

$$\frac{d}{dx}\tan^{-1} u = \frac{1}{1+u^2}\frac{du}{dx}. \tag{21}$$

$$\frac{d}{dx}\cot^{-1} u = \frac{-1}{1+u^2}\frac{du}{dx}. \tag{22}$$

$$\frac{d}{dx}\sec^{-1} u = \frac{1}{u\sqrt{u^2-1}}\frac{du}{dx}. \tag{23}$$

$$\frac{d}{dx}\csc^{-1} u = \frac{-1}{u\sqrt{u^2-1}}\frac{du}{dx}. \tag{24}$$

$$\frac{d}{dx}\text{vers}^{-1} u = \frac{1}{\sqrt{2u-u^2}}\frac{du}{dx}. \tag{25}$$

MISCELLANEOUS EXERCISES

In Ex. 1–10 find $\frac{dy}{dx}$:

1. $y = \log(e^x + e^{-x})$.
2. $y = \left(\frac{x}{n}\right)^{nx}$.
3. $y = \log \cot x$.
4. $y = (x-3)e^{2x} + 4xe^x + x + 3$.
5. $y = \dfrac{x}{e^x - 1}$.
6. $y = e^{-x^2}\cos x$.
7. $y = x^{\sin^{-1} 2x}$.
8. $y = \sec^{-1}\dfrac{1}{2x^2 - 1}$.
9. $y = \sin(2u - 7)$; $u = \log x^2$.
10. $y = e^u$; $u = \log x$.

11. $y = \log s^2 + e^s$; $s = \sec t$; find $\dfrac{dy}{dt}$.

12. $y = \dfrac{x^3}{a^2 + x^2}$. For what values of x is y an increasing function?

13. Prove that $\tan^{-1}\left(\dfrac{\sqrt{1+x^2}-1}{x}\right)$ always increases with x.

14. Show that the derivative of $\tan^{-1}\sqrt{\dfrac{1-\cos x}{1+\cos x}}$ is not a function of x.

15. Find at what points of the ellipse $\dfrac{x^2}{a^2} + \dfrac{y^2}{b^2} = 1$, the tangent cuts off equal intercepts on the axes.

16. Find $\dfrac{dy}{dx}$ from the expression $x^2 y^3 - y^2 x^5 + 6x^2 - 5y + 3 = 0$.

CHAPTER III

SUCCESSIVE DIFFERENTIATION

48. Definition of nth derivative. When a given function $y = \phi(x)$ is differentiated with regard to x by the rules of Chapter I, then the result

$$\frac{dy}{dx} = \phi'(x)$$

defines a new function which may itself be differentiated by the same rules. Thus,

$$\frac{d}{dx}\left(\frac{dy}{dx}\right) = \frac{d}{dx}\phi'(x).$$

The left-hand member is usually abbreviated to $\frac{d^2y}{dx^2}$, and the right-hand member to $\phi''(x)$; thus,

$$\frac{d^2y}{dx^2} = \phi''(x).$$

Differentiating again and using a similar notation,

$$\frac{d}{dx}\left(\frac{d^2y}{dx^2}\right) \equiv \frac{d^3y}{dx^3} \equiv \phi'''(x),$$

and so on for any number of differentiations. Thus the symbol $\frac{d^2y}{dx^2}$ expresses that y is to be differentiated with regard to x, and that the resulting derivative is to be differentiated again; or, in other words, that the operation $\frac{d}{dx}$ is to be performed upon y twice in succession. Similarly,

$\frac{d^3y}{dx^3}$ indicates the performance of the operation $\frac{d}{dx}$ three times upon y, and so on. Thus the symbol $\frac{d^ny}{dx^n}$ is equivalent to $\left(\frac{d}{dx}\right)^n y$. It is called the nth derivative of y with regard to x.

Ex. If $\quad\quad\quad y = x^4 + \sin 2x,$

$$\frac{dy}{dx} = 4x^3 + 2\cos 2x,$$

$$\frac{d^2y}{dx^2} = 12x^2 - 4\sin 2x,$$

$$\frac{d^3y}{dx^3} = 24x - 8\cos 2x,$$

$$\frac{d^4y}{dx^4} = 24 + 16\sin 2x.$$

49. Expression for the nth derivative in certain cases. For certain functions, a general expression for the nth derivative can be readily obtained in terms of n.

Ex. 1. If $\quad y = e^x$; $\frac{dy}{dx} = e^x$; $\frac{d^2y}{dx^2} = e^x$; $\ldots \frac{d^ny}{dx^n} = e^x$;

where n is any positive integer. If $y = e^{ax}$, $\frac{d^ny}{dx^n} = a^n e^{ax}$.

Ex. 2. If $\quad\quad\quad y = \sin x,$

$$\frac{dy}{dx} = \cos x = \sin\left(x + \frac{\pi}{2}\right),$$

$$\frac{d^2y}{dx^2} = \cos\left(x + \frac{\pi}{2}\right) = \sin\left(x + \frac{2\pi}{2}\right),$$

. . . .

$$\frac{d^ny}{dx^n} = \sin\left(x + \frac{n\pi}{2}\right).$$

If $\quad y = \sin ax$, $\frac{d^ny}{dx^n} = a^n \sin\left(ax + n\frac{\pi}{2}\right).$

50. Leibnitz's * theorem concerning the nth derivative of a product.

Let $y = uv$, where u, v are functions of x; then

$$\frac{dy}{dx} = u\frac{dv}{dx} + v\frac{du}{dx} = uv_1 + u_1v, \text{ where } \frac{du}{dx}, \frac{dv}{dx}$$

are replaced by u_1, v_1 for convenience;

again, $\quad \dfrac{d^2y}{dx^2} = uv_2 + 2u_1v_1 + u_2v,$

$$\frac{d^3y}{dx^3} = uv_3 + 3u_1v_2 + 3u_2v_1 + u_3v. \quad \left[u_r \equiv \frac{d^r u}{dx_r}\right]$$

These subscripts and coefficients thus far follow the same law as the exponents and coefficients of the binomial series. To test whether this law is true universally, assume its truth for some particular value of n,

$$\frac{d^n y}{dx^n} = uv_n + nu_1v_{n-1} + \cdots \frac{n!}{(n-r)!\,r!} \cdot u_r v_{n-r}$$

$$+ \frac{n!}{(n-r-1)!\,(r+1)!} \cdot u_{r+1}v_{n-r-1} + \cdots \quad (1)$$

and compare the result of differentiating once more with the result of changing n into $n+1$ in (1); if these two results are the same, it proves that if the law be true for any one value of n, it will also be true for the next higher, and so on, universally.

---

* Gottfried Wilhelm Leibnitz (1646-1716), the founder of the nomenclature and one of the chief founders of the philosophy of the differential calculus. By a remarkable coincidence Sir Isaac Newton (1642 O.S.-1727) simultaneously developed the same science, but his methods and notation are somewhat different. For the history of this remarkable discovery, see Cantor: Geschichte der Mathematik, Vol. 3, p. 150 ff.

By differentiating (1),

$$\frac{d^{n+1}y}{dx^{n+1}} = uv_{n+1} + u_1v_n + nu_1v_n + \cdots$$

$$+ \frac{n!}{(n-r)!\, r!} \{u_r v_{n-r+1} + u_{r+1} v_{n-r}\}$$

$$+ \frac{n!}{(n-r-1)!\, (r+1)!} \{u_{r+1} v_{n-r} + u_{r+2} v_{n-r-1}\} + $$

$$= uv_{n+1} + (n+1) u_1 v_n + \cdots$$

$$+ u_{r+1} v_{n-r} \left[\frac{n!}{(n-r)!\, r!} + \frac{n!}{(n-r-1)!\, (r+1)!} \right] + \cdots,$$

i.e.,
$$\frac{d^{n+1}y}{dx^{n+1}} = uv_{n+1} + (n+1) u_1 v_n + \cdots$$

$$+ \frac{(n+1)!}{(n-r)!\, (r+1)!} u_{r+1} v_{n-r} + \cdots; \qquad (2)$$

and by changing n into $n+1$ in (1), the result is seen to be the same as (2).

Now the expression for $\dfrac{d^2y}{dx^2}$ shows that the law is true for $n = 2$; hence it is universally true, and thus formula (1) is established.

It is of special value when the general expression for u_n and for v_n can be readily obtained.

Ex. Given $y = x^2 e^{ax}$; find $\dfrac{d^n y}{dx^n}$.

Let $u = x^2$, $v = e^{ax}$,

then $\dfrac{du}{dx} = u_1 = 2x$, $v_1 = ae^{ax}$,

$u_2 = 2$, $v_2 = a^2 e^{ax}$,

$u_3 = 0$, $v_3 = a^3 e^{ax}$,

. . . .

$u_n = 0$, $(n > 2)$, $v_n = a^n e^{ax}$.

Substituting these values in (1),

$$\frac{d^n}{dx^n}(x^2 e^{ax}) = x^2 a^n e^{ax} + 2 a^{n-1} x e^{ax} + n(n-1) a^{n-2} e^{ax}.$$

51. Successive x-derivatives of y when neither variable is independent. Hitherto the differentiations have always been performed with regard to the independent variable. It is, however, sometimes necessary to differentiate a function with regard to a variable which itself depends on some other variable. Let y and x be each directly given as functions of an independent variable t, and suppose it is required to express $\dfrac{dy}{dx}$ in terms of t.

From Arts. 21, 22,

$$\frac{dy}{dx} = \frac{dy}{dt} \cdot \frac{dt}{dx} = \frac{\frac{dy}{dt}}{\frac{dx}{dt}}, \qquad (1)$$

but $\dfrac{dy}{dt}, \dfrac{dx}{dt}$ can each be directly expressed in terms of t from the given expressions for x, y, hence $\dfrac{dy}{dx}$ is known in terms of t.

Thus if $y = \phi(t)$, and $x = f(t)$, then

$$\frac{dy}{dt} = \phi'(t), \ \frac{dx}{dt} = f'(t),$$

and $$\frac{dy}{dx} = \frac{\phi'(t)}{f'(t)}.$$

To obtain an expression similar to (1), for $\dfrac{d^2y}{dx^2}$, it may be put in the form

$$\frac{d^2y}{dx^2} = \frac{d}{dx}\left(\frac{dy}{dx}\right) = \frac{d}{dt}\left(\frac{dy}{dx}\right) \cdot \frac{dt}{dx} = \frac{\frac{d}{dt}\left(\frac{dy}{dx}\right)}{\frac{dx}{dt}}, \qquad (2)$$

but, by differentiating (1) with regard to t,

$$\frac{d}{dt}\left(\frac{dy}{dx}\right) = \frac{\frac{dx}{dt} \cdot \frac{d^2y}{dt^2} - \frac{dy}{dt} \cdot \frac{d^2x}{dt^2}}{\left(\frac{dx}{dt}\right)^2};$$

hence, by (2), $\quad \dfrac{d^2y}{dx^2} = \dfrac{\dfrac{dx}{dt}\cdot\dfrac{d^2y}{dt^2} - \dfrac{dy}{dt}\cdot\dfrac{d^2x}{dt^2}}{\left(\dfrac{dx}{dt}\right)^3}.$ (3)

Thus, if $y = \phi(t)$ and $x = f(t)$,

$$\dfrac{d^2y}{dx^2} = \dfrac{f'(t)\phi''(t) - \phi'(t)f''(t)}{[f'(t)]^3}.$$

Expressions similar to (1) and (3), but more complicated, can be obtained for the higher derivatives.

Next let y be given directly in terms of x, and x in terms of t; then $\dfrac{dy}{dx}$ can be first expressed in terms of x, and the result in terms of t by elimination.

Thus if $\quad y = \phi(x),\ x = f(t),$

then $\quad \dfrac{dy}{dx} = \phi'(x) = \phi'[f(t)],$

$\dfrac{d^2y}{dx^2} = \phi''(x) = \phi''[f(t)],$

$\dfrac{d^3y}{dx^3} = \phi'''(x) = \phi'''[f(t)].$

EXERCISES

1. $y = x^4 - 4x^3 + 6x^2 - 4x + 1$; find $\dfrac{d^2y}{dx^2}.$

2. $y = (x-3)e^{2x} + 4xe^x + x$; find $\dfrac{d^2y}{dx^2}.$

3. $y = x^6$; find $\dfrac{d^6y}{dx^6}.$

4. $y = x^3 \log x$; find $\dfrac{d^4y}{dx^4}.$

5. $y = \log(e^x + e^{-x})$; find $\dfrac{d^2y}{dx^2}.$

6. $y = \dfrac{x^3}{6} \log x - \dfrac{5}{x}$; find $\dfrac{d^2y}{dx^2}$.

7. $y = \tan^2 x + 8 \log \cos x + 3 x^2$; find $\dfrac{d^2y}{dx^2}$.

8. $y = e^{ax} \sin bx$; prove $\dfrac{d^2y}{dx^2} - 2a\dfrac{dy}{dx} + (a^2 + b^2) y = 0$.

9. $y = a \cos (\log x) + b \sin (\log x)$; prove $x^2 \dfrac{d^2y}{dx^2} + x \dfrac{dy}{dx} + y = 0$.

10. $y = \tan x + \sec x$; find $\dfrac{d^2y}{dx^2}$.

11. $y = (x^2 + a^2) \tan^{-1} \dfrac{x}{a}$; find $\dfrac{d^3y}{dx^3}$.

12. $y = e^{-x} \cos x$; prove $\dfrac{d^4y}{dx^4} + 4 y = 0$.

13. $y = \sqrt{\dfrac{x^3}{x-a}}$; find $\dfrac{d^2y}{dx^2}$.

14. $y = x^{n-1} \log x$, (n a positive integer); find $\dfrac{d^n y}{dx^n}$.

15. $y = \log \dfrac{1-x}{1+x}$; find $\dfrac{d^n y}{dx^n}$.

16. $y = \dfrac{x^3}{1-x}$; find $\dfrac{d^4y}{dx^4}$.

17. $y = \sec 2 x$; find $\dfrac{d^2y}{dx^2}$.

18. $y = 1 + xe^y$; find $\dfrac{d^2y}{dx^2}$ in terms of y.

19. $y = \tan (x + y)$; find $\dfrac{d^3y}{dx^3}$ in terms of y.

20. $y^2 + y = x^2$; find $\dfrac{d^3y}{dx^3}$.

21. $e^x + x = e^y + y$; find $\dfrac{d^2y}{dx^2}$.

22. $e^y + xy - e = 0$; find $\dfrac{d^2y}{dx^2}$.

23. $y^3 + x^3 - 3 axy = 0$; find $\dfrac{d^2y}{dx^2}$.

24. $y = \dfrac{1}{4x^2 - 1}$; find $\dfrac{d^n y}{dx^n}$.

25. $y = \tfrac{1}{2} \log \dfrac{x+a}{x-a}$; find $\dfrac{d^n y}{dx^n}$.

26. $y = e^x \cdot x$; find $\dfrac{d^n y}{dx^n}$.

27. $y = x^2 e^{ax}$; find $\dfrac{d^n y}{dx^n}$.

28. $y = x^2 \log x$; find $\dfrac{d^n y}{dx^n}$.

29. $y = \dfrac{x^n}{1+x}$; find $\dfrac{d^n y}{dx^n}$.

30. $y = e^x \sin x$; find $\dfrac{d^6 y}{dx^6}$.

31. Show that the members of equation (3), p. 78, become identical when t is replaced by x.

32. Replacing t by y, show that

$$\frac{d^2 y}{dx^2} = -\frac{\dfrac{d^2 x}{dy^2}}{\left(\dfrac{dx}{dy}\right)^3}.$$

Also derive this relation independently.

33. Verify this relation when $y = \sin x$.

34. Find when the slope of the curve $y = \tan x$ increases with x; and when it decreases as x increases.

35. Show that the slope of the curve $y = f(x)$ changes from increasing to decreasing when $f''(x)$ changes its sign. Apply to the curves $y = \sin x$, $y = \sin^2 x$.

CHAPTER IV

EXPANSION OF FUNCTIONS

52. It is sometimes necessary to expand a given function in a series of powers of one of its variables. For instance, in order to compute and tabulate the successive numerical values of $\sin x$ for different values of x, it is convenient to have $\sin x$ developed in a series of powers of x with coefficients independent of x.

Simple cases of such development have been seen in algebra; for example, by the binomial theorem,

$$(a+x)^n = a^n + na^{n-1}x + \frac{n(n-1)}{1\cdot 2}a^{n-2}x^2 + \cdots ; \qquad (1)$$

and again, by ordinary division,

$$\frac{1}{1-x} = 1 + x + x^2 + x^3 + \cdots . \qquad (2)$$

It is to be observed, however, that the series is a proper representative of the function only for values of x within a certain interval; for instance, it is shown in works on algebra that when n is not a positive integer, the identity in (1) holds only for values of x between $-a$ and $+a$, and that the identity in (2) holds only for values of x between -1 and $+1$. In each case, if a finite value outside of the stated limits be given to x, the sum of an infinite number of terms of the series will be infinite, while the function itself will be finite. In both of these examples the stated *interval of equivalence* of the series and its generating function is the same as the *interval of convergence* of the series itself. The general theory of the convergence and divergence of series, so far as necessary for the present purpose, is briefly outlined in the next two articles.

53. Convergence and divergence of series.* An infinite series is said to be convergent or divergent according as the sum of the first n terms of the series does or does not approach a finite limit when n is increased without limit.

For example, the sum of the first n terms of the geometric series
$$a + ax + ax^2 + ax^3 + \cdots$$
is
$$s_n = \frac{a(1-x^n)}{1-x}.$$

First let x be numerically less than unity; then when n is taken sufficiently large, the term $x^n \doteq 0$; hence
$$s_n \doteq \frac{a}{1-x}, \text{ when } n \doteq \infty.$$

Next let x be numerically greater than unity; then when $n \doteq \infty$, $x^n \doteq \infty$; hence, in this case
$$s_n \doteq \infty, \text{ when } n \doteq \infty.$$

Thus the given series is convergent or divergent according as x is numerically less or greater than unity. The condition of convergence may then be written
$$-1 < x < 1,$$
and the interval of convergence is between -1 and $+1$.

---

* For an elementary, yet comprehensive and rigorous, treatment of this subject see Professor Osgood's "Introduction to Infinite Series" (Harvard University Press, 1897).

Similarly the geometric series
$$1 - 3x + 9x^2 - 27x^3 + \ldots,$$
whose common ratio is $-3x$, is convergent or divergent according as $3x$ is numerically less or greater than unity.

The condition of convergence is $-1 < 3x < 1$, and the interval of convergence is between $-\frac{1}{3}$ and $+\frac{1}{3}$.

The definition just given and illustrated is sometimes more briefly stated as follows: An infinite series is said to be convergent or divergent according as the sum of the series to infinity is finite or infinite.

It is, however, to be carefully borne in mind that the phrase "the sum of the series to infinity" is only an abbreviation for the more precise phrase "the limit approached by the sum of the first n terms when n is made larger and larger without limit."

54. General test for interval of convergence. The following summary of algebraic principles leads up to a test that is sufficient to find the interval of convergence of a series of the most usual kind, that is, a series consisting of positive integral powers of x, in which the coefficient of x^n is a known function of n.

1. If s_n is a variable that continually increases with n, but for all values of n remains less than some fixed number k, then s_n approaches some definite limit not greater than k. [An exercise on the definition of a limit.]

2. If one series of positive terms is known to be convergent, and if the terms of another series be positive and less than the corresponding terms of the first series, then the latter series is convergent. [Use 1.]

3. If after a given term the terms of a series form a decreasing geometric progression, then (a) the successive

terms approach nearer and nearer to zero as a limit; and (b) the sum of all the terms approaches some fixed constant as a limit. [Use method of last article.]

4. If the terms of a series be positive, and if after a given term the ratio of each term to the preceding be less than a fixed proper fraction, the series is convergent. [Use 2 and 3.]

5. If there be a series A consisting of an infinite number of both positive and negative terms, and if another series B, obtained therefrom by making all the terms positive, is known to be convergent, then the series A is convergent.

For the positive terms of A must form a convergent series, otherwise the series B could not be convergent; similarly the negative terms of A must form a convergent series. Let the sums of these convergent series be u, $-v$. Let the first n terms of series A contain m positive terms and p negative terms; and let their three sums be respectively S_n, Σ_m, $-T_p$; then $S_n = \Sigma_m - T_p$. Now when $n \doteq \infty$, so does $m \doteq \infty$, and $p \doteq \infty$, hence

$$\lim_{n \doteq \infty} S_n = \lim_{m \doteq \infty} \Sigma_m - \lim_{p \doteq \infty} T_p, \ i.e., \ S = u - v;$$

therefore the series A is convergent.

DEFINITIONS. The *absolute value* of a real number x is its numerical value taken positively, and is written $|x|$. The equation $|x| = |a|$ indicates that the absolute value of x is equal to the absolute value of a. When, however, x and a are replaced by longer expressions, it is convenient to write the relation in the form $x|=|a$, in which the symbol $|=|$ is read "equals in absolute value." Similarly for the symbols $|<|, |>|$.

Any series of terms is said to be *absolutely* or *unconditionally* convergent when the series formed by their absolute values is convergent. When a series is convergent, but the series formed by making each term positive is not convergent, the first series is said to be *conditionally* convergent.*

E.g., the series $\frac{1}{1^2} - \frac{1}{2^2} + \frac{1}{3^2} - \cdots$ is absolutely convergent; but the series $1 - \frac{1}{2} + \frac{1}{3} - \cdots$ is conditionally convergent.

6. If there be any series of terms in which after some fixed term the ratio of each term to the preceding is numerically less than a fixed proper fraction; then,

(*a*) the successive terms of the series approach nearer and nearer to zero as a limit;

(*b*) the sum of all the terms approaches some fixed constant as a limit; and the series is absolutely convergent. [Use 3, 4, 5.]

Ex. 1. Find the interval of convergence of the series
$$1 + 2 \cdot 2x + 3 \cdot 4x^2 + 4 \cdot 8x^3 + 5 \cdot 16x^4 + \cdots.$$
Here the nth term is $n\, 2^{n-1} x^{n-1}$, and the $(n+1)$st term is $(n+1)\, 2^n x^n$,

hence
$$\frac{u_{n+1}}{u_n} = \frac{(n+1)\, 2^n x^n}{n\, 2^{n-1} x^{n-1}} = \frac{(n+1)\, 2x}{n},$$

therefore when $n \doteq \infty$, $\quad \dfrac{u_{n+1}}{u_n} \doteq 2x$.

It follows by (6) that the series is absolutely convergent when $-1 < 2x < 1$, and that the interval of convergence is between $-\frac{1}{2}$ and $+\frac{1}{2}$. The series is evidently not convergent when x has either of the extreme values.

Ex. 2. Find the interval of convergence of the series
$$\frac{x}{1 \cdot 3} - \frac{x^3}{x \cdot 3^3} + \frac{x^5}{5 \cdot 3^5} - \frac{x^7}{7 \cdot 3^7} + \cdots + \frac{(-1)^n\, x^{2n-1}}{(2n-1)\, 3^{2n-1}} + \cdots$$

---

* The appropriateness of this terminology is due to the fact that the terms of an absolutely convergent series can be rearranged in any way, without altering the limit of the sum of the series; and that this is not true of a conditionally convergent series. Thus the sum of the series $\frac{1}{1^2} - \frac{1}{2^2} + \frac{1}{3^2} - \cdots$ is independent of the order or grouping; but the sum of the series $1 - \frac{1}{2} + \frac{1}{3} - \frac{1}{4} + \cdots$ can be made equal to any number whatever by suitable re-arrangement. [For a simple proof see Osgood, pp. 43, 44.]

Here $\left|\dfrac{u_{n+1}}{u_n}\right| = \left|\dfrac{2n-1}{2n+1} \cdot \dfrac{3^{2n-1}}{3^{2n+1}} \cdot \dfrac{x^{2n+1}}{x^{2n-1}}\right| = \dfrac{2n-1}{2n+1} \cdot \dfrac{x^2}{3^2}$,

hence $\dfrac{u_{n+1}}{u_n} \doteq \dfrac{x^2}{3^2}$, when $n \doteq \infty$;

thus the series is absolutely convergent when $\dfrac{x^2}{3^2} < 1$, *i.e.*, when $-3 < x < 3$, and the interval of convergence is from -3 to $+3$. The extreme values of x, in the present case, render the series conditionally convergent.

Ex. 3. Show that the series $\dfrac{1}{1^2}\left(\dfrac{x}{3}\right) - \dfrac{1}{3^2}\left(\dfrac{x}{3}\right)^3 + \dfrac{1}{5^2}\left(\dfrac{x}{3}\right)^5 - \dfrac{1}{7^2}\left(\dfrac{x}{3}\right)^7 + \cdots$ has the same interval of convergence as the last; but that the extreme values of x render the series absolutely convergent.

55. Interval of equivalence. Remainder after n terms. The last article treated of the interval of convergence of a given series without reference to the question whether or not it was the development of any known function. On the other hand, the series that present themselves in this chapter are the developments of given functions, and the first question that arises is concerning the interval of equivalence of the function and its development.

When a series has such a generating function, the difference between the value of the function and the sum of the first n terms of its development is called the *remainder after n terms*.* Thus if $f(x)$ be the function, $S_n(x)$ the sum of the first n terms of the series, and $R_n(x)$ the remainder obtained by subtracting $S_n(x)$ from $f(x)$, then

$$f(x) = S_n(x) + R_n(x),$$

in which $S_n(x)$, $R_n(x)$ are functions of n as well as of x.

---

* In some discussions of convergence of series without any reference to a generating function, the phrase "remainder after n terms" is occasionally used in a sense different from that given above, which is the recognized usage in treating of the equivalence of a function and its development.

A sufficient condition for the convergence of the series is that $R_n(x)$ approach a finite limit when $n \doteq \infty$; for in that case $S_n(x), = f(x) - R_n(x), \doteq$ a finite number, when $n \doteq \infty$. Thus the interval of convergence extends over those values of x that make $\lim_{n \doteq \infty} R_n(x)$ equal to any number not infinite, and $f(x)$ itself not infinite.

On the other hand, the interval of equivalence of the series and its generating function extends only over those values of x that make $\lim_{n \doteq \infty} R_n(x) = 0$; for it is only in that case that $S_n(x) \doteq f(x)$, when $n \doteq \infty$.

Thus the interval of equivalence may possibly be narrower than the interval of convergence.

It will appear later, however, that in the case of all the ordinary functions, $\lim_{n \doteq \infty} R_n(x)$ will be zero for certain values of x, and infinite for all other values of x; and that thus the intervals of convergence and of equivalence are identical.

56. Maclaurin's expansion of a function in power-series.[*] It will now be shown that all the developments of functions in power-series which were studied in algebra and trigonometry are but special cases of one general formula of expansion.

It is proposed to find a formula for the expansion, in ascending positive integral powers of x, of any assigned function which, with its successive derivatives, is continuous in the vicinity of the value $x = 0$.

---

[*] Named after Colin Maclaurin (1698-1746), who published it in his "Treatise on Fluxions" (1742); but he distinctly says it was known by Stirling (1690-1772), who also published it in his "Methodus Differentialis" (1730), and by Taylor (see Art. 65).

The preliminary investigation will proceed on the hypothesis that the assigned function $f(x)$ has such a development, and that the latter can be treated as identical with the former for all values of x within a certain interval of equivalence that includes the value $x = 0$. From this hypothesis the coefficients of the different powers of x will be determined. It will then remain to test the validity of the result by finding the conditions that must be fulfilled, in order that the series so obtained may be a proper representation of the generating function.

Let the assumed identity be

$$f(x) \equiv A + Bx + Cx^2 + Dx^3 + Ex^4 + \cdots, \qquad (1)$$

in which $A, B, C, \cdots$ are undetermined coefficients independent of x.

Successive differentiation with regard to x supplies the following additional identities, on the hypothesis that the derivative of each series can be obtained by differentiating it term by term, and that it has some interval of equivalence with its corresponding function;

$$f'(x) = B + 2\,Cx + 3\,Dx^2 + 4\,Ex^3 + \cdots,$$
$$f''(x) = 2\,C + 3\cdot 2\,Dx + 4\cdot 3\,Ex^2 + \cdots,$$
$$f'''(x) = 3\cdot 2\,D + 4\cdot 3\cdot 2\,Ex + \cdots,$$
$$f^{\text{IV}}(x) = 4\cdot 3\cdot 2\,E + \cdots.$$

in which, by the hypothesis,* x may have any value within a certain interval including the value $x = 0$.

---

* The hypothesis here made with regard to each series would not be admissible in a process of demonstration. This preliminary investigation is for the purpose of discovering what the development is, if any exists. The validity of the development is fully tested in Arts. 60–65.

It may be of interest to refer to Professor Osgood's "Introduction to Infinite Series," pp. 54, 61, for a proof *that within its interval of convergence a power-series is a continuous and differentiable function of x, and that its*

The substitution of zero for x in each identity furnishes the following equations:

$$f(0) = A,\ f'(0) = B,\ f''(0) = 2\,C,\ f'''(0) = 3 \cdot 2\,D, \cdots;$$

hence $A = f(0),\ B = f'(0),\ C = \dfrac{f''(0)}{2!},\ D = \dfrac{f'''(0)}{3!}, \cdots.$

The unknown coefficients of (1) are thus expressed in terms of known indicated operations; and substitution in (1) gives the form of development sought;

$$f(x) = f(0) + f'(0)x$$
$$+ \frac{f''(0)}{2!}x^2 + \frac{f'''(0)}{3!}x^3 + \cdots + \frac{f^n(0)}{n!}x^n + \cdots. \quad (2)$$

Here the symbol $f^n(0)$ is used to indicate the operation of differentiating $f(x)$ with regard to x, n times in succession, and then substituting zero for x in the expression for the nth derivative.

The resulting constant, when divided by $n!$, is the required coefficient of the nth power of the variable in the assumed development of the function.

It remains to examine what are the conditions that must be fulfilled in order that the series so found may be a proper representative of the function. This question can only be fully answered when the expression for $R_n(x)$, the remainder after n terms, has been obtained. This expression will be derived after another series, which may be regarded as a generalization of (2), has been established.

There are, however, certain preliminary conditions that

---

true derivative can be obtained, within the same interval, by differentiating the series term by term.

This theorem is, however, not necessary to the demonstration of Maclaurin's or Taylor's theorem, as the series treated in Art. 60 consists of only a finite number of terms.

are easily seen to be necessary in order that the series may give the true value of the function.

First, the functions $f(x)$, $f'(x)$, $f''(x)$, $\cdots$ must all satisfy the condition of being continuous in the vicinity of $x = 0$; otherwise some of the coefficients $f(0)$, $f'(0)$, $f''(0)$, $\cdots$ would be infinite or indeterminate, and the series would have no definite sum for any value of x, showing that the given function $f(x)$ could have no development in the form prescribed.

Ex. Show that the functions $\log x$, $x^{\frac{3}{2}}$, $\dfrac{1}{e^{\frac{1}{x}}+1}$ cannot be developed in powers of x.

When this condition is satisfied, it is further necessary for the equivalence of the function and its development that the values of x be restricted to lie within a certain interval not wider than the interval of convergence of the series.

The method of computing the coefficients of the successive powers of x in the development of a given function, will be illustrated by a few examples.

Ex. 1. Expand $\sin x$ in powers of x, and find the interval of convergence of the series.

Here $f(x) = \sin x,$ $f(0) = 0,$
$f'(x) = \cos x,$ $f'(0) = 1,$
$f''(x) = -\sin x,$ $f''(0) = 0,$
$f'''(x) = -\cos x,$ $f'''(0) = -1,$
$f^{\text{IV}}(x) = \sin x,$ $f^{\text{IV}}(0) = 0,$
$f^{\text{V}}(x) = \cos x,$ $f^{\text{V}}(0) = 1,$
. . . .

Hence, by (2),

$$\sin x = 0 + 1 \cdot x + 0 \cdot x^2 - \frac{1}{3!}x^3 + 0 \cdot x^4 + \frac{1}{5!}x^5 \cdots,$$

thus the required development is

$$\sin x = x - \frac{1}{3!}x^3 + \frac{1}{5!}x^5 - \frac{1}{7!}x^7 + \cdots + \frac{(-1)^{n-1}}{(2n-1)!}x^{2n-1} + \cdots.$$

To find the interval of convergence of the series, use the method of Art. 54, then

$$\left|\frac{u_{n+1}}{u_n}\right| = \left|\frac{x^{2n+1}}{(2n+1)!} \cdot \frac{x^{2n-1}}{(2n-1)!}\right| = \frac{x^2}{(2n+1)2n},$$

and this ratio approaches the limit zero, when n becomes infinite, however large be the constant value assigned to x. This limit being less than unity, the series is convergent for any finite value of x, and hence the interval of convergence is from $-\infty$ to $+\infty$.

Assuming, for the present, that the value $x = .5$, for example, lies within the interval of equivalence of $\sin x$ and its development, the numerical value of the sine of half a radian may be computed as follows:

$$\sin(.5) = .5 - \frac{(.5)^3}{2\cdot 3} + \frac{(.5)^5}{2\cdot 3\cdot 4\cdot 5} - \frac{(.5)^7}{2\cdot 3\cdot 4\cdot 5\cdot 6\cdot 7} + \cdots,$$

$$\begin{aligned}
&= .5000000 \\
&- .0208333 \\
&+ .0002604 \\
&- .0000015 \\
&+ .0000000 \\
\hline
\sin(.5) &= .4794256 \cdots
\end{aligned}$$

Show that the ratio of u_5 to u_4 is $\frac{1}{272}$; and hence that the error in stopping at u_4 is numerically less than $u_4\left[\frac{1}{272} + \left(\frac{1}{272}\right)^2 + \cdots\right], < \frac{1}{271}u_4$.

Ex. 2. Show that the development of $\cos x$ is

$$\cos x = 1 - \frac{x^2}{2!} + \frac{x^4}{4!} - \frac{x^6}{6!} + \cdots + \frac{(-1)^{n-1}x^{2n-2}}{(2n-2)!} + \cdots,$$

and that the interval of convergence is from $-\infty$ to $+\infty$.

Ex. 3. Develop the exponential functions a^x, e^x.

Here

$f(x) = a^x,\ f'(x) = a^x \log a,\ f''(x) = a^x(\log a)^2, \cdots f^n(x) = a^x(\log a)^n,$

hence $\quad f(0) = 1,\ f'(0) = \log a,\ f''(0) = (\log a)^2, \cdots f^n(0) = (\log a)^n,$

and $\quad a^x = 1 + (\log a)x + \frac{(\log a)^2}{2!}x^2 + \cdots + \frac{(\log a)^n}{n!}x^n + \cdots.$

As a special case, putting $a = e$, the Naperian base,

then $\quad\quad\quad\quad \log a = \log e = 1,$

and $\quad\quad\quad e^x = 1 + x + \frac{x^2}{2!} + \frac{x^3}{3!} + \cdots + \frac{x^n}{n!} + \cdots.$

These series are convergent for every finite value of x.

Ex. 4. Find the development of $\tan x$.

Let $\quad f(x) = \tan x,$
then $\quad f'(x) = \sec^2 x,$
$\quad f''(x) = 2\sec^2 x \tan x,$
$\quad f'''(x) = 4\sec^2 x \tan^2 x + 2\sec^4 x,$
$\quad f^{\mathrm{IV}}(x) = 8\sec^2 x \tan^3 x + 16 \sec^4 x \tan x,$
$\quad f^{\mathrm{V}}(x) = 16 \sec^2 x \tan^4 x + 88 \sec^4 x \tan^2 x + 16 \sec^6 x,$
$\quad f^{\mathrm{VI}}(x) = 32 \sec^2 x \tan^5 x + 416 \sec^4 x \tan^3 x + 272 \sec^6 x \tan x,$
$f^{\mathrm{VII}}(x) = 64 \sec^2 x \tan^6 x + 1814 \sec^4 x \tan^4 x + 2880 \sec^6 x \tan^2 x + 272 \sec^8 x, \cdots$

Hence $\quad f(0) = 0, \; f'(0) = 1, \; f''(0) = 0, \; f'''(0) = 2, \; f^{\mathrm{IV}}(0) = 0,$
$\quad\quad f^{\mathrm{V}}(0) = 16, \; f^{\mathrm{VI}}(0) = 0, \; f^{\mathrm{VII}}(0) = 272, \cdots$

therefore $\quad \tan x = x + \dfrac{2}{3!} x^3 + \dfrac{16}{5!} x^5 + \dfrac{272}{7!} x^7 + \cdots$

$\quad\quad\quad\quad = x + \dfrac{1}{3} x^3 + \dfrac{2}{15} x^5 + \dfrac{17}{315} x^7 + \cdots.$

Here, as in many other cases, the law of succession of the coefficients is very complicated, and it is not possible to express the coefficient of x^n directly in terms of n. Thus the interval of convergence of the series cannot be obtained by simple methods.

Ex. 5. Develop $e^{\sin x}$ in powers of x.

Let $\quad f(x) = e^{\sin x},$
then $\quad f'(x) = e^{\sin x} \cos x,$
$\quad f''(x) = e^{\sin x}(\cos^2 x - \sin x),$
$\quad f'''(x) = e^{\sin x}(\cos^3 x - 3 \sin x \cos x - \cos x),$
$\quad f^{\mathrm{IV}}(x) = e^{\sin x}(\cos^4 x - 6 \sin x \cos^2 x - 4 \cos^2 x + 3 \sin^2 x + \sin x),$
$\quad f^{\mathrm{V}}(x) = e^{\sin x}$
$(\cos^5 x - 10 \sin x \cos^3 x - 10 \cos^3 x + 15 \sin^2 x \cos x + 7 \sin x \cos x + \cos x),$

hence
$\quad f(0) = 0, \; f'(0) = 1, \; f''(0) = 1, \; f'''(0) = 0, \; f^{\mathrm{IV}}(0) = -3, \; f^{\mathrm{V}}(0) = -8,$

therefore $\quad e^{\sin x} = x + \dfrac{x^2}{2!} - \dfrac{3}{4!} x^4 - \dfrac{8}{5!} x^5 + \cdots$

$\quad\quad\quad\quad = x + \dfrac{1}{2} x^2 - \dfrac{1}{8} x^4 - \dfrac{1}{15} x^5 + \cdots.$

There is no observable law of succession for the numerical coefficients, and the coefficient of x^n is not expressible as a simple function of n.

57. Development of $f(x)$ in powers of $x - a$. It was seen in Art. 56 that if $f(x)$ or any of its derivatives be discontinuous in the vicinity of $x = 0$, then $f(x)$ has no development in powers of x.

It will be shown, however, that if these successive functions be continuous in the vicinity of some other value $x = a$, then $f(x)$ will have a development in powers of $x - a$, which will be a true representative of the function for values of x within a certain interval in the vicinity of $x = a$.

First, to find the form of such development, let

$$f(x) \equiv A + B(x-a) + C(x-a)^2 \\ + D(x-a)^3 + E(x-a)^4 + \cdots \quad (1)$$

be regarded as an identity, the coefficients $A, B, C, \cdots$ being independent of x. With the same hypothesis for the vicinity of $x = a$ as was before made for the vicinity of $x = 0$, differentiation furnishes the additional identities:

$$f'(x) = B + 2C(x-a) + 3D(x-a)^2 + 4E(x-a)^3 + \cdots$$
$$f''(x) = \qquad 2C \qquad + 3 \cdot 2 D(x-a) + 4 \cdot 3 E(x-a)^2 + \cdots$$
$$f'''(x) = \qquad\qquad\qquad\quad 3 \cdot 2 D \qquad + 4 \cdot 3 \cdot 2 E(x-a) + \cdots$$

If, now, the special value a be given to x, the following equations will be obtained:

$$f(a) = A, \quad f'(a) = B, \quad f''(a) = 2C, \quad f'''(a) = 3 \cdot 2 D, \cdots$$

hence,

$$A = f(a), \quad B = f'(a), \quad C = \frac{f''(a)}{2!}, \quad D = \frac{f'''(a)}{3!}, \cdots$$

Thus the coefficients in (1) are determined, and the required development is

$$f(x) = f(a) + f'(a)(x-a) + \frac{f''(a)}{2!}(x-a)^2 + \frac{f'''(a)}{3!}(x-a)^3 \\ + \cdots + \frac{f^n(a)}{n!}(x-a)^n + \cdots. \quad (2)$$

Ex. 1. Expand $\log x$ in powers of $x - a$.

Here $f(x) = \log x, f'(x) = \dfrac{1}{x}, f''(x) = -\dfrac{1}{x^2}, f'''(x) = \dfrac{1\cdot 2}{x^3}\cdots,$

$$f^n(x) = \frac{(-1)^{n-1}(n-1)!}{x^n}$$

hence, $f(a) = \log a, f'(a) = \dfrac{1}{a}, f''(a) = -\dfrac{1}{a^2}, f'''(a) = \dfrac{1\cdot 2}{a^3}\cdots,$

$$f^n(a) = \frac{(-1)^{n-1}(n-1)!}{a^n},$$

and, by (2) the required development is

$$\log x = \log a + \frac{1}{a}(x-a) - \frac{1}{2a^2}(x-a)^2 + \frac{1}{3a^3}(x-a)^3 - \cdots$$

$$+ \frac{(-1)^{n-1}}{na^n}(x-a)^n + \cdots.$$

The condition for the convergence of this series is that

$$\lim_{n \doteq \infty}\left[\frac{(x-a)^{n+1}}{(n+1)a^{n+1}} : \frac{(x-a)^n}{na^n}\right]|<|1\,;$$

i.e.,
$$\frac{x-a}{a}|<|1,$$

$$x - a|<|a,$$

$$0 < x < 2a.$$

It will be shown presently that this is also the interval of equivalence of $\log x$ and its development. This series may be called the development of $\log x$ in the vicinity of $x = a$. Its development in the vicinity of $x = 1$ has the simpler form

$$\log x = x - 1 - \tfrac{1}{2}(x-1)^2 + \tfrac{1}{3}(x-1)^3 - \cdots,$$

which holds for values of x between 0 and 2.

Ex. 2. Show that the development of $\dfrac{1}{x}$ in powers of $x - a$ is

$$\frac{1}{x} = \frac{1}{a} - \frac{1}{a^2}(x-a) + \frac{1}{a^3}(x-a)^2 - \frac{1}{a^4}(x-a)^3 + \cdots,$$

and that the series is convergent from $x = 0$ to $x = 2a$.

Ex. 3. Develop e^x in powers of $x - 2$.

Ex. 4. Develop $x^3 - 2x^2 + 5x - 7$ in powers of $x - 1$.

Ex. 5. Develop $3y^2 - 14y + 7$ in powers of $y - 3$.

58. Remainder. The second restriction imposed upon the series in order that it may be a correct representative of the generating function, is that the remainder after n terms may be made smaller than any given number by taking n large enough.

Before getting the general form for this remainder it is necessary to prove the following lemma.

59. Rolle's theorem. If $f(x)$ and its first $n+1$ derivatives are continuous for all values of x between a and b, and $f(a), f(b)$ both vanish, then $f'(x)$ will vanish for some value of x between a and b.

By supposition $f(x)$ cannot become infinite for any value of x, such that $a < x < b$; and if $f'(x)$ does not vanish, it must always be positive or always be negative; hence, $f(x)$ must continually increase or continually decrease (Art. 20).

This is impossible, as $f(a) = 0$ and $f(b) = 0$, hence at some point x between a and b, $f(x)$ must cease to increase and begin to decrease, or cease to decrease and begin to increase.

This point x is defined by the equation $f'(x) = 0$.

To prove the same thing geometrically, let $y = f(x)$ be the equation of a continuous curve, which crosses the x-axis at distances $x = a$, $x = b$ from the origin; then at some point between a and b the tangent to the curve is parallel to the x-axis, since by supposition there is no discontinuity in the direction of the tangent. Hence

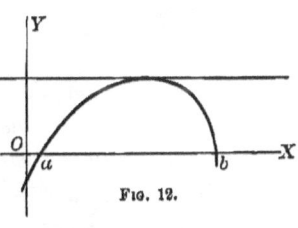

Fig. 12.

$$\frac{dy}{dx} = f'(x) = 0.$$

60. Form of remainder in development of $f(x)$ in powers of $x - a$. Let the remainder after n terms be denoted by

$R_n(x, a)$, which is a function of x and of a as well as of n. From the form of the succeeding terms, R_n may be conveniently written in the form

$$R_n(x, a) = \frac{(x-a)^n}{n!} \phi(x, a),$$

and then the problem is to determine $\phi(x, a)$, so that the following may be an algebraic identity:

$$f(x) = f(a) + f'(a)(x-a) + \frac{f''(a)}{2!}(x-a)^2 + \cdots$$

$$+ \frac{f^{n-1}(a)}{(n-1)!}(x-a)^{n-1} + \frac{\phi(x, a)}{n!}(x-a)^n, \quad (1)$$

in which the right-hand member contains only the first n terms of the series, with the remainder after n terms. Thus

$$f(x) - f(a) - f'(a)(x-a) - \frac{f''(a)}{2!}(x-a)^2 - \cdots$$

$$- \frac{f^{n-1}(a)}{(n-1)!}(x-a)^{n-1} - \frac{\phi(x, a)}{n!}(x-a)^n = 0. \quad (2)$$

Let a new function $F(z)$ be defined as follows:

$$F(z) \equiv f(x) - f(z) - f'(z)(x-z) - \frac{f''(z)}{2!}(x-z)^2 - \cdots$$

$$- \frac{f^{n-1}(z)}{(n-1)!}(x-z)^{n-1} - \frac{\phi(x, a)}{n!}(x-z)^n, \quad (3)$$

in which the right-hand member is obtained from (2) by replacing a by the variable z in every term except $\phi(x, a)$.

This function $F(z)$ vanishes when $z = x$, by inspection; and it also vanishes when $z = a$, by (2); hence, by Rolle's theorem, its derivative $F'(z)$ vanishes for some value of z between x and a, say z_1. But

$$F'(z) = -f'(z) + f'(z) - f''(z)(x-z) + f''(z)(x-z) - \cdots$$

$$- \frac{f^n(z)}{(n-1)!}(x-z)^{n-1} + \frac{\phi(x, a)}{(n-1)!}(x-a)^{n-1};$$

and these terms cancel each other off in pairs except the last two; hence
$$F'(z) = \frac{(x-z)^{n-1}}{(n-1)!}[\phi(x, a) - f^n(z)];$$
then since $F'(z)$ vanishes when $z = z_1$, it follows that
$$\phi(x, a) = f^n(z_1), \qquad (4)$$
wherein z_1 lies between x and a, and may thus be represented by
$$z_1 = a + \theta(x - a),$$
where θ is a positive proper fraction. Hence from (4)
$$\phi(x, a) = f^n[a + \theta(x - a)],$$
and
$$R_n(x, a) = \frac{f^n[a + \theta(x - a)]}{n!}(x - a)^n.*$$

The complete form of the expansion of $f(x)$ is then
$$f(x) = f(a) + f'(a)(x - a) + \frac{f''(a)}{2!}(x - a)^2 + \cdots$$
$$+ \frac{f^{n-1}(a)}{(n-1)!}(x - a)^{n-1} + \frac{f^n(a + \theta(x - a))}{n!}(x - a)^n, \quad (5)$$
in which n is any positive integer. The series may be carried to any desired number of terms by increasing n, and the last term in (5) gives the remainder (or error) after the first n terms of the series. The symbol $f^n(a + \theta(x - a))$ indicates that $f(x)$ is to be differentiated n times with regard to x, and that x is then to be replaced by $a + \theta(x - a)$.

61. Another expression for the remainder. Instead of putting $R_n(x, a)$ in the form
$$\frac{(x - a)^n}{n!}\phi(x, a),$$

---

* This form of the remainder was found by Lagrange (1736–1813), who published it in the Mémoires de l'Academie des Sciences à Berlin, 1772.

it is sometimes convenient to write it in the form

$$R_n(x, a) = (x - a)\psi(x, a).$$

Proceeding as before, the expression for $F'(z)$ will be

$$F'(z) = -\frac{f^n(z)}{(n-1)!}(x-z)^{n-1} + \psi(x, a);$$

and this is to vanish when $z = z_1$; hence

$$\psi(x, a) = \frac{f^n(z_1)}{(n-1)!}(x - z_1)^{n-1},$$

in which $z_1 = a + \theta(x - a)$, $x - z_1 = (x - a)(1 - \theta)$;

thus $\quad \psi(x, a) = \dfrac{f^n(a + \theta(x - a))}{(n-1)!}(1-\theta)^{n-1}(x-a)^{n-1},$

and $\quad R_n(x, a) = (1 - \theta)^{n-1}\dfrac{f^n(a + \theta(x - a))}{(n-1)!}(x - a)^n.$*

An example of the use of this form of remainder is furnished by the series for $\log x$ in powers of $x - a$, when $x - a$ is negative, and also in Art. 64.

Ex. 1. Find the interval of equivalence of $\log x$ and its development in powers of $x - a$, when a is a positive number.

Here, from Art. 57, Ex. 1,

$$|f^n(x)| = \left|\frac{(n-1)!}{x^n}\right|,$$

hence $\quad |f^n(x + \theta(x - a))| = \left|\dfrac{(n-1)!}{(a + \theta(x - a))^n}\right|,$

and $\quad |R_n(x, a)| = \left|\dfrac{(x-a)^n}{n(a+\theta(x-a))^n}\right| = \dfrac{1}{n}\left[\dfrac{x-a}{a + \theta(x-a)}\right]^n.$

Now the interval of equivalence is not wider than the interval of convergence; hence, by Art. 54, the first condition of equivalence is that $x - a$ be numerically less than a. First let $x - a$ be positive, then when

---

* This form of the remainder was found by Cauchy (1789–1857), and first published in his " Leçons sur le calcul infinitésimal," 1826.

it lies between 0 and a it is numerically less than $a + \theta(x - a)$, since θ is a positive proper fraction; hence when $n \doteq \infty$

$$\left[\frac{x-a}{a+\theta(x-a)}\right]^n \doteq 0, \text{ and } R_n(x, a) \doteq 0.$$

Again, when $x - a$ is negative, and numerically less than a, the second form of the remainder must be employed. As before,

$$f^n(a + \theta(x - a))| = |\frac{(n-1)!}{(a + \theta(x - a))^n},$$

hence $\quad R_n(x, a) | = |(1 - \theta)^{n-1} \cdot \frac{(x-a)^n}{[a + \theta(x - a)]^n}$

$$| = |(1 - \theta)^{n-1} \cdot \frac{(a-x)^n}{[a - \theta(a - x)]^n}$$

$$| = |\left[\frac{(a-x) - \theta(a-x)}{a - \theta(a-x)}\right]^{n-1} \cdot \frac{a-x}{a - \theta(a-x)}.$$

The factor within the brackets is always less than 1, hence the $(n-1)$st power can be made less than any given number, by taking n large enough. This is true for all values of x between 0 and a.

Therefore, $\log x$ and its development in powers of $x - a$ are equivalent within the interval of convergence of the series, that is, for all values of x between 0 and $2a$.

Ex. 2. Show that the development of $x^{-\frac{1}{2}}$ in positive powers of $x - a$ holds for all values of x that make the series convergent; that is, when x lies between 0 and $2a$.

62. Form of remainder in Maclaurin's series. The above form of remainder is at once applicable to Maclaurin's series by putting $a = 0$. The result is

$$f(x) = f(0) + f'(0)x + \frac{f''(0)}{2!}x^2 + \cdots + \frac{f^{n-1}(0)}{(n-1)!}x^{n-1} + \frac{f^n(\theta x)}{n!}x^n.$$

The remainder formula

$$R_n(x) = \frac{f^n(\theta x)}{n!}x^n$$

will now be used to show that the interval of equivalence of any one of the ordinary functions, and its development in powers of x, is co-extensive with the interval of convergence of the development itself.

The following lemma will be useful in several cases.

63. Lemma. When x has any finite value, however great, and n is a positive integer, then

$$\frac{x^n}{n!} \doteq 0 \text{ when } n \doteq \infty.$$

For, let $u_r = \frac{x^r}{r!}$, then $u_{r+1} = \frac{x^{r+1}}{(r+1)!}$,

and $\frac{u_{r+1}}{u_r} = \frac{x}{r+1}.$

Now, however large be the assigned value of x, it is possible to take r so great that

$$\frac{x}{r+1} < k,$$

where k is some proper fraction, and then for terms subsequent to u_r, the ratio of each term to the preceding term will be less than the fixed proper fraction k; hence, by 6 (*a*) of Art. 59, these successive terms approach nearer and nearer to zero as a limit.

64. Remainder in the development of a^x, sin x, cos x.

If $f(x) = a^x$, then $f^n(x) = a^x(\log a)^n$, $f^n(\theta x) = a^{\theta x}(\log a)^n$,

and $R_n(x) = a^{\theta x}(\log a)^n \cdot \frac{x^n}{n!} = a^{\theta x} \cdot \frac{(x \log a)^n}{n!};$

but $\frac{(x \log a)^n}{n!} \doteq 0$, when $n \doteq \infty$, by Art. 63; and $a^{\theta x}$ is finite, when x is any finite number, however great; hence

$$R_n(x) \doteq 0, \text{ when } n \doteq \infty.$$

Again let $f(x) = \sin x$, then $f^n(x) = \sin\left(x + \dfrac{n\pi}{2}\right)$ by Art. 49; hence

$$f^n(\theta x) = \sin\left(\theta x + \dfrac{n\pi}{2}\right), \text{ and } R_n(x) = \sin\left(\theta x + \dfrac{n\pi}{2}\right) \cdot \dfrac{x^n}{n!},$$

but $\sin\left(\theta x + \dfrac{n\pi}{2}\right)$ never exceeds unity for any value of x or of n, hence, by Art. 63, $R_n(x) \doteq 0$, when $n \doteq \infty$.

Similarly, if $f(x) = \cos x$, $f^n(x) = \cos\left(x + \dfrac{n\pi}{2}\right)$, and as before, $R^n(x) \doteq 0$, when $n \doteq \infty$.

Hence the developments of a^x, $\sin x$, $\cos x$, hold for every finite value of x.

Ex. 1. If $f(x) = \sin x$, compute $R_3(x)$, when $x = \tfrac{1}{3}\pi$ radians.

Ex. 2. Expand $\sin ax$ by Maclaurin's theorem, and determine the remainder after 7 terms, counting the terms that have zero coefficients.

Ex. 3. Show that the absolute error in stopping the series for $\sin x$, $\cos x$, at any term, is less than the next term of the series.

Ex. 4. Show that the relative error in stopping the series for e^x, at any term, is less than the next term of the series; the *relative error* being the ratio of the absolute error to the true value of the function to be computed.

Ex. 5. Prove by expansion that

$$e^{\sqrt{-1}x} + e^{-\sqrt{-1}x} = 2\cos x, \; e^{\sqrt{-1}x} - e^{-\sqrt{-1}x} = 2\sin x;$$

and, hence by addition, $e^{\sqrt{-1}x} = \cos x + \sqrt{-1}\sin x$.

Ex. 6. From the last example, prove De Moivre's theorem:

$$(\cos x + \sqrt{-1}\sin x)^m = \cos mx + \sqrt{-1}\sin mx.$$

65. Taylor's series.* It will next be shown how to write down the development for a function of the sum of two

---

* So named from its discoverer, Dr. Brook Taylor (1685-1731), who published it in his "Methodus Incrementorum," 1715; but the formula remained almost unnoticed until Lagrange completed it by finding an expression for the remainder after n terms (Art. 60). Since then it has been regarded as the most important formula in the Calculus.

numbers in ascending powers of either number, and also an expression for the remainder after n terms of the series.

If in the identity of Art. 60, which gives an expression for $f(x)$ in powers of $x - a$, the letter x be everywhere replaced by $x + a$, then $x - a$ will be replaced by x, and the identity will assume the form

$$f(x+a) = f(a) + f'(a)x + \frac{f''(a)}{2!}x^2 + \cdots$$

$$+ \frac{f^{n-1}(a)}{(n-1)!}x^{n-1} + \frac{f^n(a + \theta x)}{n!}x^n, \quad (1)$$

in which x, a are any two numbers, n is any positive integer, and θ is some positive proper fraction, which may not, however, be independent of the values of the other letters.

If the second form of remainder be used, the last term on the right will be replaced by $(1 - \theta)^{n-1}\frac{f^n(a + \theta x)}{(n-1)!}x^n$.

In the identity (1) the letters x and a may be interchanged, hence the expansion for $f(x + a)$ in powers of a is

$$f(x + a) = f(x) + f'(x)a + \frac{f''(x)}{2!}a^2 + \cdots$$

$$+ \frac{f^{n-1}(x)}{(n-1)!}a^{n-1} + \frac{f^n(x+\theta a)}{n!}a^n; \quad (2)$$

and the second form of remainder is

$$R_n(x, a) = (1 - \theta)^{n-1}\frac{f^n(x + \theta a)}{(n-1)!}a^n. \quad (3)$$

Ex. 1. Expand $(a+x)^m$ in ascending powers of x, and find the interval of equivalence.

Here $\qquad f(a + x) = (a + x)^m,$

hence $\qquad f(x) = x^m,$

EXPANSION OF FUNCTIONS

and $f'(x) = mx^{m-1}$, $f''(x) = m(m-1)x^{m-2}, \ldots$,

$$f^n(x) = m(m-1)\cdots(m-n+1)x^{m-n},$$

$f(a) = a^m$, $f'(a) = ma^{m-1}$, $f''(a) = m(m-1)a^{m-2}, \ldots$,

$$f^n(a) = m(m-1)\cdots(m-n+1)a^{m-n},$$

therefore $(a+x)^m = a^m + ma^{m-1}x + \dfrac{m(m-1)}{2!}a^{m-2}x^2 + \cdots$

$$+ \frac{m(m-1)\cdots(m-n+2)}{(n-1)!}a^{m-n+1}x^{n-1} + R_n(a,x),$$

in which, from the first form of remainder,

$$R_n(a,x) = \frac{m(m-1)\cdots(m-n+1)}{n!}(a+\theta x)^{m-n}x^n$$

$$= \frac{m(m-1)\cdots(m-n+1)}{n!}(a+\theta x)^m \left(\frac{x}{a+\theta x}\right)^n.$$

It will first be shown that the factor

$$\frac{m(m-1)\cdots(m-n+1)}{n!} \neq \infty \text{ when } n \doteq \infty ;$$

for, if it be denoted by u_n, then

$$\frac{u_{n+1}}{u_n} = \frac{m-n}{n+1},$$

and this ratio can, by taking n large enough, be made as near unity as may be desired; but it can never exceed unity, hence the successive values of u_n will approach the limit zero or a finite number, when $n \doteq \infty$ (Art. 54, 6 (a)).

Next, the expression

$$\left(\frac{x}{a+\theta x}\right)^n \doteq 0 \text{ when } n \doteq \infty,$$

if x be positive and less than a. Hence

$$R_n(a,x) \doteq 0, \text{ when } n \doteq \infty,$$

if x be positive and less than a.

Since the interval of convergence is, by Art. 54, from $x = -a$ to $x = a$, it remains to examine the value of $R_n(a, x)$ when x is negative and numerically less than a. For this purpose it is necessary to use the second form of remainder,

$$R_n(a, x) = (1-\theta)^{n-1} \frac{m(m-1)\cdots(m-n+1)}{(n-1)!} (a+\theta x)^{m-n} x^n$$

$$= \frac{m(m-1)\cdots(m-n+1)}{(n-1)!} \cdot \frac{(a+\theta x)^{m-1}}{a} \left(\frac{1-\theta}{1+\theta\frac{x}{a}}\right)^{n-1} \cdot \left(\frac{x}{a}\right)^n.$$

But when $\dfrac{x}{a}$ is negative and less than 1, the expression $\dfrac{1-\theta}{1+\theta\frac{x}{a}}$ is a proper fraction, hence its $(n-1)$st power approaches zero as a limit; and it can be shown as before that the factorial expression is not infinite. Hence

$$R_n(a, x) \doteq 0, \text{ when } n \doteq \infty,$$

if x lies between $-a$ and $+a$.

This is therefore the required interval of equivalence.

Ex. 2. Expand $\log(x + a)$ in powers of x, and find the interval of equivalence.

Here
$$f(x + a) = \log(x + a),$$
$$f(x) = \log x,$$
$$f'(x) = \frac{1}{x}, \; f''(x) = -\frac{1}{x^2}, \; f'''(x) = \frac{2!}{x^3}, \; \cdots$$
$$f^n(x) = \frac{(n-1)!}{x^n}(-1)^{n-1};$$

hence
$$\log(x + a) = \log a + \frac{x}{a} - \frac{1}{2}\frac{x^2}{a^2} + \frac{1}{3}\frac{x^3}{a^3} \cdots$$
$$\frac{(-1)^{n-2}}{(n-1)a^{n-1}} x^{n-1} + \frac{(-1)^{n-1}}{n(a+\theta x)^n} x^n.$$

This expansion could also be obtained from the development of $\log x$ in powers of $x - a$, in Art. 57.

Similarly,

$$\log(a - x) = \log a - \frac{x}{a} - \frac{1}{2a^2}x^2 - \frac{1}{3a^3}x^3 \cdots - \frac{x^{n-1}}{(n-1)a^{n-1}}$$

$$- \frac{x^n}{n(a - \theta_1 x)^n}.$$

When $a = 1$, these series become

$$\log(1 + x) = x - \frac{x^2}{2} + \frac{x^3}{3} - \frac{x^4}{4} + \cdots + \frac{(-1)^{n-2}x^{n-1}}{n-1}$$

$$+ \frac{(-1)^{n-1}x^n}{n(1 + \theta x)^n},$$

$$\log(1 - x) = -x - \frac{x^2}{2} - \frac{x^3}{3} - \frac{x^4}{4} - \cdots - \frac{x^{n-1}}{n-1} - \frac{x^n}{n(1 - \theta_1 x)^n},$$

in which, as in Ex. 1 of Art. 61, the remainder $R_n(x) \doteq 0$, when $n \doteq \infty$, if $-1 < x < 1$. Also, by subtraction,

$$\log\left(\frac{1+x}{1-x}\right) = 2\left(x + \frac{x^3}{3} + \frac{x^5}{5} + \frac{x^7}{7} + \cdots\right),$$

which can be used for computation when x is numerically less than unity.

This identity can be thrown into a form suitable for the numerical calculation of the logarithm of any number; for, put

$$\frac{1+x}{1-x} = \frac{n+h}{n}, \quad \text{then} \quad x = \frac{h}{2n+h};$$

from which

$$\log\left(\frac{n+h}{n}\right) = 2\left(\frac{h}{2n+h} + \frac{1}{3}\frac{h^3}{(2n+h)^3} + \frac{1}{5}\frac{h^5}{(2n+h)^5} + \cdots\right).$$

This is an identity for all positive values of n and h, since the original condition $x | < | 1$ is replaced by $\frac{h}{2n+h} | < | 1$.

and the latter condition is always fulfilled when n and h are positive.

Suppose it is required to find log 10. This could be done by putting $n = 1$, $h = 9$ in the last equation, but the series thus obtained would converge too slowly to be of practical value. Let log 2 be first calculated by putting both n and h equal to 1; thus

$$\log 2 = 2\left[\frac{1}{3} + \frac{1}{3} \cdot \frac{1}{3^3} + \frac{1}{5} \cdot \frac{1}{3^5} + \frac{1}{7} \cdot \frac{1}{3^7} + \cdots\right].$$

Next, put $n = 8$, $h = 2$,

$$\log 10 = 3 \log 2 + \frac{2}{3}\left[\frac{1}{3} + \frac{1}{3} \cdot \frac{1}{3^5} + \frac{1}{5} \cdot \frac{1}{3^9} + \frac{1}{7} \cdot \frac{1}{3^{13}} + \cdots\right].$$

The numerical work can be greatly facilitated by proper arrangement of terms. The result correct to 8 places of decimals is $\log 10 = 2.30258509$.

The student should bear in mind the distinction between theoretical and practical convergence. Here, only theoretical convergence has been considered. To make a series practically useful, R_n should be so small that after ten or twelve terms it could be neglected without affecting the desired numerical approximation. Sometimes, however, the expression for R_n does not lend itself easily to a numerical estimate of the error made in stopping the series at a given term. The method of comparison with a descending geometrical progression, stated in (6) of Art. 54, and illustrated in Ex. 1 of Art. 56, and in Exs. 1, 2, 3 of Art. 67, will be found very useful in practice.

Ex. 3. Expand $\sin(x + y)$ in ascending powers of y. Hence verify that $\sin(x + y) = \sin x \cos y + \cos x \sin y$.

66. Theorem of mean value. Increment of function in terms of increment of variable.

An important special case of Taylor's theorem is

$$f(x + h) = f(x) + hf'(x + \theta h), \qquad (1)$$

which is obtained by putting $n = 1$, in equation (2) of Art. 65, and replacing a by h.

If $f(x)$ be transposed, and Δx be written for h, the identity may be written

$$\Delta f(x) = \Delta x \cdot f'(x + \theta \cdot \Delta x), \qquad (2)$$

which expresses that: The increment of the function is equal to the increment of the variable multiplied by the value of the derivative taken at some intermediate value of x.

This theorem is true whether the increments be large or small. It has a simple geometrical interpretation. Since $f(x)$ is continuous, it can be represented by a curve whose equation is $y = f(x)$.

In Fig. 13, let

$$x = ON, \; x + \Delta x = OR,$$

$$f(x) = NH, \; f(x + \Delta x) = RK,$$

then $\Delta f(x) = MK$, and

$$\frac{\Delta f(x)}{\Delta x} = \frac{MK}{HM} = \tan MHK;$$

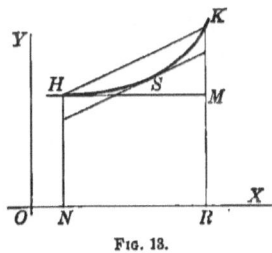

Fig. 13.

hence, $\quad f'(x + \theta \cdot \Delta x) = \tan MHK.$

But $f'(x + \theta \cdot \Delta x)$ is the slope of the tangent at some point S between H and K; thus the theorem of mean value expresses that at some point between H and K the tangent to the curve is parallel to the secant HK. This is self-evident, geometrically; and has already been mentioned in Art. 59.

Ex. 1. Verify the theorem of mean value for the function $f(x) = x^2$.
Here $f(x + h) = (x + h)^2 = x^2 + h \cdot 2(x + \theta h)$,
which is evidently a true identity when $\theta = \frac{1}{2}$. In most cases the exact numerical value of the proper fraction θ is not so apparent.

When the given increment of x is small and the increment of the function is desired, it is sometimes sufficiently accurate in practical computation to replace $f'(x + \theta \cdot \Delta x)$ in equation (2) by its approximate value $f'(x)$,

then $\qquad \Delta f(x) = \Delta x \cdot f'(x), \qquad (3)$

in which the error is, by Taylor's theorem,

$$\tfrac{1}{2}(\Delta x)^2 \cdot f''(x + \theta \cdot \Delta x),$$

a term of the second order of smallness.

A second approximation to the value of $\Delta f(x)$ is given by

$$\Delta f(x) = \Delta x \cdot f'(x) + \tfrac{1}{2}(\Delta x)^2 \cdot f''(x), \qquad (4)$$

in which the error is $\tfrac{1}{6}(\Delta x)^3 \cdot f'''(x + \theta \cdot \Delta x)$, of the third order.

The third approximation is obtained by adding the term $\tfrac{1}{6}(\Delta x)^3 f'''(x)$, and the error will then be

$$\tfrac{1}{24}(\Delta x)^4 \cdot f^{iv}(x + \theta \cdot \Delta x).$$

Ex. 2. Compute the first, second, and third approximations to the increment of $\log x$ when x changes from 10 to 10.1.

Ex. 3. Show how to compute the difference for one minute in a table of natural sines.

Increment of the increment. Let $y = f(x)$ be a function which can be developed in the vicinity of $x = x_1$; and let x have the three successive equidistant values $x_1 - h$, x_1, $x_1 + h$. When x changes from $x_1 - h$ to x_1, let y take the increment $\Delta_1 y = f(x_1) - f(x_1 - h)$
$= h \cdot f'(x_1) - \tfrac{1}{2} h^2 f''(x_1 - \theta h)$; and

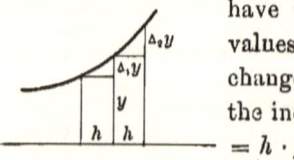

when x changes further from x_1 to $x_1 + h$, let y take the increment

$$\Delta_2 y = f(x_1 + h) - f(x_1) = hf'(x_1) + \tfrac{1}{2} h^2 f''(x_1 + \theta' h);$$

let the difference of these successive increments of y be written $\Delta(\Delta y)$ or $\Delta^2 y$;

then $\quad \Delta^2 y = \Delta_2 y - \Delta_1 y = \dfrac{h^2}{2}[f''(x_1 + \theta' h) + f''(x_1 - \theta h)]. \quad (5)$

This result may be expressed in words thus: The increment of the increment of the function, corresponding to successive equal increments of the variable, is equal to the square of the latter increment multiplied by half the sum of the values of the second derivative taken at intermediate values of the variable on each side of its middle value. This may be called the theorem of mean value for the second derivative.

Ex. 4. Prove that $\Delta^2 y$ is an infinitesimal of the same order as $(\Delta x)^2$.

Ex. 5. Show how to compute the change in the difference for one minute in exercise 3.

Limit of the ratio of $\Delta^2 y$ *to* $(\Delta x)^2$. In equation (5), replace h by Δx, divide by $(\Delta x)^2$, and take the limit of both members as $\Delta x \doteq 0$, then

$$\lim_{\Delta x \doteq 0} \frac{\Delta^2 y}{\Delta x^2} = \frac{d^2 y}{dx^2}.$$

67. To find the development of a function when that of its derivative is known. Development of the anti-trigonometric functions.

The derivative of an anti-trigonometric function being an algebraic binomial, it is easy to expand it by the binomial theorem; it is now proposed to show how to use the development of the derivative to determine the coefficients in the development of the given function, so as to avoid the labor of successive differentiation.

1. Power-series for $\tan^{-1} x$.

Assume, within an interval including $x = 0$, the identity

$$\tan^{-1} x = A + Bx + Cx^2 + Dx^3 + Ex^4 + \cdots. \qquad (1)$$

With the same preliminary hypothesis as in Art. 56, differentiation furnishes the identity

$$\frac{1}{1+x^2} = B + 2\,Cx + 3\,Dx^2 + 4\,Ex^3 + \cdots, \qquad (2)$$

but, within the interval from -1 to $+1$, the left member is identical with

$$1 - x^2 + x^4 - x^6 + \cdots,$$

hence, within a certain interval including $x = 0$, there exists the identity

$$1 - x^2 + x^4 - x^6 + \cdots = B + 2\,Cx + 3\,Dx^2 + 4\,Ex^3 + \cdots. \qquad (3)$$

therefore $B = 1$, $C = 0$, $D = -\frac{1}{3}$, $E = 0$, $F = \frac{1}{5}, \cdots$.

The first coefficient A is found to be zero by putting $x = 0$ in (1), hence

$$\tan^{-1} x = x - \tfrac{1}{3} x^3 + \tfrac{1}{5} x^5 - \tfrac{1}{7} x^7 + \cdots. \qquad (4)$$

The interval of convergence of this series, found by the usual method, is from $x = -1$ to $x = 1$.

To show that this is also the interval of equivalence of the function and the series, and thus to establish the validity of the development, let $R_n(x)$ denote the remainder obtained by subtracting the sum of the first n terms from the function, then

$$\tan^{-1} x = x - \tfrac{1}{3} x^3 + \tfrac{1}{5} x^5 - \cdots \mp \frac{x^{2n-1}}{2n-1} + R_n(x), \qquad (5)$$

EXPANSION OF FUNCTIONS

hence by differentiation with regard to x,

$$\frac{1}{1+x^2} = 1 - x^2 + x^4 - \cdots \mp x^{2n-2} + R_n'(x), \qquad (6)$$

therefore $R_n'(x)$ is the remainder after n terms obtained by dividing 1 by $1 + x^2$,

i.e.,
$$R_n'(x) = \pm \frac{x^{2n}}{1+x^2}. \qquad (7)$$

By the theorem of mean value, Art. 66,

$$R_n(x) = R_n(0) + xR_n'(\theta x), \qquad [0 < \theta < 1$$

but, from (5), $R_n(0) = 0$; and when x is less than 1, θx is less than 1, hence, by (7),

$$R_n'(\theta x) \doteq 0, \text{ when } n \doteq \infty;$$

therefore
$$\lim_{n \doteq \infty} R_n(x) = 0.$$

Thus the interval of equivalence is from -1 to $+1$.

Ex. 1. Compute $\tan^{-1}\frac{1}{2}$, $\tan^{-1}\frac{1}{3}$, $\tan^{-1}1$; and hence the value of π.

$\tan^{-1}\frac{1}{2} = \frac{1}{2} - \frac{1}{3}(\frac{1}{2})^3 + \frac{1}{5}(\frac{1}{2})^5 - \frac{1}{7}(\frac{1}{2})^7 + \cdots$

$\phantom{\tan^{-1}\frac{1}{2}} = +\ .5$
$\phantom{\tan^{-1}\frac{1}{2} = }-\ .0416667$
$\phantom{\tan^{-1}\frac{1}{2} = }+\ .0062500$
$\phantom{\tan^{-1}\frac{1}{2} = }-\ .0011162$
$\phantom{\tan^{-1}\frac{1}{2} = }+\ .0002170$
$\phantom{\tan^{-1}\frac{1}{2} = }-\ .0000444$
$\phantom{\tan^{-1}\frac{1}{2} = }+\ .0000095$
$\phantom{\tan^{-1}\frac{1}{2} = }-\ .0000023$
$\phantom{\tan^{-1}\frac{1}{2} = }+\ .0000005$
$\phantom{\tan^{-1}\frac{1}{2} = }-\ .0000001$
$\phantom{\tan^{-1}\frac{1}{2} = }\overline{\ .4636473+}$ radians.

$\tan^{-1}\frac{1}{3} = \frac{1}{3} - \frac{1}{3}(\frac{1}{3})^3 + \frac{1}{5}(\frac{1}{3})^5 - \cdots$

$\phantom{\tan^{-1}\frac{1}{3}} = +\ .3333333$
$\phantom{\tan^{-1}\frac{1}{3} = }-\ .0123457$
$\phantom{\tan^{-1}\frac{1}{3} = }+\ .0008230$
$\phantom{\tan^{-1}\frac{1}{3} = }-\ .0000653$
$\phantom{\tan^{-1}\frac{1}{3} = }+\ .0000056$
$\phantom{\tan^{-1}\frac{1}{3} = }-\ .0000005$
$\phantom{\tan^{-1}\frac{1}{3} = }\overline{\ .3217506+}$ radians.

To find $\tan^{-1} 1$, use the formula:

$$\tan^{-1}\frac{1}{2} + \tan^{-1}\frac{1}{3} = \tan^{-1}\frac{\frac{1}{2}+\frac{1}{3}}{1-\frac{1}{2}\cdot\frac{1}{3}} = \tan^{-1} 1 = \frac{\pi}{4};$$

hence $\quad \frac{\pi}{4} = .4636473 + .3217506 = 7853979 + \cdots,$

and $\quad \pi = 3.1415916 \cdots.$

In the first series, to estimate roughly the error made by stopping at the tenth term, it may be observed that the ratio of any term to the preceding is numerically less than $\frac{1}{2}$, and approaches $\frac{1}{4}$ as a limit; hence, if all the terms after u_{10} were positive, their sum would be less than the geometric series

$$u_{10}(\tfrac{1}{2} + \tfrac{1}{4} + \tfrac{1}{8} + \cdots),$$

which is less than u_{10}; moreover, since the alternate terms are negative, it follows that the error made in stopping at u_{10} is really much less than u_{10}, and is thus less than one unit in the seventh decimal place.

Similarly, in the second series, the error is much less than

$$u_6(\tfrac{1}{3} + \tfrac{1}{9} + \tfrac{1}{27} + \cdots),$$

i.e., less than $\frac{u_6}{2}$, or less than $2\frac{1}{4}$ units in the seventh place. Thus the error in the value of $\frac{\pi}{4}$ is less than $3\frac{1}{4}$ such units, and the error in the value of π is less than $1\frac{1}{2}$ units in the sixth place.

Therefore the numerical value of π lies between 3.1415916 and 3.1415941.

2. Power-series for $\sin^{-1} x$.

Proceed as before, and use the development

$$(1-x^2)^{-\frac{1}{2}} = 1 + \tfrac{1}{2}x^2 + \tfrac{1}{2}\cdot\tfrac{3}{4}x^4 + \tfrac{1}{2}\cdot\tfrac{3}{4}\cdot\tfrac{5}{6}\cdot x^6 + \cdots, \quad (1)$$

then $\quad \sin^{-1} x = x + \frac{1}{2}\cdot\frac{x^3}{3} + \frac{1}{2}\cdot\frac{3}{4}\cdot\frac{x^5}{5} + \frac{1}{2}\cdot\frac{3}{4}\cdot\frac{5}{6}\cdot\frac{x^7}{7} + \cdots, \quad (2)$

in which the interval of convergence is from -1 to 1.

Let $R_n(x)$ be the remainder after n terms in (2), then by differentiation, $R_n'(x)$ is the remainder after n terms in (1),

but $\quad R_n(x) = R_n(0) + xR_n'(\epsilon x) = xR_n'(\epsilon x), \quad [0 < \epsilon < 1$

67.]   EXPANSION OF FUNCTIONS   113

and, by Art. 66,

$R'_n(\epsilon x) \doteq 0$ when $n \doteq \infty$, if $x |<| 1$; hence $R_n(x) \doteq 0$;

and the interval of equivalence in (2) is from -1 to 1.

Ex. 2. Compute $\sin^{-1}(\tfrac{1}{2})$, and hence obtain the numerical value of π.

$\dfrac{\pi}{6} = \sin^{-1}(\tfrac{1}{2}) = \tfrac{1}{2} + \tfrac{1}{2}\cdot\tfrac{1}{3}(\tfrac{1}{2})^3 + \tfrac{1}{2}\cdot\tfrac{3}{4}\cdot\tfrac{1}{5}(\tfrac{1}{2})^5 + \tfrac{1}{2}\cdot\tfrac{3}{4}\cdot\tfrac{5}{6}\cdot\tfrac{1}{7}(\tfrac{1}{2})^7 + \cdots$

= .5000000000
+ .0208333333
+ .0023437500
+ .0003487723
+ .0000593397
+ .0000109239
+ .0000021183
+ .0000004262
+ .0000000881
+ .0000000186
  .5235987704 ;   hence $\pi = 3.1415926224+$.

Here each term may be used to obtain the next by applying as a factor the corresponding term in the series of ratios:

$\dfrac{1}{2}$, $\dfrac{1\cdot 1}{2\cdot 3\cdot 4}$, $\dfrac{3\cdot 3}{4\cdot 5\cdot 4}$, $\dfrac{5\cdot 5}{6\cdot 7\cdot 4}$, $\dfrac{7\cdot 7}{8\cdot 9\cdot 4}$, $\dfrac{9\cdot 9}{10\cdot 11\cdot 4}$, $\dfrac{11\cdot 11}{12\cdot 13\cdot 4}$, $\cdots$.

To determine the maximum error made by stopping at the tenth term, it is evident that the ratio of each term to the preceding is less than $\tfrac{1}{4}$, and approaches $\tfrac{1}{4}$ as a limit; therefore the sum of the remaining terms is less than

$$u_{10}(\tfrac{1}{4} + \tfrac{1}{16} + \tfrac{1}{64} + \cdots),$$

that is, less than $\tfrac{1}{3} u_{10}$. Hence the error in the value of $\tfrac{1}{6}\pi$ is less than 63 units in the tenth place, and the error in the above value of π is less than 378 units in the tenth place. Thus the numerical value of π lies between 3.1415926224 and 3.1415926602.*

Ex. 3. Show that the error made by stopping at any term in the series for log 10, Ex. 2, Art. 65, is less than $\tfrac{1}{10}$ of the last term used.

---

* Both of these formulas for π were found by Montferier. The correct value to ten places is 3.1415926536. By various methods mathematicians have carried the approximation to a much larger number of places. Mr. Shanks, of Durham, England, published the value of π to 607 places in 1853. No other constant has so much engaged the attention of mathematicians. See "Famous Problems in Elementary Geometry," by Professor Klein, translated by Professors Beman and Smith, 1897.

EXERCISES

Derive the following expansions:

1. $\sec x = 1 + \dfrac{x^2}{2!} + \dfrac{5\,x^4}{4!} + \dfrac{61}{6}x^6 + R.$

2. $\log \sec x = \dfrac{x^2}{2} + \dfrac{x^4}{12} + \dfrac{x^6}{45} + R.$

3. $\cos^2 x = 1 - x^2 + \dfrac{2^3\,x^4}{4!} - \dfrac{2^5\,x^6}{6!} + R.$

4. $e^x \cos x = 1 + x - \dfrac{2\,x^3}{3!} - \dfrac{4\,x^5}{5!} + R.$

5. $\log \dfrac{x}{x-1} = \dfrac{1}{x-1} - \dfrac{1}{2(x-1)^2} + \dfrac{1}{3(x-1)^3} + R.$

6. $\log(1 + \sin x) = x - \dfrac{x^2}{2} + \dfrac{x^3}{6} - \dfrac{x^4}{12} + R.$

7. $\sin^2 x = x^2 - \dfrac{x^4}{3} + \dfrac{2\,x^6}{3^2 \cdot 5} + R.$

8. $\sqrt{1 + 4x + 12\,x^2} = 1 + 2x + 4x^2 + R.$

9. $\cos(x + a) = \cos x - a \sin x - \dfrac{a^2}{2!}\cos x + \dfrac{a^3}{3!}\sin x + R.$

10. $\log \sin(x + a) = \log \sin x + a \cot x - \dfrac{a^2}{2}\csc^2 x + \dfrac{a^3}{3}\dfrac{\cos x}{\sin^3 x} + R.$

11. $e^x \sec x = 1 + x + x^2 + \dfrac{2\,x^3}{3} + R.$

12. $\log(1 + e^x) = \log 2 + \dfrac{x}{2} + \dfrac{x^2}{2^3} - \dfrac{x^4}{2^3 \cdot 4!} + R.$

13. $\cot^{-1} x = \tfrac{1}{2}\pi - x + \tfrac{1}{3}x^3 - \tfrac{1}{5}x^5 + \cdots,$ $\quad [|x|<|1]$

14. $\cot^{-1} x = \dfrac{1}{x} - \dfrac{1}{3\,x^3} + \dfrac{1}{5\,x^5} - \cdots,$ $\quad [|x|>|1]$

15. $\tan^{-1} x = \dfrac{\pi}{2} - \dfrac{1}{x} + \dfrac{1}{3\,x^3} - \dfrac{1}{5\,x^5} + \cdots.$ $\quad [|x|>|1]$

16. $\csc^{-1} x = \sin^{-1}\dfrac{1}{x} = \dfrac{1}{x} + \dfrac{1}{2}\cdot\dfrac{1}{3\,x^3} + \dfrac{1}{2}\cdot\dfrac{3}{4}\cdot\dfrac{1}{5\,x^5} + \cdots.$

17. Expand $\cos^{-1} x$ in powers of x; $\sec^{-1} x$ in powers of x^{-1}.

CHAPTER V

INDETERMINATE FORMS

68. Hitherto the values of a given function $f(x)$, corresponding to assigned values of the variable x, have been obtained by direct substitution. The function may, however, involve the variable in such a way that for certain values of the latter the corresponding values of the function cannot be found by mere substitution.

For example, the function

$$\frac{e^x - e^{-x}}{\sin x},$$

for the value $x = 0$, assumes the form $\frac{0}{0}$, and the corresponding value of the function is thus not directly determined. In such a case the expression for the function is said to assume an *indeterminate form* for the assigned value of the variable.

The example just given illustrates the indeterminateness of most frequent occurrence; namely, that in which the given function is the quotient of two other functions that vanish for the same value of the variable.

Thus if $$f(x) = \frac{\phi(x)}{\psi(x)},$$

and if, when x takes the special value a, the functions $\phi(x)$ and $\psi(x)$ both vanish, then

$$f(a) = \frac{\phi(a)}{\psi(a)} = \frac{0}{0}$$

is indeterminate in form, and cannot be rendered determinate without further transformation.

69. Indeterminate forms may have determinate values. A case has already been seen (Art. 16) in which an expression that assumes the form $\dfrac{0}{0}$ for a certain value of its variable takes a definite value, dependent upon the law of variation of the function in the vicinity of the assigned value of the variable.

As another example, consider the function

$$y = \frac{x^2 - a^2}{x - a}.$$

If this relation between x and y be written in the forms

$$y(x - a) = x^2 - a^2, \qquad (x - a)(y - x - a) = 0,$$

it will be seen that it can be represented graphically, as in the figure (Fig. 14), by the pair of lines

$$x - a = 0,$$
$$y - x - a = 0.$$

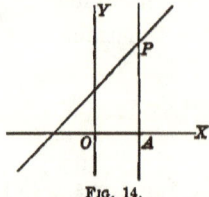

Fig. 14.

Hence when x has the value a there is an indefinite number of corresponding points on the locus, all situated on the line $x = a$; and thus for this value of x the function y may have any value whatever, and is then indeterminate.

When x has any value different from a, the corresponding value of y is determined from the equation $y = x + a$. Now, of the infinite number of different values of y corresponding to $x = a$, there is one particular value AP which is continuous with the series of values taken by y when x takes successive values in the vicinity of $x = a$. This may be called the *determinate* or *singular value* of y when $x = a$. It is obtained by putting $x = a$ in the equation $y = x + a$, and is therefore $y = 2a$.

This result may be stated without a locus as follows: When $x = a$, the function

$$\frac{x^2 - a^2}{x - a}$$

is indeterminate, and has an infinite number of different values; but among these values there is one determinate value which is continuous with the series of values taken by the function as x increases through the value a; this determinate or singular value may then be defined by

$$\lim_{x \doteq a} \frac{x^2 - a^2}{x - a}.$$

In evaluating this limit the infinitesimal factor $x - a$ may be removed from numerator and denominator, since this factor is not zero, while x is different from a; hence the determinate value of the function is

$$\lim_{x \doteq a} \frac{x + a}{1} = 2a.$$

Ex. 1. Find the determinate value, when $x \doteq 1$, of the function

$$\frac{x^{\frac{3}{2}} - 1 + (x - 1)^{\frac{3}{2}}}{(x^2 - 1)^{\frac{3}{2}} - x + 1},$$

which at the limit takes the indeterminate form $\frac{0}{0}$.

This expression may be written in the form

$$\frac{(x^{\frac{1}{2}} - 1)(x + x^{\frac{1}{2}} + 1) + (x^{\frac{1}{2}} - 1)^{\frac{3}{2}}(x^{\frac{1}{2}} + 1)^{\frac{3}{2}}}{(x^{\frac{1}{2}} - 1)^{\frac{3}{2}}(x^{\frac{1}{2}} + 1)^{\frac{3}{2}}(x + 1)^{\frac{1}{2}} - (x^{\frac{1}{2}} - 1)(x^{\frac{1}{2}} + 1)},$$

from which the infinitesimal factor $x^{\frac{1}{2}} - 1$ may be removed, giving

$$\frac{x + x^{\frac{1}{2}} + 1 + (x^{\frac{1}{2}} - 1)^{\frac{1}{2}}(x^{\frac{1}{2}} + 1)^{\frac{3}{2}}}{(x^{\frac{1}{2}} - 1)^{\frac{1}{2}}(x^{\frac{1}{2}} + 1)^{\frac{3}{2}}(x + 1)^{\frac{1}{2}} - (x^{\frac{1}{2}} + 1)},$$

which, when $x \doteq 1$, approaches the determinate value $-\frac{3}{2}$.

Ex. 2. Find the determinate value, when $x \doteq a$, of the expression

$$\frac{\sqrt{x} - \sqrt{a} + \sqrt{x-a}}{\sqrt{x^3 - a^3} + \sqrt{x^3 - a^2 x}},$$

by removing the infinitesimal factor $\sqrt{\sqrt{x} - \sqrt{a}}$.

70. Evaluation by transformation and removal of common factor. Sometimes a transformation must be made, before the common vanishing factor can be discovered and removed.

For instance, to evaluate, when $x \doteq 0$, the expression

$$\frac{a - \sqrt{a^2 - x^2}}{x^2},$$

which takes the form $\dfrac{0}{0}$. On multiplying numerator and denominator by $a + \sqrt{a^2 - x^2}$, the fraction becomes

$$\frac{x^2}{x^2(a + \sqrt{a^2 - x^2})},$$

which, by the removal of the common vanishing factor x^2, reduces to

$$\frac{1}{a + \sqrt{a^2 - x^2}},$$

and has therefore, when x is replaced by zero, the determinate value $\dfrac{1}{2a}$.

Ex. 1. Evaluate, when $x \doteq 0$, the function

$$\frac{1 - \sqrt{1-x}}{\sqrt{1+x} - \sqrt{1+x^2}}.$$

[Multiply numerator and denominator by

$$(1 + \sqrt{1-x})(\sqrt{1+x} + \sqrt{1+x^2}).]$$

Ex. 2. Evaluate, when $x \doteq 1$, the function

$$\frac{1 - x^{\frac{1}{3}}}{1 - (\sqrt{2x - x^2})^{\frac{1}{3}}}.$$

71. Evaluation by development. In some cases the common vanishing factor can be best removed after expansion in series.

Ex. 1. Consider the function mentioned in Art. 67,

$$\frac{e^x - e^{-x}}{\sin x}.$$

On developing numerator and denominator in powers of x, it becomes

$$\frac{1 + x + \frac{x^2}{2!} + \frac{x^3}{3!} + \cdots - \left(1 - x + \frac{x^2}{2!} - \frac{x^3}{3!} + \cdots\right)}{x - \frac{x^3}{3!} + \cdots},$$

$$= \frac{2x + \frac{2}{3!}x^3 + \cdots}{x - \frac{x^3}{3!} + \cdots} = \frac{2 + \frac{x^2}{3} + \cdots}{1 - \frac{x^2}{6} + \cdots},$$

which has the determinate value 2, when x takes the value zero.

Ex. 2. As another example, evaluate, when $x \doteq 0$, the function

$$\frac{x - \sin^{-1}x}{\sin^3 x}.$$

By development it becomes

$$\frac{x - \left(x - \frac{1}{2} \cdot \frac{x^3}{3} + \cdots\right)}{\left(x - \frac{x^3}{3!} + \cdots\right)} = \frac{\frac{x^3}{6} + \cdots}{x^3 + \cdots}.$$

Removing the common factor, and then putting $x = 0$, the result is $\frac{1}{6}$.

In these two examples the assigned value of x, for which the indeterminateness occurs, is zero, and the developments are made in powers of x. If the assigned value of x be some other number a, then the development should be made in powers of $x - a$.

Ex. 3. Evaluate, when $x \doteq a$, the expression
$$\frac{\log x - \log a}{\tan (x - a)},$$
which then takes the form $\frac{0}{0}$.

Developing $\log x$ in powers of $x - a$ by Art. 57 and $\tan (x - a)$ in powers of $x - a$ by Art. 56, the expression becomes
$$\frac{x - a - \frac{1}{2}(x - a)^2 + \cdots}{x - a + \frac{1}{3}(x - a)^3 + \cdots},$$
which, on removing the common vanishing factor $x - a$, is
$$\frac{1 - \frac{1}{2}(x - a) + \cdots}{1 + \frac{1}{3}(x - a)^2 + \cdots},$$
and reduces to unity, when x takes the assigned value a.

In such a case it is usually convenient to write for $x - a$ a single letter h, and then x is replaced by $a + h$.

Ex. 4. Evaluate, when $x \doteq 1$, the function,
$$\frac{1 - x + \log x}{1 - \sqrt{2x - x^2}}.$$

Let $x - 1 = h$, $x = 1 + h$, then by developing in powers of h, the expression becomes
$$\frac{-h + \log(1 + h)}{1 - \sqrt{1 + h^2}} = \frac{-h + \left(h - \frac{h^2}{2} + \cdots\right)}{1 - \left(1 + \frac{h^2}{2} + \cdots\right)} = \frac{-\frac{1}{2}h^2 + \cdots}{-\frac{1}{2}h^2 + \cdots},$$
which, on removing the common vanishing factor h^2, and then putting $x = 1$ (that is, $h = 0$), reduces to the value unity.

Ex. 5. Evaluate, when $x = \dfrac{\pi}{2}$, the function

$$\frac{\cos x}{1 - \sin x}.$$

Putting $x - \dfrac{\pi}{2} = h$, $x = \dfrac{\pi}{2} + h$, the expression becomes

$$\frac{\cos\left(\dfrac{\pi}{2} + h\right)}{1 - \sin\left(\dfrac{\pi}{2} + h\right)} = \frac{-\sin h}{1 - \cos h} = \frac{-h + \dfrac{h^3}{6} - \cdots}{\dfrac{h^2}{2} - \dfrac{h^4}{24} + \cdots} = \frac{-1 + \dfrac{h^2}{6} - \cdots}{\dfrac{h^2}{2} - \dfrac{h^3}{24} + \cdots},$$

which becomes infinite when $h = 0$, that is, when $x = \dfrac{\pi}{2}$,

hence $\quad\displaystyle\lim_{x \doteq 0} \frac{\cos x}{1 - \sin x} = \infty.$

72. Evaluation by differentiation. Let the given function be of the form $\dfrac{f(x)}{\phi(x)}$, and suppose that $f(a) = 0$, $\phi(a) = 0$. It is required to find $\displaystyle\lim_{x \doteq a} \frac{f(x)}{\phi(x)}$.

As before, let $f(x)$, $\phi(x)$ be developed in the vicinity of $x = a$, by expanding them in powers of $x - a$, then

$$\frac{f(x)}{\phi(x)} = \frac{f(a) + f'(a)(x-a) + \dfrac{f''(a + \theta(x-a))}{2!}(x-a)^2}{\phi(a) + \phi'(a)(x-a) + \dfrac{\phi''(a + \theta_1(x-a))}{2!}(x-a)^2}$$

$$= \frac{f'(a)(x-a) + \dfrac{f''(a + \theta(x-a))}{2!}(x-a)^2}{\phi'(a)(x-a) + \dfrac{\phi''(a + \theta_1(x-a))}{2!}(x-a)^2}.$$

By dividing by $x - a$ and then letting $x \doteq a$, it follows that

$$\lim_{x \doteq a} \frac{f(x)}{\phi(x)} = \frac{f'(a)}{\phi'(a)}.$$

The functions $f'(a)$, $\phi'(a)$ will in general both be finite.

If $f'(a) = 0$, $\phi'(a) \neq 0$, then $\dfrac{f(a)}{\phi(a)} = 0$.

If $f'(a) \neq 0$, $\phi'(a) = 0$, then $\dfrac{f(a)}{\phi(a)} = \infty$.

If $f'(a)$ and $\phi'(a)$ are both zero, the limiting value of $\dfrac{f(x)}{\phi(x)}$ is to be obtained by carrying Taylor's development one term further, removing the common factor $(x - a)^2$, and then letting $x \doteq a$. The result is $\dfrac{f''(a)}{\phi''(a)}$.

Similarly, if $f(a)$, $f'(a)$, $f''(a)$; $\phi(a)$, $\phi'(a)$, $\phi''(a)$ all vanish, it is proved in the same manner that

$$\lim_{x \doteq a} \frac{f(x)}{\phi(x)} = \frac{f'''(a)}{\phi'''(a)},$$

and so on, until a result is obtained that is not indeterminate in form.

Hence the rule:

To evaluate an expression of the form $\dfrac{0}{0}$, *differentiate numerator and denominator separately; substitute the critical value of x in their derivatives, and equate the quotient of the derivatives to the indeterminate form.*

Ex. 1. Evaluate $\dfrac{1-\cos\theta}{\theta^2}$ when $\theta = 0$.

Put $\quad\quad f(\theta) = 1 - \cos\theta, \quad \phi(\theta) = \theta^2;$
then $\quad\quad f'(\theta) = \sin\theta, \quad\quad \phi'(\theta) = 2\theta,$
and $\quad\quad f'(0) = 0, \quad\quad\quad\quad \phi'(0) = 0.$

gain, $f''(\theta) = \cos\theta,\quad \phi''(\theta) = 2,$
$f''(0) = 1,\quad \phi''(0) = 2,$

hence $\lim\limits_{\theta \doteq 0} \dfrac{1-\cos\theta}{\theta^2} = \dfrac{1}{2}.$

Ex. 2. Find $\lim\limits_{x \doteq 0} \dfrac{e^x + e^{-x} + 2\cos x - 4}{x^4},$

$= \lim\limits_{x \doteq 0} \dfrac{e^x - e^{-x} - 2\sin x}{4\,x^3},$

$= \lim\limits_{x \doteq 0} \dfrac{e^x + e^{-x} - 2\cos x}{12\,x^2},$

$= \lim\limits_{x \doteq 0} \dfrac{e^x - e^{-x} + 2\sin x}{24\,x},$

$= \lim\limits_{x \doteq 0} \dfrac{e^x + e^{-x} + 2\cos x}{24},$

$= \dfrac{1}{4}.$

Ex. 3. Find $\lim\limits_{x \doteq 0} \dfrac{x - \sin x \cos x}{x^3}.$

Ex. 4. Find $\lim\limits_{x \doteq 1} \dfrac{x^5 - 2x^3 - 4x^2 + 9x - 4}{x^4 - 2x^3 + 2x - 1}.$

In this example, show that $x-1$ is a factor of both numerator and denominator.

Ex. 5. Find $\lim\limits_{x \doteq 0} \dfrac{3\tan x - 3x - x^3}{x^5}$

In applying this process to particular problems, the work can often be shortened by evaluating a non-vanishing factor in either numerator or denominator before performing the differentiation.

Ex. 6. Find $\lim\limits_{x \doteq 0} \dfrac{(x-4)^2 \tan x}{x}$

$= \lim\limits_{x \doteq 0} \dfrac{(x-4)^2 \sec^2 x + 2(x-4)\tan x}{1}$

$= 16.$

The example shows that it is unnecessary to differentiate the factor $(x-4)^2$, as the coefficient of its derivative vanishes.

In general, if $f(x) = \psi(x)\chi(x)$, and if $\psi(a) = 0$, $\chi(a) \neq 0$, $\phi(a) = 0$, then

$$\lim_{x \doteq a} \frac{f(x)}{\phi(x)} = \chi(a) \frac{\psi'(a)}{\phi'(a)},$$

for $\lim_{x \doteq a} \frac{f(x)}{\phi(x)} = \lim_{x \doteq a} \frac{\psi(x)\chi(x)}{\phi(x)} = \frac{\psi(a)\chi'(a) + \psi'(a)\chi(a)}{\phi'(a)}$

$$= \chi(a) \frac{\psi'(a)}{\phi'(a)}, \text{ since } \psi(a) = 0.$$

Otherwise thus:

$$\lim_{x \doteq a} \frac{\psi(x)\chi(x)}{\phi(x)} = \lim_{x \doteq a} \chi(x) \cdot \lim_{x \doteq a} \frac{\psi(x)}{\phi(x)} = \chi(a) \cdot \frac{\psi'(a)}{\phi'(a)}.$$

Ex. 7. Find $\lim_{x \doteq \frac{\pi}{2}} \frac{\sin x \cos^2 x}{(2x - \pi)^2}$.

Ex. 8. Find $\lim_{x \doteq 1} \frac{(x-3)^2 \log(2-x)}{\sin(x-1)}$.

There are other indeterminate forms than $\frac{0}{0}$; they are $\frac{\infty}{\infty}$, $\infty - \infty$, 0^0, 1^∞, ∞^0. The form $0 - 0$ is not indeterminate, the value of the function being evidently zero.

The form $\infty - \infty$ may be finite, zero, or infinite.

For instance, consider $\sqrt{x^2 + ax} - x$ for the value $x = \infty$; it is of the form $\infty - \infty$, but by multiplying and dividing by $\sqrt{x^2 + ax} + x$ it becomes $\dfrac{ax}{\sqrt{x^2 + ax} + x}$, which has the form $\dfrac{\infty}{\infty}$ when $x = \infty$.

Again, by dividing both terms by x, it takes the form $\dfrac{a}{\sqrt{1 + \dfrac{a}{x}} + 1}$, and this becomes $\dfrac{a}{2}$ when $x = \infty$.

73. Evaluation of the indeterminate form $\frac{\infty}{\infty}$.

Let the function $\dfrac{f(x)}{\phi(x)}$ become $\dfrac{\infty}{\infty}$ when $x = a$; it is required to find $\lim\limits_{x \doteq a} \dfrac{f(x)}{\phi(x)}$.

This function can be written

$$\frac{f(x)}{\phi(x)} = \frac{\dfrac{1}{\phi(x)}}{\dfrac{1}{f(x)}},$$

which takes the form $\dfrac{0}{0}$ when $x = a$, and can therefore be evaluated by the preceding rule.

When $x \doteq a$,

$$\frac{f(x)}{\phi(x)} = \frac{\dfrac{1}{\phi(x)}}{\dfrac{1}{f(x)}} = \frac{-\dfrac{\phi'(x)}{[\phi(x)]^2}}{-\dfrac{f'(x)}{[f(x)]^2}} = \left[\frac{f(x)}{\phi(x)}\right]^2 \cdot \frac{\phi'(x)}{f'(x)}. \qquad (1)$$

Dividing through by $\dfrac{f(x)}{\phi(x)}$, it becomes

$$1 = \frac{f(x)}{\phi(x)} \cdot \frac{\phi'(x)}{f'(x)},$$

therefore
$$\left[\frac{f(x)}{\phi(x)}\right]_{x \doteq a} = \frac{f'(a)}{\phi'(a)}. \qquad (2)$$

This is exactly the same result as was obtained for the form $\dfrac{0}{0}$; hence the procedure for evaluating the indeterminate forms $\dfrac{0}{0}, \dfrac{\infty}{\infty}$, is the same in both cases.

When the true value of $\dfrac{f(a)}{\phi(a)}$ is 0 or ∞, equation (1) is satisfied, independent of the value of $\dfrac{f'(a)}{\phi'(a)}$; but (2) still gives the correct form; for suppose $\lim\limits_{x \doteq a} \dfrac{f(x)}{\phi(x)} = 0$; and consider the function

$$\frac{f(x)}{\phi(x)} + c, = \frac{f(x) + c\phi(x)}{\phi(x)},$$

which has the form $\dfrac{\infty}{\infty}$ when $x = a$, and has the determinate value c, which is not zero; hence by (2)

$$\lim_{x \doteq a} \frac{f(x) + c\phi(x)}{\phi(x)} = \frac{f'(a) + c\phi'(a)}{\phi'(a)} = \frac{f'(a)}{\phi'(a)} + c;$$

therefore, by subtracting c,

$$\lim_{x \doteq a} \frac{f(x)}{\phi(x)} = \frac{f'(a)}{\phi'(a)}.$$

If $\lim\limits_{x \doteq a} \dfrac{f(x)}{\phi(x)} = \infty$, then $\lim\limits_{x \doteq a} \dfrac{\phi(x)}{f(x)} = 0$, which can be treated as the previous case.

74. Evaluation of the form $\infty \cdot 0$.

Let the function be $\phi(x) \cdot \psi(x)$, such that $\phi(a) = \infty$, $\psi(a) = 0$.

This may be written $\dfrac{\psi(x)}{\dfrac{1}{\phi(x)}}$, which takes the form $\dfrac{0}{0}$ when a is substituted for x, and therefore comes under the above rule. (Art. 72.)

75. Evaluation of the form $\infty - \infty$. — There is here no general rule of procedure as in the previous cases, but by means of transformations and proper grouping of terms it is often possible to bring it into one of the forms $\dfrac{0}{0}$, $\dfrac{\infty}{\infty}$. Frequently a function which becomes $\infty - \infty$ for a critical value of x can be put in the form

$$\frac{u}{v} - \frac{t}{w},$$

in which v, w becomes zero; and this equals

$$\frac{uw - vt}{vw},$$

which is then of the form $\frac{0}{0}$.

Ex. 1. Find $\lim\limits_{x \doteq \frac{\pi}{2}} (\sec x - \tan x)$.

This expression assumes the form $\infty - \infty$, but can be written

$$\frac{1}{\cos x} - \frac{\sin x}{\cos x} = \frac{1 - \sin x}{\cos x},$$

which is of the form $\frac{0}{0}$, and gives, when evaluated,

$$\lim\limits_{x \doteq \frac{\pi}{2}} (\sec x - \tan x) = 0.$$

Ex. 2. Prove $\lim\limits_{x \doteq \frac{\pi}{2}} (\sec^n x - \tan^n x) = \infty, 1, 0$ according as
$n > 2, \ = 2, \ < 2$.

EXERCISES

Evaluate the following expressions, both by expansion and also by differentiation:

1. $\dfrac{\log x}{x - 1}$ when $x = 1$.

2. $\dfrac{e^x - e^{-x}}{\sin x}$ $\quad x = 0$.

3. $\dfrac{\log (2x^2 - 1)}{\tan (x - 1)}$ $\quad x = 1$.

4. $\dfrac{\log \sin x}{(\pi - 2x)^2}$ $\quad x = \dfrac{\pi}{2}$.

5. $\dfrac{x^4 - 2x^3 + 2x - 1}{x^6 - 15x^2 + 24x - 10}$ $\quad x = 1$.
   (Evaluate also without the use of derivatives.)

6. $\dfrac{e^x - e^{-x} - 2x}{x - \sin x}$ $\quad x = 0$.

7. $\dfrac{x - 2}{(x - 1)^n - 1}$ $\quad x = 2$.

8. $\dfrac{x - \sin^{-1}x}{\sin^3 x}$ when $x = 0$.

9. $\dfrac{a^x - b^x}{x}$ $x = 0$.

10. $\dfrac{\tan x - x}{x - \sin x}$ $x = 0$.

11. $\dfrac{1 - x + \log x}{1 - \sqrt{2x - x^2}}$ $x = 1$.

12. $\dfrac{m \sin x - \sin mx}{x(\cos x - \cos mx)}$ $x = 0$.

13. $\dfrac{x^2}{1 - \cos mx}$ $x = 0$.

14. $\dfrac{\sqrt{a^2 - x^2} + a - x}{\sqrt{a^2 - \dfrac{x^3}{a}} + \sqrt{ax - x^2}}$ $x = a$.

15. $\dfrac{x^{\frac{3}{2}} - 1 + (x-1)^{\frac{3}{2}}}{(x^2 - 1)^{\frac{1}{2}} - x + 1}$ $x = 1$.

16. $\dfrac{x^3 \cot^2 x + \sin x}{x}$ $x = 0$.

17. $\dfrac{(e^x - e^{-x})^2 - 2x^2(e^x + e^{-x})}{x^4}$ $x = 0$.

18. $\dfrac{1 - \sqrt{1-x}}{\sqrt{1+x} - \sqrt{1+x^2}}$ $x = 0$.

19. $\dfrac{\sqrt{2} - \sin x - \cos x}{\log \sin 2x}$ $x = \dfrac{\pi}{4}$.

20. $\dfrac{e^x + \log\left(\dfrac{1-x}{e}\right)}{\tan x - x}$ $x = 0$.

21. $\dfrac{\sec x}{\sec 3x}$ $x = \dfrac{\pi}{2}$.

22. $\dfrac{a^x}{\csc(ma^{-x})}$ $x = \infty$.

23. $\dfrac{\log(1+x)}{x}$ $x = \infty$.

24. $e^x \sin \dfrac{1}{x}$ when $x = \infty$.

25. $\dfrac{\tan x}{\tan 5x}$ $x = \dfrac{\pi}{2}$.

26. $\dfrac{x^2 - a^2}{a^2} \tan \dfrac{\pi x}{2a}$ $x = a$.

27. $e^{-\frac{1}{x}}(1 - \log x)$ $x = 0$.

28. $\log(x - a) \tan(x - a)$ $x = a$.

29. $\dfrac{\sec^n x}{e^{\tan x}}$ $x = \dfrac{\pi}{2}$.

30. $(1 - x) \tan \dfrac{\pi x}{2}$ $x = 1$.

31. $\dfrac{x(e^{2x} + e^x)}{(e^x - 1)^3} - \dfrac{2 e^x}{(e^x - 1)^2}$ $x = 0$.

32. $\dfrac{2}{x} - \cot \dfrac{x}{2}$ $x = 0$.

33. $\dfrac{x}{x - 1} - \dfrac{1}{\log x}$ $x = 1$.

34. $\dfrac{x}{\sin^3 x} - \cot^2 x$ $x = 0$.

35. $\dfrac{\pi}{4x} - \dfrac{\pi}{2x(e^{\pi x} + 1)}$ $x = 0$.

36. Prove that if $f(a) = 1$, $\phi(a) = 1$,
$$\lim_{x \doteq a} \dfrac{\log f(x)}{\log \phi(x)} = \dfrac{f'(a)}{\phi'(a)}.$$

37. $2^x \sin \dfrac{a}{2^x}$ $x = \infty$.

38. $\sqrt{a^2 - x^2} \cot \left\{ \dfrac{\pi}{2} \sqrt{\dfrac{a - x}{a + x}} \right\}$ $x = a$.

39. $\dfrac{\left(x + \sin x - 4 \sin \dfrac{x}{2}\right)^4}{\left(3 + \cos x - 4 \cos \dfrac{x}{2}\right)^3}$ $x = 0$.

76. Evaluation of the form 1^∞.

Let the function $u = [\phi(x)]^{\psi(x)}$ assume the form 1^∞ when $x = a$.

To make the exponent a multiplier, take the logarithm of both sides; then

$$\log u = \psi(x) \cdot \log \phi(x) = \frac{\log \phi(x)}{\frac{1}{\psi(x)}}.$$

This expression assumes the form $\frac{0}{0}$ when $x = a$, and can be evaluated by the method of Art. 71.

If the reduced value of this fraction be denoted by m, then $\log u = m$ and $u = e^m$.

NOTE. The form 1^0 is not indeterminate, but is equal to 1. For, let $[\phi(x)]^{\psi(x)}$ assume the form 1^0 when $x = a$.

Put $\qquad u = [\phi(x)]^{\psi(x)},$
then $\qquad \log u = \psi(x) \log [\phi(x)],$
which equals zero when $x = a$;
hence $\qquad \log u = 0, \quad u = e^0 = 1.$

77. Evaluation of the forms 0^0, ∞^0.

Let $[\phi(x)]^{\psi(x)}$ become ∞^0 when $x = a$.

Put $\qquad u = [\phi(x)]^{\psi(x)},$
then $\qquad \log u = \psi(x) \log \phi(x) = \frac{\log \phi(x)}{\frac{1}{\psi(x)}}.$

This is of the form $\frac{\infty}{\infty}$, and can be evaluated by the method of Art. 72. Similarly for the form 0^0.

NOTE. The form 0^∞ is not indeterminate, but is equal to 0. For let $u = [\phi(x)]^{\psi(x)}$ become 0^∞ when $x = a$, then $\log u = \psi(x) \log \phi(x) = -\infty$, and $u = e^{-\infty} = 0$.

This completes the list of ordinary indeterminate forms.

The evaluation of all of them depends upon the same principle, namely, that each form (or its logarithm) may be brought to the form $\frac{0}{0}$, and then evaluated by differentiating numerator and denominator separately.

This principle will be subsequently employed in the application of the Differential Calculus to Geometry.

EXERCISES

Evaluate the following indeterminate forms:

1. $(\cos x)^{\cot^2 x}$ when $x = 0$.
2. $(\cos \alpha x)^{\csc^2 \beta x}$ $x = 0$.
3. $\left(\dfrac{1}{x}\right)^{\tan x}$ $x = 0$.
4. $(1-x)^{\frac{1}{x}}$ $x = 0$.
5. $x^{\frac{1}{1-x}}$ $x = 1$.
6. $\left(\dfrac{\log x}{x}\right)^{\frac{1}{x}}$ $x = \infty$.
7. $(1-x)^{\frac{1}{x}}$ $x = \infty$.
8. $\left(\dfrac{a_1^{\frac{1}{x}} + a_2^{\frac{1}{x}} + \cdots a_n^{\frac{1}{x}}}{n}\right)^{nx}$ $x = \infty$.

CHAPTER VI

MODE OF VARIATION OF FUNCTIONS OF ONE VARIABLE

78. In this chapter methods of exhibiting the march or mode of variation of functions, as the variable takes all values in succession from $-\infty$ to $+\infty$, will be discussed. Simple examples have been given in Art. 19 of the use that can be made of the derivative function $\phi'(x)$ for this purpose.

The fundamental principle employed is that when x increases through the value a, $\phi(x)$ increases through the value $\phi(a)$ if $\phi'(a)$ is positive, and that $\phi(x)$ decreases through the value $\phi(a)$ if $\phi'(a)$ is negative. Thus the question of finding whether $\phi(x)$ increases or decreases through an assigned value $\phi(a)$, is reduced to determining the sign of $\phi'(a)$.

Ex. 1. Find whether the function

$$\phi(x) \equiv x^2 - 4x + 5$$

increases or decreases through the values $\phi(3) = 2$, $\phi(0) = 5$, $\phi(2) = 1$, $\phi(-1) = 10$, and state at what value of x the function ceases to increase and begins to decrease, or conversely.

79. Turning values of a function. It follows that the values of x, at which $\phi(x)$ ceases to increase and begins to decrease are those at which $\phi'(x)$ changes sign from positive to negative; and that the values of x, at which $\phi(x)$ ceases to decrease and begins to increase, are those at which $\phi'(x)$ changes its sign from negative to positive. In the former case, $\phi(x)$ is said to pass through a *maximum*, in the latter, a *minimum* value.

Ex. 2. Find the turning values of the function
$$\phi(x) \equiv 2x^3 - 3x^2 - 12x + 4,$$
and exhibit the general march of the function by sketching the curve $y = \phi(x)$.

Here $\phi'(x) = 6x^2 - 6x - 12, = 6(x+1)(x-2)$, hence $\phi'(x)$ is negative when x lies between -1 and $+2$, and positive for all other values of x. Thus $\phi(x)$ increases from $x = -\infty$ to $x = -1$, decreases from $x = -1$ to $x=2$ and increases from $x = 2$ to $x = \infty$. Hence $\phi(-1)$ is a maximum value of $\phi(x)$, and $\phi(2)$ a minimum.

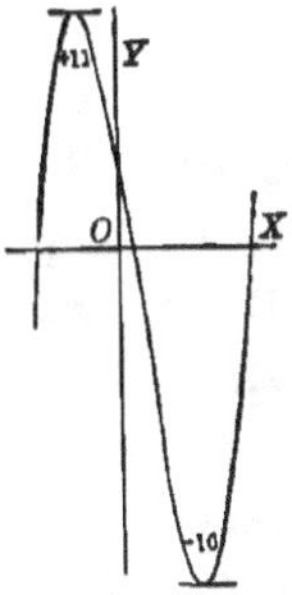

Fig. 15.

The general form of the curve $y = \phi(x)$ (Fig. 15) may be inferred from the last statement, and from the following simultaneous values of x and y:

$$x = -\infty, \; -2, \; -1, \; 0, \quad 1, \quad 2, \quad 3, \; 4, \; \infty.$$
$$y = -\infty, \quad 0, \quad 11, \; 4, \; -9, \; -16, \; -5, \; 36, \; \infty.$$

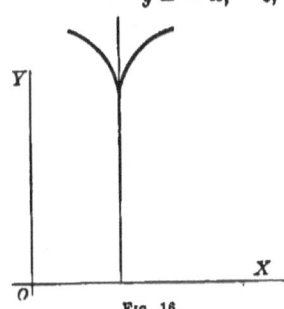

Fig. 16.

Ex. 3. Exhibit the march of the function
$$\phi(x) \equiv (x-1)^{\frac{2}{3}} + 2,$$
especially its turning values.

Since $\phi'(x) = \dfrac{2}{3} \dfrac{1}{(x-1)^{\frac{1}{3}}}$,

hence $\phi'(x)$ changes sign at $x = 1$; being negative when $x < 1$, infinite when $x = 1$, and positive when $x > 1$. Thus $\phi(1) = 2$ is a minimum turning value of $\phi(x)$; and the graph of the function is as shown in Fig. 16, with a vertical tangent at the point $(1, 2)$.

Ex. 4. Examine for maxima and minima the function
$$\phi(x) \equiv (x-1)^{\frac{1}{3}} + 1.$$

Here $\phi'(x) = \dfrac{1}{3} \dfrac{1}{(x-1)^{\frac{2}{3}}}$,

hence $\phi'(x)$ never changes sign, but is always positive. Thus there is no turning value. The curve $y = \phi(x)$ has a vertical tangent at the point $(1, 1)$, since $\dfrac{dy}{dx} = \phi'(x)$ is infinite when $x = 1$. (Fig. 17.)

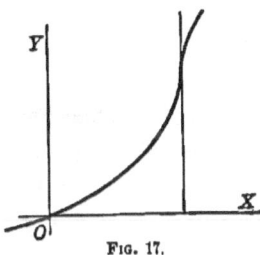

Fig. 17.

80. Critical values of the variable. It has been shown that the necessary and sufficient condition for a turning value of $\phi(x)$ is that $\phi'(x)$ shall change its sign. Now a function can only change its sign either when it passes through zero, as in Ex. 2, or when its reciprocal passes through zero, as in Exs. 3, 4. In the latter case it is usual to say that the function passes through infinity. It is not true, conversely, that a function always changes its sign in passing through zero or infinity, e.g., $y = x^2$.

Nevertheless all the values of x, at which $\phi'(x)$ passes through zero or infinity, are called *critical* values of x, because they are to be further examined to determine whether $\phi'(x)$ actually changes sign as x passes through these values; and whether, in consequence, $\phi(x)$ passes through a turning value.

For instance, in Ex. 2, the derivative $\phi'(x)$ vanishes when $x = -1$, and when $x = 2$, and it does not become infinite for any finite value of x. Thus the critical values are $-1, 2$; and it is found that both give turning values to $\phi(x)$. Again, in Exs. 3, 4, the critical value is $x = 1$, since it makes $\phi'(x)$ infinite, and it gives a turning value to $\phi(x)$ in Ex. 3, but not in Ex. 4.

81. Method of determining whether $\phi'(x)$ changes its sign in passing through zero or infinity.

Let a be a critical value of x, in other words let $\phi'(a)$ be either zero or infinite, and let h be a very small positive number; then $a - h$ and $a + h$ are two numbers very close to a, and on opposite sides of it; thus in order to determine whether $\phi'(x)$ changes sign as x increases through the value a, it is only necessary to compare the signs of $\phi'(a + h)$ and $\phi'(a - h)$. If it is possible to take h so small that $\phi'(a - h)$

is positive and $\phi'(a+h)$ negative, then $\phi'(x)$ changes sign as x passes through the value a, and $\phi(x)$ passes through a maximum value $\phi(a)$. Similarly, if $\phi'(a-h)$ is negative and $\phi'(a+h)$ positive, then $\phi(x)$ passes through a minimum value $\phi(a)$.

If $\phi'(a-h)$ and $\phi'(a+h)$ have the same sign, however small h may be, then $\phi(a)$ is not a turning value of $\phi(x)$.

Ex. 5. Find the turning values of the function
$$\phi(x) \equiv (x-1)^2 (x+1)^3.$$
Here $\phi'(x) = 2(x-1)(x+1)^3 + 3(x-1)^2(x+1)^2$
$= (x-1)(x+1)^2(5x-1),$

hence $\phi'(x)$ passes through zero at $x = -1, \tfrac{1}{5}$, and 1; and it does not become infinite for any finite value of x.

Thus, the critical values are $-1, \tfrac{1}{5}, 1$.

When $x = -1 - h$, the three factors of $\phi'(x)$ take the signs $- + -$, and when $x = -1 + h$, they become $- + -$; thus $\phi'(x)$ does not change sign as x increases through -1; hence $\phi(-1) = 0$ is not a turning value of $\phi(x)$.

When $x = \tfrac{1}{5} - h$, the three factors of $\phi'(x)$ are $- + -$, and when $x = \tfrac{1}{5} + h$, they become $- + +$; thus $\phi'(x)$ changes sign from $+$ to $-$ as x increases through $\tfrac{1}{5}$, and $\phi(\tfrac{1}{5}) = 1 \cdot 1 \cdots$ is a maximum value of $\phi(x)$.

Finally, when $x = 1 - h$, the three factors of $\phi'(x)$ have the signs $- + +$, and when $x = 1 + h$ they become $+ + +$; thus $\phi'(x)$ changes sign from $-$ to $+$ as x increases through 1, and $\phi(1) = 0$ is a minimum value of $\phi(x)$.

The deportment of the function and its first derivative in the vicinity of the critical values may be tabulated thus:

x	$-1-h$	-1	$-1+h$	$\tfrac{1}{5}-h$	$\tfrac{1}{5}$	$\tfrac{1}{5}+h$	$1-h$	1	$1+h$
$\phi'(x)$	+	0	+	+	0	−	−	0	+
$\phi(x)$	inc.	infl. 0	inc.	inc.	max. 1·1	dec.	dec.	min. 0	inc.

The general march of the function may be exhibited graphically by tracing the curve $y = \phi(x)$ (Fig. 18), using the foregoing result and also the following simultaneous values of x and y:

$$x = -\infty,\ -2,\ -1,\ 0,\ \tfrac{1}{5},\ 1,\ 2,\ \infty.$$
$$y = -\infty,\ -9,\ 0,\ 1,\ 1\tfrac{1}{10},\ 0,\ 27,\ \infty.$$

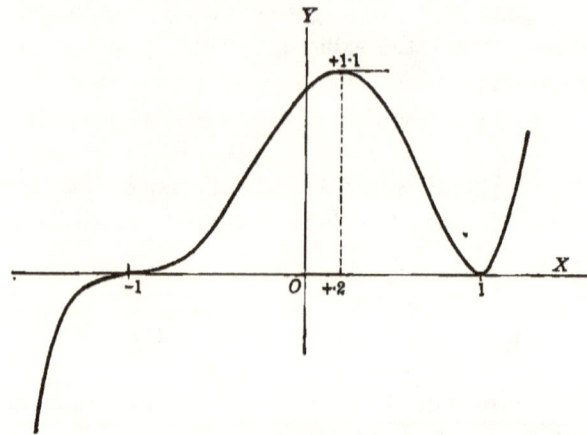

Fig. 18.

Ex. 6. Show the march of the function

$$\phi(x) \equiv \sin^2 x \cdot \cos x.$$
$$\phi'(x) = 2 \sin x \cos^2 x - \sin^3 x$$
$$= \sin x (2 \cos^2 x - \sin^2 x),$$

hence the critical values of x are found from the equations

$$\sin x = 0, \text{ and } 2\cos^2 x - \sin^2 x = 0, \text{ or } \tan x = \pm \sqrt{2}.$$

Thus the critical values of x are $x = 0$, $x = \pi$, $x = 2\pi \cdots$ and $x = \pm a$, $\pi \pm a$, $2\pi \pm a$, $\cdots$ where $a = \tan^{-1} \sqrt{2} = .85 \cdots$ radians.

When $x = -h$, the factors of $\phi'(x)$ are $-$, $+$.
$x = 0,$ $\quad\quad\quad\quad\quad\quad\quad\quad$ $0, +,$
$x = +h,$ $\quad\quad\quad\quad\quad\quad\quad$ $+, +;$

thus $\phi'(x)$ changes from $-$ to $+$ as x increases through zero, and $\phi(0) = 0$ is a minimum value of $\phi(x)$.

When $x = \pi - h$, the factors of $\phi'(x)$ are $+, +,$
$x = \pi,$ $\quad\quad\quad\quad\quad\quad\quad\quad$ $0, +,$
$x = \pi + h,$ $\quad\quad\quad\quad\quad\quad\quad$ $-, +;$

thus $\phi'(x)$ changes from $+$ to $-$ at $x = \pi$, and $\phi(\pi) = 0$ is a maximum value of $\phi(x)$.

Similarly, $\phi'(x)$ changes from $-$ to $+$ at $x = 2\pi$, and $\phi(2\pi) = 0$ is a minimum value of $\phi(x)$.

VARIATION OF FUNCTIONS

Again, when $x = a - h$, the factors of $\phi'(x)$ are $+$, $+$,
  $x = a$, $+$, 0,
  $x = a + h$, $+$, $-$.

(Observe that when x increases to $a + h$, $\cos 2$ diminishes, and $\sin x$ increases; thus the zero factor at $x = a$ becomes negative at $x = a + h$. Similarly, it becomes positive at $x = a - h$.)

Thus $\phi'(x)$ changes from $+$ to $-$ at $x = a$, and $\phi(x)$ has a maximum value at $\phi(a)$.

When $x = \pi - a - h$, the factors of $\phi'(x)$ are $+$, $-$,
  $x = \pi - a$, $+$, 0,
  $x = \pi - a + h$, $+$, $+$.

(Observe that since $\pi - a$ is in the second quarter, diminishing $\pi - a$ increases the sine and diminishes the cosine numerically, and thus changes the zero factor to negative.)

Thus $\phi'(x)$ changes from $-$ to $+$ as x increases through $\pi - a$, and $\phi(\pi - a)$ is a minimum value of $\phi(x)$.

It may be shown in the same manner that $\phi(\pi + a)$ is a minimum, $\phi(2\pi - a)$ a maximum, and so on.

Combining the two sets of results, the form of the curve is found to be that of the accompanying figure (Fig. 19).

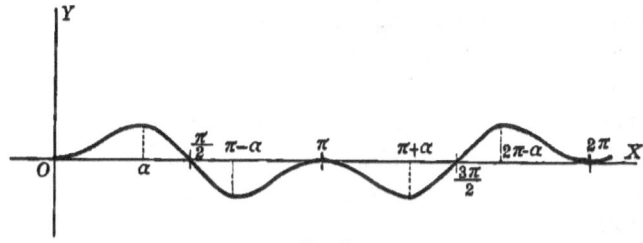

Fig. 19.

82. Second method of determining whether $\phi'(x)$ changes sign in passing through zero. The following method may be employed when the function and its derivatives are continuous in the vicinity of the critical value $x = a$.

Suppose, when x increases through the value a, that $\phi'(x)$ changes sign from positive through zero to negative. Its change from positive to zero is a decrease, and so is the change from zero to negative; thus $\phi'(x)$ is a decreasing function at $x=a$, and hence its derivative, $\phi''(x)$, is negative at $x=a$.

On the other hand, if $\phi'(x)$ changes sign from negative through zero to positive, it is an increasing function, and $\phi''(x)$ is positive at $x = a$; hence:

The function $\phi(x)$ has a maximum value $\phi(a)$, when $\phi'(a) = 0$ and $\phi''(a)$ is negative; $\phi(x)$ has a minimum value $\phi(a)$, when $\phi'(a) = 0$ and $\phi''(a)$ is positive.

It may happen, however, that $\phi''(a)$ is also zero.

In this case, to determine whether $\phi(x)$ has a turning value, it is necessary to proceed to the higher derivatives. If $\phi(x)$ is a maximum, $\phi''(x)$ is negative just before vanishing, and negative just after, for the reason given above; but the change from negative to zero is an increase, and the change from zero to negative is a decrease; thus $\phi''(x)$ changes from increasing to decreasing as x passes through a. Hence $\phi'''(x)$ changes sign from positive through zero to negative, and it follows, as before, that its derivative, $\phi^{\text{IV}}(x)$, is negative.

Thus $\phi(a)$ is a maximum value of $\phi(x)$ if $\phi'(a) = 0$, $\phi''(a) = 0$, $\phi'''(a) = 0$, $\phi^{\text{IV}}(a)$ negative. Similarly, $\phi(a)$ is a minimum value of $\phi(x)$ if $\phi'(a) = 0$, $\phi''(a) = 0$, $\phi'''(a) = 0$, and $\phi^{\text{IV}}(a)$ positive.

If it happen that $\phi^{\text{IV}}(a) = 0$, it is necessary to proceed to still higher derivatives to test for turning values. The result may then be generalized thus:

The function $\phi(x)$ has a maximum (or minimum) value at $x=a$ if one or more of the derivatives $\phi'(a)$, $\phi''(a)$, $\phi'''(a)$ vanish and if the first one that does not vanish is of even order, and negative (or positive).

Ex. 7. Find the critical values of Ex. 5 by the second method.

$\phi''(x) = (x+1)^2(5x-1) + 2(x-1)(x+1)(5x-1) + 5(x-1)(x+1)^2$,

$\phi''(1) = 16$, hence $\phi(1)$ is a minimum value of $\phi(x)$,

$\phi''(-1) = 0$, hence it is necessary to find $\phi'''(-1)$,

$\phi'''(x) = 2(x+1)(5x-1) + 5(x+1)^2 + 2(x+1)(5x-1) + 2(x-1)(5x-1)$
$\qquad + 10(x-1)(x+1) + 5(x+1)^2 + 10(x-1)(x+1)$.

$\phi'''(-1) = 24$, hence $\phi(-1)$ is neither a maximum nor a minimum value of $\phi(x)$.

Again, $\phi''\left(\dfrac{1}{5}\right) = 5\left(\dfrac{1}{5} - 1\right)\left(\dfrac{1}{5} + 1\right)^2$ is negative, hence $\phi\left(\dfrac{1}{5}\right)$ is a maximum value of $\phi(x)$.

Ex. 8. Examine similarly the critical values of Ex. 6.

In this case the second derivative reduces to

$$\phi''(x) = \cos x(2\cos^2 x - 7\sin^2 x),$$

hence $\phi''(0)$ is positive, $\phi'(\pi)$ is negative; thus $\phi(0)$ is a minimum and $\phi(\pi)$ a maximum value of $\phi(x)$.

Again, $\qquad \phi''(a) = \cos a(2\cos^2 a - 7\sin^2 a)$,

but a satisfies the equation $2\cos^2 a - \sin^2 a = 0$, hence $\phi''(a)$ is negative and $\phi(a)$ is a minimum value of $\phi(x)$.

Also $\phi''(\pi - a) = -\cos a(2\cos^2 a - 7\sin^2 a)$ is positive, and $\phi(\pi - a)$ a minimum value of $\phi(x)$. Similarly for the other critical values of Ex. 6.

83. Conditions for maxima and minima derived from Taylor's theorem.

In this article, as in the preceding, the function and its derivatives are supposed to be continuous in the vicinity of $x = a$; otherwise the method of Art. 81 must be used.

Let $\phi(a)$ be a maximum value of $\phi(x)$; then it follows from the definition that $\phi(a)$ is greater than either of the neighboring values, $\phi(a+h)$, $\phi(a-h)$, when h is taken small enough. Hence $\phi(a+h) - \phi(a)$ and $\phi(a-h) - \phi(a)$ are both negative.

Similarly, these expressions are both positive if $\phi(a)$ is a minimum value of $\phi(x)$.

Let $\phi(x+h)$, $\phi(x-h)$ be expanded in powers of h by Taylor's theorem;

then $\phi(x+h) = \phi(x) + \phi'(x)h + \dfrac{\phi''(x)}{2!}h^2 + \dfrac{\phi'''(x+\theta h)}{3!}h^3$,

$\phi(x-h) = \phi(x) - \phi'(x)h + \dfrac{\phi''(x)}{2!}h^2 - \dfrac{\phi'''(x-\theta_1 h)}{3!}h^3$.

If x be replaced by a, and $\phi(a)$ transposed, there results

$\phi(a+h) - \phi(a) = \phi'(a)h + \dfrac{\phi''(a)}{2!}h^2 + \dfrac{\phi'''(a+\theta h)}{3!}h^3$,

$\phi(a-h) - \phi(a) = -\phi'(a)h + \dfrac{\phi''(a)}{2!}h^2 - \dfrac{\phi'''(a-\theta_1 h)}{3!}h^3$.

The increment h can now be taken so small that $h\phi'(a)$ will be numerically larger than the sum of the remaining terms in the second member of either of the last two equations. Thus $\phi(a+h) - \phi(a)$ and $\phi(a-h) - \phi(a)$ cannot have the same sign unless $\phi'(a)$ be zero, hence the first condition for a turning value is $\phi'(a) = 0$.

In this case

$\phi(a+h) - \phi(a) = \dfrac{\phi''(a)}{2!}h^2 + \dfrac{\phi'''(a+\theta h)}{3!}h^3$,

$\phi(a-h) - \phi(a) = \dfrac{\phi''(a)}{2!}h^2 - \dfrac{\phi'''(a-\theta_1 h)}{3!}h^3$,

and h can be taken so small that the first term on the right is numerically larger than either of the second terms, hence $\phi(a+h) - \phi(a)$ and $\phi(a-h) - \phi(a)$ are both negative when $\phi''(a)$ is negative, and both positive when $\phi''(a)$ is positive.

Thus $\phi(a)$ is a maximum (or minimum) value of $\phi(x)$ when $\phi'(a)$ is zero and $\phi''(a)$ is negative (or positive).

If it should also happen that $\phi''(a)$ is also zero, then

$$\phi(a+h) - \phi(a) = \frac{\phi'''(a)}{3!}h^3 + \frac{\phi^{IV}(a+\theta h)}{4!}h'',$$

$$\phi(a-h) - \phi(a) = -\frac{\phi'''(a)}{3!}h^3 + \frac{\phi^{IV}(a-\theta_1 h)}{4!}h'',$$

and by the same reasoning as before, it follows that for a maximum (or minimum) there are the further conditions that $\phi'''(a)$ equals zero, and that $\phi^{IV}(a)$ is negative (or positive).

Proceeding in this way, the general conclusion stated in the last article is evident.

Ex. 1. Which of the preceding examples can be solved by the general rule here referred to?

Ex. 2. Why was the restriction imposed upon $\phi'(x)$ that it should change sign by passing through zero, rather than by passing through infinity?

84. Application to rational polynomials. When $\phi(x)$ is a rational polynomial, its derivative $\phi'(x)$ is of similar form. Let the real roots of the equation $\phi'(x) = 0$ be $a, b, c, \cdots l$, arranged in descending order of algebraic magnitude; suppose, first, that no two of them are equal; then $\phi'(x)$ has the form

$$\phi'(x) = (x-a)(x-b)(x-c)\cdots(x-l)P, \quad (1)$$

in which P is the product of the imaginary factors of the polynomial $\phi'(x)$. This product will have the same sign for all values of x, and by giving the coefficient of the highest power of x in $\phi'(x)$ a positive value, P will always be positive, by the theory of equations.

Differentiating (1) with regard to x, and putting $x = a$, it follows that

$$\phi''(a) = (a-b)(a-c)\cdots(a-l)P,$$

but $a-b$, $a-c$, $\cdots$ are all positive, hence $\phi''(a)$ is positive, and therefore $\phi(a)$ is a minimum value of $\phi(x)$.

Again, $\phi''(b) = (b-a)(b-c) \cdots (b-l)P$,

but $b-a$ is negative, and the remaining factors are positive; hence $\phi''(b)$ is negative, and $\phi(b)$ is a maximum value of $\phi(x)$.

Also $\phi''(c) = (c-a)(c-b) \cdots (c-l)P$,

in which the only negative factors are $c-a$, $c-b$; hence $\phi''(c)$ is positive and $\phi(c)$ is a minimum value of $\phi(x)$.

Similarly, the fourth root (in descending order) gives a maximum, and so do the sixth, eighth, $\cdots$, while the first, third, fifth, $\cdots$ correspond to minima.

Thus, if the equation $\phi'(x) = 0$ has $2n$ real roots, all of which are distinct, the function $\phi(x)$ has n maxima and n minima occurring alternately; if $\phi'(x) = 0$ has $2n+1$ distinct real roots, then $\phi(x)$ has $n+1$ minima and n maxima, the latter being situated, respectively, between successive minima.

Next, suppose that two of the roots are each equal to a;

then $\phi'(x) = (x-a)^2 \psi(x)$,

and $\phi''(x) = (x-a)^2 \psi'(x) + 2(x-a)\psi(x)$,

$\phi'''(x) = (x-a)^2 \psi''(x) + 4(x-a)\psi'(x) + 2\psi(x)$;

hence $\phi'(a) = 0$, $\phi''(a) = 0$, $\phi'''(a) = 2\psi(a)$;

therefore $\phi'(a)$ is neither a maximum nor a minimum.

If three of the roots of $\phi'(x)$ are each equal to a, it is proved similarly that $\phi(a)$ is a maximum or minimum according as $\psi(a)$ is negative or positive.

These conclusions may be extended to the cases of n equal roots, in which n is even or odd, respectively.

An illustrative example was given in Art. 81.

85. The maxima and minima of any continuous function occur alternately. It has been seen that the maximum and minimum values of a rational polynomial occur alternately when the variable is continually increased or diminished.

This principle is also true in the case of every continuous function of a single variable; for, let $\phi(a)$, $\phi(b)$ be two maximum values of $\phi(x)$, in which a is supposed less than b; then when $x = a + h$, the function is decreasing; when $x = b - h$, the function is increasing, h being taken sufficiently small, and positive. But in passing from a decreasing to an increasing state, a continuous function must, at some intermediate value of x, change from decreasing to increasing, that is, must pass through a minimum. Hence, between two maxima there must be at least one minimum.

It can be similarly proved that between two minima there must be at least one maximum.

86. Simplifications that do not alter critical values. The work of finding the critical values of the variable, in the case of any given function, may often be simplified by means of the following self-evident principles.

1. Any value of x that gives a turning value to $c\phi(x)$ gives also a turning value to $\phi(x)$, and conversely, when c is independent of x. These two turning values are of the same or opposite kind according as c is positive or negative.

2. Any value of x that gives a turning value to $c + \phi(x)$ gives also a turning value of the same kind to $\phi(x)$, and conversely, provided c is independent of x.

3. Any value of x that gives a turning value to $[\phi(x)]^n$ gives also a turning value to $\phi(x)$, and conversely, when n is independent of x. Whether these turning values are of the same or opposite kind depends on the sign of n, and also on the sign of $[\phi(x)]^n$.

EXERCISES

Find the critical values of x in the following examples, and determine the nature of the function at each, and obtain the graph of the function.

1. $u = x^3 + 18 x^2 + 105 x.$ 2. $u = (x - 1)^3 (x - 2)^2.$

3. $u = x (x - 1)^2 (x + 1)^3.$

4. $u = Ax^2 + Bx + C$; show that u cannot have both a maximum and a minimum value, for any values of A, B, C.

5. $u = 3 x^3 - 2 x + 4$. Show that a cubic function has in general both a maximum and a minimum value.

6. $u = 2 x + 4 - x^3$. Compare the graph of this function with that of exercise 4.

7. $u = x^x.$ 9. $u = \dfrac{(a - x)^3}{a - 2x}.$

8. $u = \dfrac{\log x}{x}.$ 10. $u = \sin 2 x - x.$

11. Show that the function $b + c (x - a)^{\frac{5}{3}}$ has neither a maximum nor a minimum.

12. $u = \sin^2 x \cos^3 x.$ 14. $u = x + \tan x.$

13. $u = \sin x + \cos 2 x.$ 15. $u = \dfrac{e^x}{x} + e^{-2x}.$

87. Geometric problems in maxima and minima. The theory of the turning values of a function has important applications in solving problems concerning geometric maxima or minima, *i.e.*, the determination of the largest or the smallest value a magnitude may have while satisfying certain stated geometric conditions.

The first step is to express the magnitude in question algebraically. If the resulting expression contains more than one variable, the stated conditions will furnish enough relations between these variables, so that all the others may be expressed in terms of one. The expression to be maximized or minimized can then be made a function of a single variable, and can be treated by the preceding rules.

Ex. 1. Find the largest rectangle whose perimeter is 100. Let x, y denote the dimensions of any of the rectangles whose perimeter is 100. The magnitude to be maximized is the area

$$u = xy, \tag{1}$$

in which the variables x, y are subject to the stated condition

$$2x + 2y = 100,$$

i.e.,
$$y = 50 - x, \tag{2}$$

hence the function to be maximized, expressed in terms of the single variable x, is

$$u = \phi(x) = x(50 - x) = 50x - x^2. \tag{3}$$

The critical value of x is found from the equation

$$\phi'(x) = 50 - 2x = 0,$$

and is $x = 25$. When x increases through this value, $\phi'(x)$ changes sign from positive to negative, and hence $\phi(x)$ is a maximum when $x = 25$. Equation (2) shows that the corresponding value of y is 25. Thus the maximum rectangle whose perimeter is 100, is the square whose side is 25.

Ex. 2. The sum of the three dimensions of a rectangular box is 10, the total surface is 34; find its dimensions so that its volume may be a maximum.

Here the function

$$u = xyz \tag{1}$$

is to be maximized, the three variables being subject to the two conditions

$$x + y + z = 10, \tag{2}$$

$$xy + xz + yz = 17. \tag{3}$$

Equation (2) multiplied by z, subtracted from (3) and transposed, gives

$$xy = 17 - 10z + z^2,$$

by means of which the variables x and y can be eliminated from (1), giving

$$u = (17 - 10z + z^2)z.$$

Hence the function to be maximized by varying z is

$$\phi(z) = z^3 - 10z^2 + 17z,$$

then
$$\phi'(z) = 3z^2 - 20z + 17 = (z - 1)(3z - 17),$$

$$\phi''(z) = 6z - 20;$$

hence the critical value $z=1$, which makes $\phi'(z)$ zero and $\phi''(z)$ negative, gives to $\phi(z)$ the maximum value 8. The other two dimensions, found from (2) and (3), are 8 and 1. The second critical value, $z = 5\frac{2}{3}$, makes $\phi''(z)$ positive, and $\phi(z)$ an algebraic minimum. The corresponding dimensions are $5\frac{2}{3}$, $-1\frac{1}{3}$, $5\frac{2}{3}$, a result not applicable to the special problem in question. Thus the required dimensions are 8, 1, 1. Any change of these dimensions subject to the given conditions will lessen the volume.

Ex. 3. If, from a square piece of tin whose side is a, a square be cut out at each corner, find the side of the latter square in order that the remainder may form a box of maximum capacity, with open top.

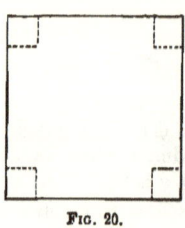

Fig. 20.

Let x be a side of each square cut out, then the bottom of the box will be a square whose side is $a - 2x$, and the depth of the box will be x, hence the volume is

$$v = x(a - 2x)^2,$$

which is to be made a maximum by varying x.

Here $\dfrac{dv}{dx} = (a - 2x)^2 - 4x(a - 2x)$

$= (a - 2x)(a - 6x).$

This derivative vanishes when $x = \dfrac{a}{2}$, and when $x = \dfrac{a}{6}$. It will be found by applying the usual test, that $x = \dfrac{a}{2}$ gives the minimum value zero, and that $x = \dfrac{a}{6}$ gives it a maximum value $\dfrac{4a^3}{27}$, hence the side of the square to be cut out is one sixth the side of the given square.

Ex. 4. Find the area of the greatest rectangle that can be inscribed in a given ellipse.

An inscribed rectangle will evidently be symmetric with regard to the principal axes of the ellipse.

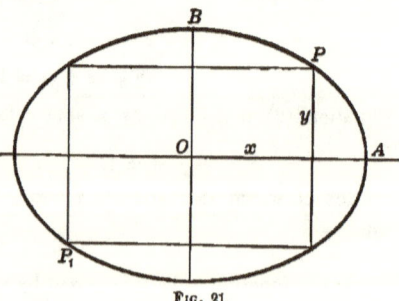

Fig. 21.

Let a, b denote the lengths of the semi-axes OA, OB (Fig. 21); let $2x$, $2y$ be the dimensions of an inscribed rectangle; then the area is

$$u = 4xy, \qquad (1)$$

87.]   *VARIATION OF FUNCTIONS*   147

in which the variables x, y may be regarded as the coördinates of the vertex P, on the curve, and are therefore subject to the equation of the ellipse

$$\frac{x^2}{a^2} + \frac{y^2}{b^2} = 1. \qquad (2)$$

It is geometrically evident that there is some position of P for which the inscribed rectangle is a maximum; for let P be supposed to take in succession all positions between A and B; then just as P moves away from A the rectangle begins by increasing from zero, and when P comes to B the rectangle ends by decreasing back to zero; hence there must be a change from increasing to decreasing, *i.e.*, a maximum, for at least one intermediate position.

The elimination of y from (1), by means of (2), gives the function of x to be maximized,

$$u = \frac{4b}{a} x \sqrt{a^2 - x^2}. \qquad (3)$$

By Art. 86, the critical values of x are not altered if this function be divided by the constant $\frac{4b}{a}$, and then squared. Hence, the values of x which render u a maximum, give also a maximum value to the function

$$\phi(x) = x^2(a^2 - x^2) = a^2x^2 - x^4.$$

Here $\phi'(x) = 2a^2x - 4x^3 = 2x(a^2 - 2x^2),$

$\phi''(x) = 2a^2 - 12x^2;$

hence, by the usual tests, the critical values $x = \pm \dfrac{a}{\sqrt{2}}$ render $\phi(x)$, and therefore the area u, a maximum. The corresponding values of y are given by (2), and the vertex P may be at any of the four points denoted by

$$x = \pm \frac{a}{\sqrt{2}}, \quad y = \pm \frac{b}{\sqrt{2}},$$

giving in each case the same maximum inscribed rectangle, whose dimensions are $a\sqrt{2}, b\sqrt{2}$, and whose area is $2ab$, or half that of the circumscribed rectangle.

Ex. 5. Find the cylinder of maximum volume that can be cut from a given prolate spheroid.

Let the spheroid and inscribed cylinder be generated by the figure of Ex. 4 revolving about OA; then the volume of the cylinder is

$$v = 2\pi xy^2, \qquad (1)$$

and this is to be maximized subject to the condition

$$\frac{x^2}{a^2}+\frac{y^2}{b^2}=1;\qquad (2)$$

hence
$$v=\frac{2\pi b^2}{a^2}x(a^2-x^2),$$

and by Art. 86, when this function is a maximum, so is the function

$$x(a^2-x^2),$$

which, according to the usual tests, has its maximum when $x=\dfrac{a}{\sqrt{3}}$. The corresponding value of y, from (2), is $\dfrac{b\sqrt{2}}{\sqrt{3}}$; hence, from (1), the maximum volume is

$$v=\frac{4\pi ab^2}{3\sqrt{3}},$$

or $\dfrac{1}{\sqrt{3}}$ of the volume of the prolate spheroid.

Ex. 6. Find the greatest cylinder that can be cut from a given right cone, whose height is h, and the radius of whose base is a.

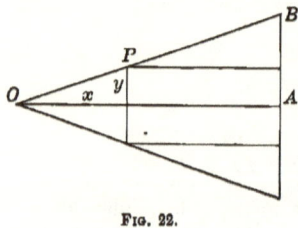

Fig. 22.

Let the cone be generated by the revolution of the triangle OAB (Fig. 22); and the inscribed cylinder by that of the rectangle AP.

Let $OA = h$, $AB = a$, and let the coördinates of P be (x, y); then the function to be maximized is $\pi y^2(h-x)$ subject to the relation $\dfrac{y}{x}=\dfrac{a}{h}$.

Ex. 7. Find the area of the greatest rectangle that can be inscribed in the segment of the parabola $y^2 = px$, cut off by the line $x = a$.

Ex. 8. What is the altitude of the maximum cylinder that can be inscribed in a given segment of a paraboloid of revolution?

Ex. 9. Find the greatest right-angled triangle that can be constructed on a given line as hypothenuse.

Ex. 10. Given the vertical angle of a triangle, and its area. Find when its base is a minimum.

Ex. 11. A Norman window consists of a rectangle surmounted by a semicircle. Given the perimeter; required the height and breadth of window when the quantity of light admitted is a maximum.

VARIATION OF FUNCTIONS

Ex. 12. The diameter of a cylindrical tree is a. Find the strongest beam that may be cut from it, assuming that the strength is proportional to the breadth multiplied by the square of the thickness.

Ex. 13. An open tank is to be constructed with a square base and vertical sides. Show that the area of the entire inner surface will be least if the depth is half the width.

Ex. 14. The sum of the perimeters of a circle and a square is fixed. Show that when the sum of the areas is least, the side of the square is double the radius of the circle.

Ex. 15. What should be the ratio between the diameter of the base and the height of a cylindrical fruit can in order that the amount of tin used in constructing it may be the least possible? Solve the same problem when the top is open.

Ex. 16. The top of a pedestal which sustains a statue c feet in height is b feet above the level of a man's eyes. Find his horizontal distance from the pedestal when the statue subtends the greatest angle.

Ex. 17. A high vertical wall is to be braced by a beam which must pass over a parallel wall a feet high, and b feet distant from the other. Find the length of the shortest beam that can be used for the purpose.

Ex. 18. Determine the cone of minimum volume that can be described about a given sphere.

Ex. 19. Find the shortest distance from the point (2, 1) to the parabola $y^2 = 4x$.

Ex. 20. The lower corner of a leaf, whose width is a, is folded over so as just to reach the inner edge of the page; find the width of the part folded over when the length of the crease is a minimum.

Ex. 21. A tangent is drawn to the ellipse whose semi-axes are a and b, such that the part intercepted by the axes is a minimum; show that its length is $a + b$.

Ex. 22. A person being in a boat 3 miles from the nearest point on the beach, wishes to reach in the shortest time a place 5 miles from that point along the shore; supposing he can walk 5 miles an hour, but row only at the rate of 4 miles an hour, find the place where he must land.

Ex. 23. A slip noose in a rope is thrown around a large square post, and the rope drawn tight in the direction as shown in the figure. At what angle does the rope leave the post?

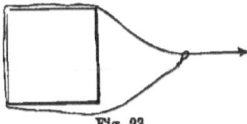

Fig. 23.

Ex. 24. Show that just before and after a turning value the function passes through equal values. Apply this principle to give geometrical solutions to Exs. 22, 23.

Ex. 25. Show that in the vicinity of a turning value $\Delta f(x)$ is an infinitesimal of an even order when Δx is of the first order. When is $\Delta f(x)$ of the third order?

Ex. 26. A rectangular court is to be built so as to contain a given area, and a wall already constructed is available for one of the sides; find its dimensions so that the least expense may be incurred.

Ex. 27. The work of driving a steamer through the water being proportional to the cube of her speed, find the most economical rate per hour against a current running a knots per hour.

Ex. 28. Assuming that the current in a voltaic cell is $C = \dfrac{E}{r + R}$, E being electromotive force, r internal resistance, R external resistance, and that the power given out is $P = RC^2$, prove that P is a maximum when $r = R$. Trace the curve that shows the variation of P, as R varies.

CHAPTER VII

RATES AND DIFFERENTIALS

88. Rates. Time as independent variable. Suppose a particle P is moving in any path, straight or curved, and let s be the number of space-units passed over in t seconds; then s may be taken as the dependent variable, and t as the independent variable.

Let Δs be the number of space-units described in the additional time Δt seconds; then the *average velocity* of P during the time Δt is $\dfrac{\Delta s}{\Delta t}$, the average number of space-units described per second during the interval.

The velocity of P is said to be *uniform* if its average velocity, $\dfrac{\Delta s}{\Delta t}$, is the same for all intervals Δt. The *actual velocity* of P at any instant denoted by t is the limit which the average velocity, for the interval between the time t and the time $t + \Delta t$, approaches as Δt is made to approach zero as a limit.

Thus
$$v = \lim_{\Delta t \doteq 0} \frac{\Delta s}{\Delta t} = \frac{ds}{dt}$$

is the actual velocity of P at the time denoted by t. It is evidently the number of space-units that would be passed over in the next second if the velocity remained uniform from the time t to the time $t + 1$.

It may be observed that if, for the word "velocity," the more general term, "rate of change," be used, the above statements will apply to any quantity that varies with the time, whether it be length, volume, strength of current, etc. For instance, let the quantity of an electric current be C at time t, and $C + \Delta C$ at time $t + \Delta t$; then the *average rate of change* of current in the interval Δt is $\dfrac{\Delta C}{\Delta t}$, the average increase in current units per second; and, as before, the actual rate of change at the instant denoted by t is

$$\lim_{\Delta t \doteq 0} \frac{\Delta C}{\Delta t} = \frac{dC}{dt}.$$

This is the number of current-units that would be gained in the next second if the rate of gain were uniform from the time t to the time $t + 1$. Since, by Art. 21,

$$\frac{dy}{dx} = \frac{dy}{dt} : \frac{dx}{dt},$$

hence $\dfrac{dy}{dx}$ measures the ratio of the rates of change of y and of x.

It follows that the result of differentiating

$$y = f(x) \tag{1}$$

may be written in either of the forms

$$\frac{dy}{dx} = f'(x), \tag{2}$$

$$\frac{dy}{dt} = f'(x)\frac{dx}{dt}. \tag{3}$$

The latter form is often convenient, and may also be obtained directly from (1) by differentiating both sides with regard to t. It may be read: the rate of change of y is $f'(x)$ times the rate of change of x.

Returning to the illustration of a moving point P, let its coördinates at time t be x and y; then $\dfrac{dx}{dt}$ measures the rate

88.] RATES AND DIFFERENTIALS 153

of change of the x-coördinate, and may be called the velocity of P resolved parallel to the x-axis, or the x-component of the velocity.

Similarly, $\dfrac{dy}{dt}$ is the y-component of velocity.

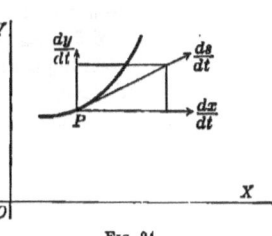

Fig. 24.

These three rates of change are connected by the equation

$$\left(\frac{ds}{dt}\right)^2 = \left(\frac{dx}{dt}\right)^2 + \left(\frac{dy}{dt}\right)^2. \tag{4}$$

Ex. 1. If a point describe the straight line $3x + 4y = 5$, and if x increase h units per second, find the rates of increase of y and of s.

Since $\quad y = \tfrac{5}{4} - \tfrac{3}{4}x,$

hence $\quad \dfrac{dy}{dt} = -\dfrac{3}{4}\dfrac{dx}{dt},$

and when $\quad \dfrac{dx}{dt} = h, \ \dfrac{dy}{dt} = -\tfrac{3}{4}h,$

$$\frac{ds}{dt} = \sqrt{h^2 + \tfrac{9}{16}h^2} = \tfrac{5}{4}h.$$

Ex. 2. A point describes the parabola $y^2 = 12x$, in such a way that when $x = 3$, the abscissa is increasing at the rate of 2 feet per second: at what rate is y then increasing? Find also the rate of increase of s.

Since $\quad y^2 = 12x,$

$$2y\frac{dy}{dt} = 12\frac{dx}{dt},$$

$$\frac{dy}{dt} = \frac{6}{y}\frac{dx}{dt} = \frac{6}{\sqrt{12x}}\frac{dx}{dt};$$

hence, when $x = 3$, and $\dfrac{dx}{dt} = 2, \ \dfrac{dy}{dt} = \pm\sqrt{2}.$

Again, $\left(\dfrac{ds}{dt}\right)^2 = \left(\dfrac{dx}{dt}\right)^2 + \left(\dfrac{dy}{dt}\right)^2$, hence $\dfrac{ds}{dt} = 2\sqrt{2}$ feet per second.

Ex. 3. A person is walking towards the foot of a tower on a horizontal plane at the rate of 5 miles per hour; at what rate is he approaching the top, which is 60 feet high, when he is 80 feet from the bottom?

Let x be the distance from foot of tower at time t, and y the distance from the top at the same time; then

$$x^2 + 60^2 = y^2,$$

$$x\frac{dx}{dt} = y\frac{dy}{dt}.$$

When x is 80 feet, y is 100 feet; hence if $\frac{dx}{dt}$ is 5 miles per hour, $\frac{dy}{dt}$ is 4 miles per hour.

89. Abbreviated notation for rates. When, as in the above examples, a time derivative is a factor of each member of an equation, it is usually convenient to write, instead of the symbols $\frac{dx}{dt}$, $\frac{dy}{dt}$, the abbreviations dx and dy, for the rates of change of the variables x and y. Thus the result of differentiating

$$y = f(x) \qquad (1)$$

may be written in either of the forms

$$\frac{dy}{dx} = f'(x), \qquad (2)$$

$$\frac{dy}{dt} = f'(x)\frac{dx}{dt}, \qquad (3)$$

$$dy = f'(x)\,dx. \qquad (4)$$

It is to be observed that the last form is not to be regarded as derived from equation (2) by separation of the symbols dy, dx; for the derivative $\frac{dy}{dx}$ has been defined as the result of performing upon y an indicated operation represented by the symbol $\frac{d}{dx}$; and thus the dy and dx of the symbol $\frac{dy}{dx}$ have been given no separate meaning.

The dy and dx of equation (4) stand for the rates or time derivatives $\dfrac{dy}{dt}$ and $\dfrac{dx}{dt}$ in (3), which is itself obtained from (1) by differentiation with regard to t, by Art. 21.

In case the dependence of y upon x be not indicated by a functional operation f, equations (3), (4) take the form

$$\frac{dy}{dt} = \frac{dy}{dx}\frac{dx}{dt},$$

$$dy = \frac{dy}{dx} dx.$$

In the abbreviated notation, equation (4) of the last article is written
$$ds^2 = dx^2 + dy^2.$$

Ex. 1. A point that is describing the parabola $y^2 = 2px$ is moving at time t with a velocity of v feet per second; find the rate of increase of the coördinates x and y at the same instant.

Differentiating the given equation with regard to t,

$$ydy = pdx,$$

but dx, dy also satisfy the relation

$$dx^2 + dy^2 = v^2;$$

hence, by solving these simultaneous equations,

$$dx = \frac{y}{\sqrt{y^2+p^2}} v, \quad dy = \frac{p}{\sqrt{y^2+p^2}} v, \text{ in feet per second.}$$

Ex. 2. A vertical wheel of radius 10 ft. is making 5 revolutions per second about a fixed axis. Find the horizontal and vertical velocities of a point on the circumference situated 30° from the horizontal.

Since $\quad x = 10 \cos \theta, \quad y = 10 \sin \theta,$

$\quad\quad\quad dx = -10 \sin \theta d\theta, \quad dy = 10 \cos \theta d\theta,$

but $\quad d\theta = 100\pi = 314.16$ feet per second,

hence $\quad dx = -3141.6 \sin \theta = -1570.8$ feet per second,

$\quad\quad\quad dy = 3141.6 \cos \theta = 2720.6$ feet per second.

Ex. 3. Trace the changes in the horizontal and vertical velocity in a complete revolution.

90. Differentials often substituted for rates. The symbols dx, dy have been defined above as the rates of change of x and y per second.

They may sometimes, however, be conveniently allowed to stand for any two numbers, large or small, that are proportional to these rates; and the equations, being homogeneous in them, will not be affected. It is usual in such cases to speak of the numbers dx and dy by the more general name of *differentials*, and they may then be either the rates themselves, or any two numbers in the same ratio.

This will be especially convenient in problems in which the time variable is not explicitly mentioned.

Ex. 1. When x increases from 45° to 45° 15', find the increase of $\log_{10} \sin x$, assuming that the ratio of the rates of change of the function and the variable remains sensibly constant throughout the short interval.

Here $\quad\quad\quad dy = -.4343 \cot x\, dx = .4343\, dx$;

let $\quad\quad\quad dx = 15' = .004363$ radians;

then $\quad\quad\quad dy = .001895$,

which is the approximate increment of $\log_{10} \sin x$,

but $\quad\quad \log_{10} \sin 45° = -\tfrac{1}{2} \log 2 = -.150515$,

∴ $\quad\quad \log_{10} \sin 45° 15' = -.148612$.

Ex. 2. Expanding $\log_{10} \sin(x+h)$ as far as h^2 by Taylor's theorem, and then putting $x = .785398$, $h = .004363$, show what is the error made by neglecting the third term, as was done in Ex. 1.

Ex. 3. When x varies from 60° to 60° 10', find the increase in $\sin x$.

Ex. 4. Show that $\log_{10} x$ increases more slowly than x, when $x > \log_{10} e$, that is, $x > .4343$.

Ex. 5. Two sides, a, b, of a triangle are measured, and also the included angle C; find the error in the computed length of the third side c due to a small error in the observed angle C.

[Differentiate the equation $c^2 = a^2 + b^2 - 2ab \cos C$, regarding a, b as constant.]

Ex. 6. In a tangent galvanometer the tangent of the deflection of the needle is proportional to the current. Show that the relative error in the computed value of the current, due to a given error of reading, is least when the angle of deflection is 45°.

Ex. 7. The error in the area A of an ellipse, due to small errors in the semi-axes, is approximately given by $\dfrac{\Delta A}{A} = \dfrac{\Delta a}{a} + \dfrac{\Delta b}{b}$.

Ex. 8. The side of an equilateral triangle is 24 inches long and is increasing at the rate of two inches per day; how fast is the area of the triangle increasing?

Ex. 9. Find the rate of change in the area of a square when the side b is increasing at a ft. per second.

Ex. 10. In the function $y = 2x^3 + 6$, what is the value of x at the point where y increases 24 times as fast as x?

Ex. 11. A circular plate of metal expands by heat so that its diameter increases uniformly at the rate of 2 inches per second; at what rate is the surface increasing when the diameter is 5 inches?

Ex. 12. What is the value of x at the point at which $x^3 - 5x^2 + 17x$ and $x^3 - 3x$ change at the same rate?

Ex. 13. Find the points at which the rate of change of the ordinate $y = x^3 - 6x^2 + 3x + 5$ is equal to the rate of change of the slope of the tangent to the curve.

Ex. 14. The relation between s, the space through which a body falls, and t, the time of falling, is $s = 16t^2$; show that the velocity is equal to $32t$.

The rate of change of velocity is called *acceleration;* show that the acceleration of the falling body is a constant.

Ex. 15. A body moves according to the law $s = \cos(nt + c)$; show that its acceleration is negative and proportional to the space through which it has moved.

CHAPTER VIII

DIFFERENTIATION OF FUNCTIONS OF MORE THAN ONE VARIABLE

In the previous chapters the dependence of one variable upon another, called the independent variable, has been discussed. The mode of dependence of one variable upon two others will next be considered; and the relation between the dependent variable z and the independent variables x and y will be expressed in the form

$$z = f(x, y). \qquad (1)$$

Examples of such dependence have been seen in coördinate geometry of three dimensions; for instance, from the equation of a sphere referred to its center as origin.

$$x^2 + y^2 + z^2 = a^2,$$

any one of the variables may be expressed as a function of the other two; thus

$$z = \sqrt{a^2 - x^2 - y^2}.$$

Conversely, any relation of the form (1) can be exhibited graphically by taking x, y as coördinates of a point on a horizontal plane, and drawing at the point an ordinate to the plane to represent the corresponding value of the function z; the form of the surface of which (1) is the equation will represent the mode of variation of the function.

91. Definition of continuity. A function $f(x, y)$ is said to be continuous in the vicinity of the values $x = a$, $y = b$; when

$f(a, b)$ is real, finite, and determinate (whether unique or multiple-valued); and when the difference $f(a + h, b + k) - f(a, b)$ can be made less than any assigned number η, by taking h, k small enough, independent of the ratio of k to h; in other words, when

$$\underset{h\doteq 0,\ k\doteq 0}{\text{limit}} f(a+h,\ b+k) = f(a, b),$$

no matter in what way h and k approach their limits.

It is implied that, when the function is multiple-valued, attention is to be paid to the correspondence of the multiple values in the two members of this limit-relation.

In geometrical language the function $f(x, y)$ is continuous at $x=a$, $y=b$, when the ordinate of the surface $z = f(x, y)$ drawn at the point $(a+h, b+k)$ approaches as a limit the ordinate drawn at the point (a, b) irrespective of the direction in which the point $(a + h, b + k)$ moves to coincidence with the point (a, b).

92. Rate of variation. Partial derivatives. The most important question concerning the variation of a continuous function z is: what is the rate of change of z when x and y vary at given rates? It is convenient to consider first the simpler question: what is the rate of change of z when x varies at a given rate, and y remains constant?

In this case z is a function of the single variable x, and its rate of change is

$$\frac{dz}{dt} = \frac{dz}{dx}\frac{dx}{dt}, \qquad (1)$$

in which it is to be understood that the operation $\dfrac{dz}{dx}$ is performed on the supposition that y is a constant, and that $\dfrac{dz}{dt}$ is the rate of change of z in so far as it depends on the change of x. To indicate these facts without the qualifying verbal statements, equation (1) will be written in the form

$$\frac{d_x z}{dt} = \frac{\partial z}{\partial x} \cdot \frac{dx}{dt}, \qquad (2)$$

in which $\frac{\partial z}{\partial x}$ stands for the x-derivative of z when y is kept constant, and is called the *partial derivative* of z with regard to x, and $\frac{d_x z}{dt}$ denotes the rate of change of z in so far as it depends on the change of x.

Thus, by Art. 18, the partial derivative is the result of the indicated operation

$$\frac{\partial z}{\partial x} = \lim_{\Delta x \doteq 0} \frac{\Delta_x z}{\Delta x} = \lim_{\Delta x \doteq 0} \frac{f(x+\Delta x, y) - f(x, y)}{\Delta x},$$

Similarly, the rate of change of z when x is kept constant and y varies at a given rate is measured by

$$\frac{d_y z}{dt} = \frac{\partial z}{\partial y} \cdot \frac{dy}{dt}, \qquad (3)$$

in which $\frac{d_y z}{dt}$ is the rate of change of z in so far as it depends upon the change of y, and $\frac{\partial z}{\partial y}$ denotes the partial derivative of z taken with regard to y, that is, the result of the operation indicated by

$$\frac{\partial z}{\partial y} = \lim_{\Delta y \doteq 0} \frac{\Delta_y z}{\Delta y} = \lim_{\Delta y \doteq 0} \frac{f(x, y+\Delta y) - f(x, y)}{\Delta y}.$$

93. Geometric illustration. Let the function $f(x, y)$ be represented graphically by the ordinate of a surface whose equation is $z = f(x, y)$ and let a vertical section be taken parallel to the plane (z, x) at a given distance $y = y_1$ from that plane; then if a point P be supposed to describe on the surface the contour of the section, the y-coördinate will remain constant, and the value of the varying ordinate z will be given by the equation $z = f(x, y_1)$. If the rate of variation of x at any instant be known, the corresponding rate of variation of z is given by

$$\frac{d_x z}{dt} = \frac{\partial z}{\partial x}\frac{dx}{dt} = \frac{\partial f(x, y_1)}{\partial x}\frac{dx}{dt},$$

which may be called the rate of variation of the ordinate in the x-direction.

The partial derivative $\frac{\partial z}{\partial x}$ is the ratio of the rates of increase of z and x, and is represented geometrically by the slope of the tangent drawn to the contour at P.

Ex. 1. A point P on the surface $z = x^2 y + 2xy^2$ moves in the plane $y = 2$ at the rate of 10 feet per second; find the rate of change of z, when P is passing through the point for which $x = 3$, and also the direction and velocity of the motion of P.

Differentiating the given identity with regard to t, y being kept constant, $\frac{d_x z}{dt} = (2xy + 2y^2)\frac{dx}{dt} = 20\frac{dx}{dt} = 200$ feet per second, and the slope of the tangent at P in the plane of motion is 20.

The velocity of P in the curve is $\sqrt{10^2 + 200^2} = 200.25$ feet per second.

Similarly, if P move on the surface in the plane $x = x_1$, the rate of change of z will be given by

$$\frac{d_y z}{dt} = \frac{\partial z}{\partial y}\frac{dy}{dt} = \frac{\partial f(x, y)}{\partial y}\frac{dy}{dt},$$

and $\frac{\partial z}{\partial y}$, the ratio of the rates of change of z and y, will measure the slope of the tangent at P in the plane of motion.

Ex. 2. Find for the same surface as before, at the point for which $x = 3$, $y = 2$ the rate of change of z in the y-direction, if y be changing at the rate of 5 feet per second, x being kept constant.

Here $\frac{d_y z}{dt} = (x^2 + 4xy)\frac{dy}{dt} = 33\frac{dy}{dt} = 165$ feet per second, and the slope in the direction of motion is 33.

94. Simultaneous variation of x and y; total rate of variation of z. It will now be shown that when x and y vary simultaneously, the total rate of change of z is the sum of its separate rates of change as x and y vary alone; that is,

$$\frac{dz}{dt} = \frac{d_x z}{dt} + \frac{d_y z}{dt}, \qquad (1)$$

or
$$\frac{dz}{dt} = \frac{\partial z}{\partial x}\frac{dx}{dt} + \frac{\partial z}{\partial y}\frac{dy}{dt}. \qquad (2)$$

For, let $z = f(x, y)$, and let x, y start at the values x_1, y_1, and take increments Δx, Δy; then the initial value of z is $f(x_1, y_1)$, and its final value is $f(x_1 + \Delta x, y_1 + \Delta y)$; hence the total increment of z is

$$f(x_1 + \Delta x, \ y_1 + \Delta y) - f(x_1, \ y_1).$$

By subtracting and adding the intermediate value

$$f(x_1 + \Delta x, \ y_1),$$

in which x alone has varied from its original value, the total increment of z may be written as the sum of two partial increments in the form

$$\Delta z = [f(x_1 + \Delta x, \ y_1 + \Delta y) - f(x_1 + \Delta x, \ y_1)]$$
$$+ [f(x_1 + \Delta x, \ y_1) - f(x_1, \ y_1)];$$

the latter being the increment of $f(x, y)$ as x changes from x_1 to $x_1 + \Delta x$, y remaining constant, and the former being the further increment of the function as x remains at the value $x_1 + \Delta x$ while y changes from y_1 to $y_1 + \Delta y$.

The result of dividing by Δt, the increment of t, may be written

$$\frac{\Delta z}{\Delta t} = \frac{f(x_1 + \Delta x, \ y_1 + \Delta y) - f(x_1 + \Delta x, \ y_1)}{\Delta y} \frac{\Delta y}{\Delta t}$$
$$+ \frac{f(x_1 + \Delta x, \ y_1) - f(x_1, \ y_1)}{\Delta x} \frac{\Delta x}{\Delta t}.$$

Taking limits as Δt, Δx, Δy, Δz, all approach the limit zero, and remembering that by Art. 92,

$$\frac{f(x_1+\Delta x, y_1)-f(x_1, y_1)}{\Delta x} \doteq \frac{\partial z}{\partial x}, \text{ taken at } x=x_1, y=y_1,$$

$$\frac{f(x_1+\Delta x, y_1+\Delta y)-f(x_1+\Delta x, y_1)}{\Delta y} \doteq \frac{\partial z}{\partial y}, \text{ taken at } x=x_1+\Delta x, y=y_1,$$

$$\doteq \frac{\partial z}{\partial y}, \text{ taken at } x=x_1, y=y_1,$$

it follows that, at any values of x and y, for which the function and its partial derivatives are continuous,

$$\frac{dz}{dt} = \frac{\partial z}{\partial x}\frac{dx}{dt} + \frac{\partial z}{\partial y}\frac{dy}{dt}.$$

In the abbreviated rate notation, equations (2), (3) of Art. 92, and (1), (2) of Art. 94, are respectively,

$$d_x z = \frac{\partial z}{\partial x} dx, \qquad d_y z = \frac{\partial z}{\partial y} dy,$$

$$dz = d_x z + d_y z = \frac{\partial z}{\partial x} dx + \frac{\partial z}{\partial y} dy.$$

Ex. 1. A particle moves on the spherical surface $x^2 + y^2 + z^2 = a^2$ in a vertical meridian plane inclined at an angle of 60° to the plane (zx).

If the x-component of its velocity be $\frac{1}{10}a$ per second, when $x = \frac{1}{4}a$, find the y-component, the z-component, and the resultant velocity.

Since $\qquad z = \sqrt{a^2 - x^2 - y^2},$

$$dz = -\frac{x\,dx}{\sqrt{a^2-x^2-y^2}} - \frac{y\,dy}{\sqrt{a^2-x^2-y^2}};$$

but since $dx = \frac{1}{10}a$, and the equation of the given meridian plane is $y = x \tan 60°$, hence $dy = dx\sqrt{3} = \frac{a}{10}\sqrt{3}$, and $y = \frac{a\sqrt{3}}{4}$. Therefore

$$dz = -\frac{dx}{2\sqrt{3}} - \frac{dy}{2} = -\frac{a\sqrt{3}}{15} = -.115\,a \text{ in feet per second.}$$

Also, $\quad ds = \sqrt{dx^2 + dy^2 + dz^2} = \frac{a\sqrt{3}}{75} = .023\,a$ in feet per second.

95. Language of differentials. The results of the preceding articles may be stated thus:

The partial z-differential due to the change of x is equal to the x-differential multiplied by the partial x-derivative.

The partial z-differential due to the change of y is equal to the y-differential multiplied by the partial y-derivative.

The total z-differential is equal to the sum of the partial z-differentials.

One advantage in keeping the equation in the differential form is that it may be divided when necessary by the differential of any other variable s, to which x and y are related, and then, remembering that the ratio of two differentials (or rates) may be expressed as a derivative, the equation becomes

$$\frac{dz}{ds} = \frac{\partial z}{\partial x}\frac{dx}{ds} + \frac{\partial z}{\partial y}\frac{dy}{ds}.$$

Ex. 1. Given $\quad z = axy^2 + bx^2y + cx^3 + ey,$
$\quad dz = (ay^2 + 2bxy + 3cx^2)dx + (2axy + bx^2 + e)dy.$

Ex. 2. Given $\quad z = x^y, \quad d_xz = yx^{y-1}dx, \quad d_yz = x^y \log x \, dy,$
$\quad dz = yx^{y-1}dx + x^y \log x \, dy.$

Ex. 3. Given $\quad u = \tan^{-1}\frac{y}{x}, \quad du = \frac{x\,dy - y\,dx}{x^2 + y^2}.$

Ex. 4. Assuming the characteristic equation of a perfect gas, $vp = Rt$, in which v is volume, p pressure, t absolute temperature, and R a constant; express each of the differentials dv, dp, dt, in terms of the other two.

Ex. 5. Being given that in the case of air, $R = 96$, when p is measured in pounds per square foot, v in cubic feet, and t is centigrade; and letting $t = 300$, $p = 2000$, $v = 14.4$; find the change in p when t changes to 301, and v to 14.5, supposing that p changes uniformly in the interval. [Perry's Calculus for Engineers, p. 138.]

Since $\quad\quad vdp + pdv = Rdt, \quad dv = .1, \quad dt = 1;$
hence $\quad\quad\quad\quad dp = -7.22.$

The actual increment of p will be a little different from this, and is easily found by direct computation to be -7.17.

The difference in the results is analogous to the difference between the ordinate of a surface and the ordinate of its tangent plane, taken near the common ordinate of the point of contact.

96. One variable a function of the other. When there is a definite relation connecting the variables x and y, the equation

$$dz = \frac{\partial z}{\partial x} dx + \frac{\partial z}{\partial y} dy$$

may be divided by the differential of either variable,

then $\qquad \dfrac{dz}{dx} = \dfrac{\partial z}{\partial x} + \dfrac{\partial z}{\partial y}\dfrac{dy}{dx}.\qquad$ (1)

It is here well to note the difference between $\dfrac{\partial z}{\partial x}$ and $\dfrac{dz}{dx}$. The former is the partial derivative of the functional expression for z with regard to x, on the supposition that y is constant. The latter is the total derivative of z with regard to x, when account is taken of the fact that y varies with x.

It is to be observed that the implied assumption in Art. 94, that the variables x and y have at any instant some definite numerical rate of change, is only equivalent to assuming that they vary in some continuous manner. They need not on that account be expressible as definite functions of the time, or have any fixed relation of dependence upon each other. On the other hand, a fixed relation of dependence is not precluded, for Art. 94 only assumes that x, y take the increments Δx, Δy in the time Δt, without inquiring whether one of the increments may not be determined by the other, or whether they may not both arise from the increment of some other hidden variable. The supposition that the letters x, y are independent in forming the partial derivatives is only a convenient algebraic rule or artifice for obtaining the coefficients of the differentials dx, dy; and does not imply the physical independence of the magnitudes denoted by these letters. Thus the "independence" is formal, or operational, rather than physical.

The value of $\dfrac{dy}{dx}$ is to be obtained by differentiating the functional relation between x and y. If this relation expresses y as an explicit function of x, the right hand member of (1) can then be expressed in terms of x alone; and the result will be the same as if z had been first reduced to the form of a function of the single variable x, and then differentiated with regard to this variable.

Ex. 1. A point moves on the surface $z = f(x, y)$ in the curve determined by the cylindrical surface $y = \phi(x)$; express dz in terms of dx.

Ex. 2. If $z = \tan^{-1}\dfrac{y}{2x}$, and $4x^2 + y^2 = 1$, find $\dfrac{dz}{dx}$.

Ex. 3. A point moves on the curve of intersection of the surfaces $z = \phi(x, y)$, $z = f(x, y)$; find the mutual ratio of the rates $dx : dy : dz$ at any point x, y, z.

For shortness, denoting partial derivatives by subscripts,
$$dz = f_1 dx + f_2 dy = \phi_1 dx + \phi_2 dy,$$
hence $\quad dx : dy : dz = \phi_2 - f_2 : f_1 - \phi_1 : f_1\phi_2 - f_2\phi_1.$

97. Differentiation of implicit functions; relative variation that keeps z constant. An important special question is how to vary x and y so as to keep $f(x, y)$ from varying. If $z = f(x, y) = $ constant,

then $\qquad dz = \dfrac{\partial f}{\partial x} dx + \dfrac{\partial f}{\partial y} dy = 0,$

hence the relative rates of change of x and y are given by the equation

$$\frac{dy}{dx} = -\frac{\dfrac{\partial f}{\partial x}}{\dfrac{\partial f}{\partial y}}.$$

Ex. 1. If x pass through the value 2 at the rate of 5 units per second, at what rate must y pass through the value 3 in order to keep the function $x^2y + 3xy^2$ constant?

Since $\quad d(x^2y + 3xy^2) = (2xy + 3y^2)dx + (x^2 + 6xy)dy = 0,$
hence $\quad 39\,dx + 40\,dy = 0, \quad dy = -4\tfrac{7}{8}$ units per second.

Ex. 2. Defining the elasticity of a gas as the limit of the ratio of an increment of pressure to the corresponding relative decrement of volume, find e, the elasticity of a perfect gas under constant temperature.

By def. $\quad e = \lim\limits_{\Delta v \doteq 0}\left(\Delta p : -\dfrac{\Delta v}{v}\right) = -v\dfrac{dp}{dv}$,

and by differentiating $vp = Rt$, keeping t constant,

$$vdp + pdv = 0, \quad \dfrac{dp}{dv} = -\dfrac{p}{v}, \quad \text{hence } e = p.$$

As a geometrical illustration, let a section of the surface $z = f(x, y)$ be made by the plane $z = c$, then for all points on the contour of the section

$$z = f(x, y) = c,$$

and if a point describe the contour, the x-rate and the y-rate will be in the ratio $-\dfrac{\partial z}{\partial y} : \dfrac{\partial z}{\partial x}$; and this ratio will measure the slope of the tangent to the contour with reference to the plane zx.

Ex. A particle is moving on the ellipsoid $\dfrac{x^2}{a^2} + \dfrac{y^2}{b^2} + \dfrac{z^2}{c^2} = 1$ at the point $x = \dfrac{a}{2}$, $y = \dfrac{b}{2}$, $z = \dfrac{c}{\sqrt{2}}$; find the relative rates of x and of y so that the rate along z may be zero.

Since $\quad \dfrac{x\,dx}{a^2} + \dfrac{y\,dy}{b^2} = 0$, hence $\dfrac{dy}{dx} = -\dfrac{b^2 a}{a^2 b} = -\dfrac{b}{a}$.

Similarly, if a point whose coördinates are x, y move in a plane so as to keep the function $f(x, y)$ constant, then it describes a curve whose equation is $f(x, y) = c$, hence the differentials dx, dy are connected by

$$\dfrac{\partial f}{\partial x}dx + \dfrac{\partial f}{\partial y}dy = 0, \tag{1}$$

and the slope of the direction of motion is given by

$$\dfrac{dy}{dx} = -\dfrac{\partial f}{\partial x} : \dfrac{\partial f}{\partial y}. \tag{2}$$

In all such cases either variable is an implicit function of the other, and thus the last equation furnishes a rule for finding the derivative of an implicit function. In many examples in practice it is preferable to equate the total differential to zero, as in (1), and then solve for $\dfrac{dy}{dx}$.

Ex. 1. Given $x^3 + y^3 + 3\,axy = c$, find $\dfrac{dy}{dx}$.

Since $(3x^2 + 3\,ay)\,dx + (3y^2 + 3\,ax)\,dy = 0$, $\dfrac{dy}{dx} = -\dfrac{x^2 + ay}{y^2 + ax}$

Ex. 2. $f(ax+by) = c$; $\dfrac{\partial f}{\partial x} = af'(ax+by)$; $\dfrac{\partial f}{\partial y} = bf'(ax+by)$; $\dfrac{dy}{dx} = -\dfrac{a}{b}$

Ex. 3. If $ax^2 + 2\,hxy + by^2 + 2\,gx + 2\,fy + c = 0$, find $\dfrac{dy}{dx}$.

Ex. 4. Given $x^4 - y^4 = c$, find $\dfrac{dy}{dx}$.

98. Functions of more than two variables. All of the methods of this chapter are applicable to functions of three or more variables.

Let $\qquad u = f(x, y, z)$,

then it can be shown, as in Art. 94, that

$$\frac{du}{dt} = \frac{\partial u}{\partial x}\frac{dx}{dt} + \frac{\partial u}{\partial y}\frac{dy}{dt} + \frac{\partial u}{\partial z}\frac{dz}{dt}, \qquad (1)$$

or in the abbreviated notation,

$$du = \frac{\partial u}{\partial x}dx + \frac{\partial u}{\partial y}dy + \frac{\partial u}{\partial z}dz. \qquad (2)$$

No simple geometric representation can be given of a function of three variables, but there are many examples in physics of functions of the three coördinates of a point; for instance, the potential u produced at a point (x, y, z) by a given distribution of fixed attracting bodies, is a definite function of the variables x, y, z, and equation (2) gives the rate of change of the potential as the point (x, y, z) changes its position in any direction.

First let it be required to vary (x, y, z) so that the potential $u = f(x, y, z)$ shall remain constant; then the point must remain on the equipotential surface whose equation is $f(x, y, z) = c$, and the differentials of x, y, z are connected by the relation

$$\frac{\partial f}{\partial x} dx + \frac{\partial f}{\partial y} dy + \frac{\partial f}{\partial z} dz = 0. \qquad (3)$$

In such cases z is an implicit function of the two variables x, y; and its differential is expressed in terms of their differentials by the last equation.

Ex. 1. If the "characteristic equation" of a substance be

$$f(p, v, t) = \text{constant}$$

prove $\qquad \left(\dfrac{dp}{dt}\right)_{v \text{ con.}} \cdot \left(\dfrac{dt}{dv}\right)_{p \text{ con.}} \cdot \left(\dfrac{dv}{dp}\right)_{t \text{ con.}} = -1.$

Since $\qquad \dfrac{\partial f}{\partial p} dp + \dfrac{\partial f}{\partial v} dv + \dfrac{\partial f}{\partial t} dt = 0,$

hence, if $dv = 0$, $\dfrac{dp}{dt} = -\dfrac{\partial f}{\partial t} : \dfrac{\partial f}{\partial v}$. Similarly for $dp = 0$, etc.

Ex. 2. In the equation $c^2 = a^2 + b^2 - 2ab \cos C$, referred to in exercise 6 of Art. 90, find the error in c arising from given small errors in a, b, C, all the errors being supposed so small that their squares can be neglected; within the required degree of accuracy.

99. One or two relations between the three variables x, y, z. Again, if the point, instead of moving on the surface $u = $ con, move on some other surface defined by

$$z = \phi(x, y), \qquad (1)$$

then $\qquad dz = \dfrac{\partial \phi}{\partial x} dx + \dfrac{\partial \phi}{\partial y} dy, \qquad (2)$

and (2) of Art. 98 becomes, by elimination of dz,

$$du = \left(\frac{\partial f}{\partial x} + \frac{\partial u}{\partial z} \frac{\partial \phi}{\partial x}\right) dx + \left(\frac{\partial f}{\partial y} + \frac{\partial u}{\partial z} \frac{\partial \phi}{\partial y}\right) dy. \qquad (3)$$

The point has then only two degrees of freedom, indicated by the independent differentials dx, dy. If the point be further restricted to the curve determined on the latter surface by the cylinder

$$y = \psi(x), \tag{4}$$

then $\qquad dy = \psi'(x)\,dx,$

and (3) becomes by elimination of dy, and division by the single independent differential dx,

$$\frac{du}{dx} = \frac{\partial f}{\partial x} + \frac{\partial u}{\partial z}\frac{\partial \phi}{\partial x} + \left(\frac{\partial f}{\partial y} + \frac{\partial u}{\partial z}\frac{\partial \phi}{\partial y}\right)\psi'(x). \tag{5}$$

This derivative could also be obtained by eliminating z and y before differentiation. The function u in terms of the single variable x is then

$$u = f(x, \psi(x), \phi(x, \psi(x))).$$

The latter method is usually longer, and is not applicable at all when equations (1) and (4) are replaced by implicit relations that cannot be solved for one of the variables.

Ex. 1. $u = e^{\frac{x}{a}}(y - z)$, $z = a \sin \frac{y}{x}$, $y = a \log \frac{x}{a}$; find $\frac{du}{dx}$.

Ex. 2. If $u = f(x, y, z)$; and if x, y, z are connected by the two relations $\phi(x, y, z) = c_1$, $\psi(x, y, z) = c_2$; find du in terms of dx.

Differentiating, and denoting partial derivatives by subscripts for shortness,

$$du = f_1\,dx + f_2\,dy + f_3\,dz,$$
$$0 = \phi_1\,dx + \phi_2\,dy + \phi_3\,dz,$$
$$0 = \psi_1\,dx + \psi_2\,dy + \psi_3\,dz;$$

hence, by elimination of dy, dz,

$$du(\phi_2\psi_3 - \phi_3\psi_2) = dx\,[f_1(\phi_2\psi_3 - \phi_3\psi_2) + f_2(\phi_3\psi_1 - \phi_1\psi_3) + f_3(\phi_1\psi_2 - \phi_2\psi_1)].$$

Geometrically speaking, the point (x, y, z) moves on the curve of intersection of two surfaces and has therefore only one degree of freedom.

Thus the variation of a single independent coördinate is sufficient to determine the variation of the other coördinates, and of the function u itself.

100. Euler's theorem. Relation between a homogeneous function and its partial derivatives. Let $u = f(x, y, z)$ be a homogeneous function of x, y, z, of degree n; then

$$x\frac{\partial u}{\partial x} + y\frac{\partial u}{\partial y} + z\frac{\partial u}{\partial z} = nu.$$

For, let $\quad u = Ax^\alpha y^\beta z^\gamma + Bx^{\alpha'}y^{\beta'}z^{\gamma'} + \cdots$
where $\quad \alpha + \beta + \gamma = \alpha' + \beta' + \gamma' = \cdots = n.$

$$\frac{\partial u}{\partial x} = \alpha A x^{\alpha-1} y^\beta z^\gamma + \alpha' B x^{\alpha'-1} y^{\beta'} z^{\gamma'} + \cdots,$$

$$x\frac{\partial u}{\partial x} = \alpha A x^\alpha y^\beta z^\gamma + \alpha' B x^{\alpha'} y^{\beta'} z^{\gamma'} + \cdots.$$

Similarly,

$$y\frac{\partial u}{\partial y} = \beta A x^\alpha y^\beta z^\gamma + \beta' B x^{\alpha'} y^{\beta'} z^{\gamma'} + \cdots,$$

$$z\frac{\partial u}{\partial z} = \gamma A x^\alpha y^\beta z^\gamma + \gamma' B x^{\alpha'} y^{\beta'} z^{\gamma'} + \cdots.$$

Adding these three equations,

$$x\frac{\partial u}{\partial x} + y\frac{\partial u}{\partial y} + z\frac{\partial u}{\partial z}$$
$$= (\alpha + \beta + \gamma)Ax^\alpha y^\beta z^\gamma + (\alpha' + \beta' + \gamma')Bx^{\alpha'}y^{\beta'}z^{\gamma'} + \cdots$$
$$= n(Ax^\alpha y^\beta z^\gamma + Bx^{\alpha'}y^{\beta'}z^{\gamma'} + \cdots)$$
$$= nu.$$

The theorem can be extended to functions of any number of variables.

If a function, homogeneous in several variables, be differentiated partially with regard to each of them; then each partial derivative be multiplied by the variable with regard to which the derivative was taken; and all these products added; the result is n times the original function; where n is the degree of the function.

This is known as Euler's theorem.*

EXERCISES

Verify Euler's theorem for the following expressions:

1. $x^4 + 3x^2y^2 - 7xy^3$.

2. $(x^{\frac{1}{2}} + y^{\frac{1}{2}})(x^n + y^n)$.

3. $\sin\frac{y}{x}$.

4. $\dfrac{1}{x^3 - 3x^2y - y^3}$.

5. $\tan^{-1}\frac{y}{x}$.

6. $e^{\frac{y}{x}}\dfrac{x^2}{y^2} + \dfrac{y^2}{x^2}\sin\frac{y}{x}$.

Prove the following identities:

1. $u = \log(e^x + e^y)$, $\quad \dfrac{\partial u}{\partial x} + \dfrac{\partial u}{\partial y} = 1$.

2. $u = \log(x^3 + y^3 + z^3 - 3xyz)$, $\quad \dfrac{\partial u}{\partial x} + \dfrac{\partial u}{\partial y} + \dfrac{\partial u}{\partial z} = \dfrac{3}{x+y+z}$.

3. $u = x^y y^x$, $\quad x\dfrac{\partial u}{\partial x} + y\dfrac{\partial u}{\partial y} = (x + y + \log u)u$.

4. $u = \sin^{-1}(xyz)$, $\quad \dfrac{\partial u}{\partial x}\dfrac{\partial u}{\partial y}\dfrac{\partial u}{\partial z} = \tan^2 u \sec u$.

5. $u = \log(\tan x + \tan y + \tan z)$, $\quad \sin 2x \dfrac{\partial u}{\partial x} + \sin 2y \dfrac{\partial u}{\partial y} + \sin 2z \dfrac{\partial u}{\partial z} = 2$.

6. $u = e^x \sin y + e^y \sin x$, $\quad \left(\dfrac{\partial u}{\partial x}\right)^2 + \left(\dfrac{\partial u}{\partial y}\right)^2 = e^{2x} + e^{2y} + 2e^{x+y}\sin(x+y)$.

7. $u = \log(x + \sqrt{x^2 + y^2})$, $\quad x\dfrac{\partial u}{\partial x} + y\dfrac{\partial u}{\partial y} = 1$.

8. $u = \log^y x$, $\quad du = \dfrac{dx}{x \log y} - \dfrac{\log x \, dy}{y(\log y)^2}$.

9. $u = \log y^x$, $\quad du = \log y \, dx + \dfrac{x}{y} dy$.

10. $u = y^{\sin x}$, $\quad du = y^{\sin x}\cos x \log y \, dx + \dfrac{\sin x}{y^{\text{covers } x}} dy$.

---

* Leonard Euler (1707–1783), one of the most eminent mathematicians of the eighteenth century.

CHAPTER IX

SUCCESSIVE PARTIAL DIFFERENTIATION

101. Successive differentiation of functions of two variables. Let $z = f(x, y)$, in which x, y are functions of another variable t, which may conveniently be thought of as time. As the rate of change of z is usually variable, it is sometimes useful to have an expression for the rate of change of this rate. Just as $\dfrac{dz}{dt}$ is the rate of change of z, so $\dfrac{d}{dt}\left(\dfrac{dz}{dt}\right)$ is the rate of change of $\dfrac{dz}{dt}$, and it is written $\dfrac{d^2z}{dt^2}$. This rate of the rate may conveniently be called the acceleration.

It will now be shown that the expression for the z-acceleration involves the x-acceleration, the y-acceleration and also the squares and products of the x-rate and the y-rate, each with a certain coefficient.

It was proved in Art. 94 that the z-rate is

$$\frac{dz}{dt} = \frac{\partial z}{\partial x}\frac{dx}{dt} + \frac{\partial z}{\partial y}\frac{dy}{dt}. \quad (1)$$

Differentiating each term of this identity with regard to t,

$$\frac{d^2z}{dt^2} = \frac{d}{dt}\left(\frac{\partial z}{\partial x}\right)\frac{dx}{dt} + \frac{\partial z}{\partial x}\frac{d^2x}{dt^2} + \frac{d}{dt}\left(\frac{\partial z}{\partial y}\right)\frac{dy}{dt} + \frac{\partial z}{\partial y}\frac{d^2y}{dt^2}; \quad (2)$$

but, since $\dfrac{\partial z}{\partial x}$ is itself a function of two variables, hence, by Art. 94,

$$\frac{d}{dt}\left(\frac{\partial z}{\partial x}\right) = \frac{\partial}{\partial x}\left(\frac{\partial z}{\partial x}\right)\frac{dx}{dt} + \frac{\partial}{\partial y}\left(\frac{\partial z}{\partial y}\right)\frac{dy}{dt},$$

also
$$\frac{d}{dt}\left(\frac{\partial z}{\partial y}\right) = \frac{\partial}{\partial x}\left(\frac{\partial z}{\partial y}\right)\frac{dx}{dt} + \frac{\partial}{\partial y}\left(\frac{\partial z}{\partial y}\right)\frac{dy}{dt},$$

hence (2) becomes, by substitution and slight re-arrangement,

$$\frac{d^2z}{dx^2} = \frac{\partial}{\partial x}\left(\frac{\partial z}{\partial x}\right)\left(\frac{dx}{dt}\right)^2 + \left[\frac{\partial}{\partial y}\left(\frac{\partial z}{\partial x}\right) + \frac{\partial}{\partial x}\left(\frac{\partial z}{\partial y}\right)\right]\frac{dx}{dt}\frac{dy}{dt}$$
$$+ \frac{\partial}{\partial y}\left(\frac{\partial z}{\partial y}\right)\left(\frac{dy}{dt}\right)^2 + \frac{\partial z}{\partial x}\frac{d^2x}{dt^2} + \frac{\partial z}{\partial y}\frac{d^2y}{dt^2}. \qquad (3)$$

The successive partial x- and y-derivatives

$$\frac{\partial}{\partial x}\left(\frac{\partial z}{\partial x}\right), \quad \frac{\partial}{\partial x}\left(\frac{\partial z}{\partial y}\right), \quad \frac{\partial}{\partial y}\left(\frac{\partial z}{\partial x}\right), \quad \frac{\partial}{\partial y}\left(\frac{\partial z}{\partial y}\right),$$

which appear as coefficients of the squares and product of the rates will be denoted by the symbols

$$\frac{\partial^2 z}{\partial x^2}, \quad \frac{\partial^2 z}{\partial x\, \partial y}, \quad \frac{\partial^2 z}{\partial y\, \partial x}, \quad \frac{\partial^2 z}{\partial y^2}.$$

In other words, $\dfrac{\partial^2 z}{\partial x^2}$ will stand for the operation of differentiating z twice in succession with regard to x, on the supposition that y is constant; and $\dfrac{\partial^2 z}{\partial x\, \partial y}$ will denote the operation of differentiating z first with regard to y, on the supposition that x is constant, and then differentiating the result with regard to x on the supposition that y is constant; and similarly for the other expressions.

With this notation, eq. (3) is

$$\frac{d^2z}{dt^2} = \frac{\partial^2 z}{\partial x^2}\left(\frac{dx}{dt}\right)^2 + \left[\frac{\partial^2 z}{\partial y\, \partial x} + \frac{\partial^2 z}{\partial x\, \partial y}\right]\frac{dx}{dt}\frac{dy}{dt}$$
$$+ \frac{\partial^2 z}{\partial y^2}\left(\frac{dy}{dt}\right)^2 + \frac{\partial z}{\partial x}\frac{d^2x}{dt^2} + \frac{\partial z}{\partial y}\frac{d^2y}{dt^2}.$$

The coefficient of the product of the rates will be further simplified by the theorem of the next article.

102. Order of differentiation indifferent.

THEOREM. The successive partial derivatives

$$\frac{\partial^2 z}{\partial y\, \partial x}, \qquad \frac{\partial^2 z}{\partial x\, \partial y}$$

are equal for any values of x and y in the vicinity of which z and its first and second partial x- and y-derivatives are continuous.

For, let $z = f(x, y)$; and first change x into $x + h$, keeping y constant, then by the theorem of mean value, the increment of the function is equal to the increment of the variable multiplied by the derivative taken for some value intermediate between x and $x + h$; that is,

$$f(x + h, y) - f(x, y) = h \frac{\partial}{\partial x} f(x + \theta h, y) \qquad [0 < \theta < 1.$$

Now let y change to $y + k$, x remaining constant, and take the increment of the function on the left; then by the theorem of mean value applied to $\frac{\partial}{\partial x} f(x + \theta h, y)$ as a function of y, for the increment k,

$$[f(x + h, y + k) - f(x, y + k)] - [f(x + h, y) - f(x, y)]$$
$$= kh \frac{\partial}{\partial y} \frac{\partial}{\partial x} f(x + \theta h, y + \theta_1 k).$$

Next let these increments be given in reversed order; then

$$[f(x + h, y + k) - f(x + h, y)] - [f(x, y + k) - f(x, y)]$$
$$= hk \frac{\partial}{\partial x} \frac{\partial}{\partial y} f(x + \theta_2 h, y + \theta_3 k);$$

hence

$$\frac{\partial}{\partial y} \frac{\partial}{\partial x} f(x + \theta h, y + \theta_1 k) = \frac{\partial}{\partial x} \frac{\partial}{\partial y} f(x + \theta_2 h, y + \theta_3 k)$$

for any values of h and k within the range around the point (x, y) within which all the functions mentioned are continuous.

When h, k approach zero,

$$x + \theta h, \quad y + \theta_1 k, \quad \text{and} \quad x + \theta_2 h, \quad y + \theta_3 k$$

approach (x, y), and

$$f(x + \theta h, y + \theta_1 k), \quad f(x + \theta_2 h, y + \theta_3 k)$$

approach $f(x, y)$; hence

$$\frac{\partial}{\partial y}\frac{\partial}{\partial x}f(x, y) = \frac{\partial}{\partial x}\frac{\partial}{\partial y}f(x, y),$$

or, since $f(x, y) = z$,

$$\frac{\partial^2 z}{\partial y \, \partial x} = \frac{\partial^2 z}{\partial x \, \partial y}.$$

Cor. 1. It follows directly that under corresponding conditions the order of differentiation in the higher partial derivatives is indifferent. In other words, if u and all its partial derivatives are continuous, the operations $\frac{\partial}{\partial x}$, $\frac{\partial}{\partial y}$ are commutative.

E.g.,
$$\frac{\partial^3 u}{\partial x \, \partial y \, \partial x} = \frac{\partial^3 u}{\partial x^2 \, \partial y} = \frac{\partial^3 u}{\partial y \, \partial x^2}.$$

Cor. 2. Equation (3) may now be written in the simple form

$$\frac{d^2 u}{dt^2} = \frac{\partial^2 u}{\partial x^2}\left(\frac{dx}{dt}\right)^2 + 2\frac{\partial^2 u}{\partial x \, \partial y}\frac{dx}{dt}\frac{dy}{dt} + \frac{\partial^2 u}{\partial y^2}\left(\frac{dy}{dt}\right)^2$$
$$+ \frac{\partial u}{\partial x}\frac{d^2 x}{dt^2} + \frac{\partial u}{\partial y}\frac{d^2 y}{dt^2}, \qquad (4)$$

or, if the independent variable t is not expressed,

$$d^2 u = \frac{\partial^2 u}{\partial x^2}(dx)^2 + 2\frac{\partial^2 u}{\partial x \, \partial y}dx \, dy + \frac{\partial^2 u}{\partial y^2}(dy)^2$$
$$+ \frac{\partial u}{\partial x}d^2 x + \frac{\partial u}{\partial y}d^2 y. \qquad (5)$$

102.] SUCCESSIVE PARTIAL DIFFERENTIATION

If x be taken as independent variable, then t is to be replaced by x; and since $\dfrac{d^2x}{dx^2} = 0$, the equation becomes

$$\frac{d^2u}{dx^2} = \frac{\partial^2 u}{\partial x^2} + 2\frac{\partial^2 u}{\partial x\,\partial y}\frac{dy}{dx} + \frac{\partial^2 u}{\partial y^2}\left(\frac{dy}{dx}\right)^2 + \frac{\partial u}{\partial y}\frac{d^2y}{dx^2}. \qquad (6)$$

Similarly, if y be taken as independent variable, and x be a function of y, then

$$\frac{d^2u}{dy^2} = \frac{\partial^2 u}{\partial x^2}\left(\frac{dx}{dy}\right)^2 + 2\frac{\partial^2 u}{\partial x\,\partial y}\frac{dx}{dy} + \frac{\partial^2 u}{\partial y^2} + \frac{\partial u}{\partial x}\frac{d^2x}{dy^2}. \qquad (7)$$

EXERCISES

1. Verify that $\dfrac{\partial^2 u}{\partial x\,\partial y} = \dfrac{\partial^2 u}{\partial y\,\partial x}$, when $u = x^2 y^3$.

2. Verify that $\dfrac{\partial^3 u}{\partial x\,\partial y^2} = \dfrac{\partial^3 u}{\partial y^2\,\partial x}$, when $u = x^2 y + xy^3$.

3. Verify that $\dfrac{\partial^2 u}{\partial x\,\partial y} = \dfrac{\partial^2 u}{\partial y\,\partial x}$, when $u = y\log(1+xy)$.

4. In Ex. 3 are there any exceptional values of x, y for which the relation is not true?

5. Given $u = (x^2 + y^2)^{\frac{1}{2}}$, verify the formula

$$x^2 \frac{\partial^2 u}{\partial x^2} + 2xy\frac{\partial^2 u}{\partial x\,\partial y} + y^2 \frac{\partial^2 u}{\partial y^2} = 0.$$

6. Given $u = (x^3 + y^3)^{\frac{1}{2}}$, show that the expression in the left member of the equation in Ex. 5 is equal to $\dfrac{3\,u}{4}$.

7. Given $u = (x^2 + y^2 + z^2)^{-\frac{1}{2}}$; prove that $\dfrac{\partial^2 u}{\partial x^2} + \dfrac{\partial^2 u}{\partial y^2} + \dfrac{\partial^2 u}{\partial z^2} = 0$.

8. Given $u = \sec(y+ax) + \tan(y-ax)$; prove that $\dfrac{\partial^2 u}{\partial x^2} = a^2 \dfrac{\partial^2 u}{\partial y^2}$.

9. Given $u = \sin x \cos y$; verify that $\dfrac{\partial^4 u}{\partial y^2\,\partial x^2} = \dfrac{\partial^4 u}{\partial x\,\partial y\,\partial x\,\partial y} = \dfrac{\partial^4 u}{\partial x^2\,\partial y^2}$.

10. Given $u = (4\,ab - c^2)^{-\frac{1}{2}}$; prove that $\dfrac{\partial^2 u}{\partial c^2} = \dfrac{\partial^2 u}{\partial a\,\partial b}$.

11. If $u = \sin v$, v being a homogeneous function of degree n in x and y, determine the value of $x\dfrac{\partial u}{\partial x} + y\dfrac{\partial u}{\partial y}$.

12. If $u = \tan^{-1}\dfrac{xy}{\sqrt{1 + x^2 + y^2}}$, show that $\dfrac{\partial^2 u}{\partial x\,\partial y} = (1 + x^2 + y^2)^{-\frac{3}{2}}$ and that $\dfrac{\partial^4 u}{\partial x^2\,\partial y^2} = \dfrac{15\,xy}{(1 + x^2 + y^2)^{\frac{7}{2}}}$.

103. Extension of Taylor's theorem to expansion of functions of two variables.

Taylor's theorem, as developed in Chapter IV, relates only to functions of one variable, but it can be readily extended to functions of any number of variables in a manner first shown by Lagrange.

Let $f(x, y)$ be a function of the two variables x, y, which, with its first $2n$ partial derivatives, as to x and y, is finite and continuous for all values of x, y within a certain portion of the coördinate plane. It is required to expand $f(x + h, y + k)$ in a series of ascending powers of h and of k.

Using an auxiliary variable t, let

$$x' = x + ht, \quad y' = y + kt, \qquad (1)$$

then $\quad f(x', y') = f(x + ht, y + kt) = F(t)$, say; $\qquad (2)$

the development of $F(t)$ in powers of t is, by Maclaurin's theorem,

$$F(t) = F(0) + F'(0)\cdot t + \dfrac{F''(0)\cdot t^2}{2!} + \cdots \dfrac{F^{n-1}(0)}{(n-1)!}\cdot t^{n-1}$$

$$+ \dfrac{F^n(\theta t)}{n!} t^n, \qquad [0 < \theta < 1. \quad (3)$$

whence, putting $t = 1$,

$$f(x + y, y + k) = F(0) + F'(0) + \cdots \dfrac{F^{n-1}(0)}{(n-1)!} + \dfrac{F^n(\theta)}{n!}. \quad (4)$$

102-103.] SUCCESSIVE PARTIAL DIFFERENTIATION

To express $F(0)$, $F'(0)$ ... in terms of h, k, first find $F'(t)$, $F''(t)$, $F'''(t)$, ... by successive differentiation of (2); then

$$F'(t) = \frac{dF(t)}{dt} = \frac{d}{dt}f(x', y') = \frac{\partial f}{\partial x'} \cdot \frac{dx'}{dt} + \frac{\partial f}{\partial y'} \cdot \frac{dy'}{dt},$$

but, from (1), $\dfrac{dx'}{dt} = h$, $\dfrac{dy'}{dt} = k$, hence

$$F'(t) = h\frac{\partial f}{\partial x'} + k\frac{\partial f}{\partial y'};$$

likewise,

$$F''(t) = h\frac{d}{dt}\left(\frac{\partial f}{\partial x'}\right) + k\frac{d}{dt}\left(\frac{\partial f}{\partial y'}\right)$$

$$= h\left(\frac{\partial^2 f}{\partial x'^2}\frac{dx'}{dt} + \frac{\partial^2 f}{\partial x'\partial y'}\frac{dy'}{dt}\right) + k\left(\frac{\partial^2 f}{\partial x'\partial y'}\frac{dx'}{dt} + \frac{\partial^2 f}{\partial y'^2}\frac{dy'}{dt}\right);$$

then putting $\dfrac{dx'}{dt} = h$, $\dfrac{dy'}{dt} = k$,

$$F''(t) = h^2\frac{\partial^2 f}{\partial x'^2} + 2hk\frac{\partial^2 f}{\partial x'\partial y'} + k^2\frac{\partial^2 f}{\partial y'^2}.$$

Similarly,

$$F'''(t) = h^3\frac{\partial^3 f}{\partial x'^3} + 3h^2k\frac{\partial^3 f}{\partial x'^2\partial y'} + 3hk^2\frac{\partial^3 f}{\partial x'\partial y'^2} + k^3\frac{\partial^3 f}{\partial y'^3}.$$

. . . .

Now when t is replaced by zero in these derivatives, x', y' reduce back to x, y; hence

$$F(0) = f(x, y),$$

$$F'(0) = h\frac{\partial f}{\partial x} + k\frac{\partial f}{\partial y},$$

$$F''(0) = h^2\frac{\partial^2 f}{\partial x^2} + 2hk\frac{\partial^2 f}{\partial x\,\partial y} + k^2\frac{\partial^2 f}{\partial y^2},$$

$$F'''(0) = h^3\frac{\partial^3 f}{\partial x^3} + 3h^2k\frac{\partial^3 f}{\partial x^2\,\partial y} + 3hk^2\frac{\partial^3 f}{\partial x\,\partial y^2} + k^3\frac{\partial^3 f}{\partial y^3},$$

and when t is replaced by θ in $F^n(t)$, x' and y' become $x + \theta h$ and $y + \theta k$, hence

$$F^n(\theta) = \sum_{r=0}^{n} \binom{n}{r} h^{n-r} k^r \frac{\partial^n f(x + \theta h, y + \theta k)}{\partial x^{n-r} \partial y^r},$$

in which $\binom{n}{r}$ stands for the binomial coefficient $\dfrac{n!}{r!(n-r)!}$.

Therefore (4) becomes

$$f(x + h, y + k) = f(x, y) + h\frac{\partial f}{\partial x} + k\frac{\partial f}{\partial y}$$

$$+ \frac{1}{2!}\left(h^2 \frac{\partial^2 f}{\partial x^2} + 2hk\frac{\partial^2 f}{\partial x \partial y} + k^2 \frac{\partial^2 f}{\partial y^2}\right) + \cdots$$

$$+ \frac{1}{(n-1)!}\sum_{r=0}^{n-1} \binom{n-1}{r} h^{n-r-1} k^r \frac{\partial^{n-1} f(x, y)}{\partial x^{n-r-1} \partial y^r}$$

$$+ \frac{1}{n!}\sum_{r=0}^{n} \binom{n}{r} h^{n-r} k^r \frac{\partial^n f(x + \theta h, y + \theta k)}{\partial x^{n-r} \partial y^r}, \quad (5)$$

which is the desired form of expansion.

104. Significance of remainder. This expansion is useful only for those values of x, y, h, k, for which the last term, called the remainder after the $(n-1)$st powers of h and k, can be made as small as desired by taking n large enough.

105. Form corresponding to Maclaurin's theorem. The expansion of a function of two variables in a series of powers of these variables can be readily obtained from the last equation.

If it be desired to expand $f(x, y)$ in the vicinity of a, b, let $x = a$, $y = b$; $h = x - a$, $k = y - b$, then equation (5) of Art. 103 becomes

$$f(x, y) = f(a, b) + (x - a)\frac{\partial f}{\partial a} + (y - b)\frac{\partial f}{\partial b}$$

$$+ \frac{1}{2!}\left((x - a)^2 \frac{\partial^2 f}{\partial a^2} + 2(x - a)(y - b)\frac{\partial^2 f}{\partial a \partial b} + (y - b)^2 \frac{\partial^2 f}{\partial b^2}\right)$$

$$+ \cdots + \frac{1}{(n-1)!}\sum_{r=0}^{n}\binom{n-1}{r}(x - a)^{n-r-1}(y - b)^r \frac{\partial^{n-1} f}{\partial a^{n-r-1} \partial b^r}$$

$$+ \frac{1}{n!}\sum_{r=0}^{n}\binom{n}{r}(x-a)^{n-r}(y-b)^r \frac{\partial^n f(a+\theta(x-a),\ y+\theta(y-b))}{\partial a^{n-r} \partial b^r},$$

in which $\frac{\partial f}{\partial a}$ denotes that $f(x, y)$ is to be differentiated with regard to x, and that $x = a$, $y = b$ are then to be substituted in the derivative; and so for the other symbols.

In particular, if $a = 0$, $b = 0$, the expansion of $f(x, y)$ in powers of x, y becomes

$$f(x, y) = f(0, 0) + x\left[\frac{\partial f}{\partial x}\right]_{\substack{x=0\\y=0}} + \left[y\frac{\partial f}{\partial y}\right]_{\substack{x=0\\y=0}}$$

$$+ \frac{1}{2!}\left(x^2 \frac{\partial^2 f(0, 0)}{\partial x^2} + 2xy \frac{\partial^2 f(0, 0)}{\partial x \partial y} + y^2 \frac{\partial^2 f(0, 0)}{\partial y^2}\right)$$

$$+ \cdots + \frac{1}{(n-1)!}\sum_{r=0}^{n-1}\binom{n-1}{r}x^{n-r-1}y^r \frac{\partial^{n-1} f(0, 0)}{\partial x^{n-r-1} \partial y^r}$$

$$+ \frac{1}{n!}\sum_{r=0}^{n}\binom{n}{r}x^{n-r}y^r \frac{\partial^n f(\theta x, \theta y)}{\partial x^{n-r} \partial y^r}.$$

These theorems for expansion can be readily extended to functions of any number of variables.

EXERCISES

1. Expand $e^x \sin(x+y)$ in powers of x and y.

2. Expand $(x+y)^4$ in powers of x and y.

3. Expand $\sqrt{x+h} \tan(y+k)$ in powers of h and k, and express the form of the remainder after two terms, when $h=1$, $k=1$.

4. Expand the function $x^2 + y^2 + z^2 - 4x + 6y - 2z - 11$ in powers of $x-2$, $y+3$, $z-1$.

5. Arrange the function
$$3x^2 - 5y^2 + 4x - 7y + 11$$
in powers of $x-2$, $y+3$.

6. Transform the equation
$$x^3 + y^3 - 3xy = 1$$
to parallel axes with the point $(1, 2)$ as origin.

7. What kind of discontinuity has the function $\dfrac{2x+6y}{x+2y}$ at the point $(0,0)$? Show that it may have any value between 2 and 3, depending on the ratio of y to x as they approach zero. Illustrate geometrically.

8. Write down an expression for the error in the approximate equation
$$\Delta f(x, y, z) = \frac{\partial f}{\partial x}\Delta x + \frac{\partial f}{\partial y}\Delta y + \frac{\partial f}{\partial z}\Delta z$$
(cf. Ex. 5, p. 156; Ex. 5, p. 164; Ex. 2, p. 169).

CHAPTER X

MAXIMA AND MINIMA OF FUNCTIONS OF TWO VARIABLES

106. Definition of maximum and minimum of a function of two variables. A continuous function $z = \phi(x, y)$ has a maximum value $\phi(a, b)$ for $x = a, y = b$ if, as the variables pass in any manner through the values a and b, the function hitherto increasing, ceases to increase and begins to decrease; the function has a minimum value $\phi(a, b)$ if it ceases to decrease, and begins to increase, for every variation of x and y through the values a and b.

This fact is expressed analytically thus: $\phi(a, b)$ is a maximum or minimum value of $\phi(x, y)$ according as the increment

$$\phi(a + h, b + k) - \phi(a, b)$$

preserves a negative or a positive sign for all values of the increments h and k which are numerically less than a given small number m.

If the function be represented by the ordinate of a surface, then a maximum (or minimum) ordinate $\phi(a, b)$ is greater (or less) than every neighboring ordinate $\phi(a + h, b + k)$ drawn at any point $(a + h, b + k)$, irrespective of the signs and relative magnitudes of h and k.

107. Determination of maxima and minima. It was shown in Art. 79 that the necessary and sufficient condition that a function of one variable may have a maximum or a minimum for a given value of the variable is that its first derivative change its sign as the variable increases through the given value. Similarly for a function of two variables, its differential must change its sign at a maximum or minimum, independent of the mode of variation of the variables through these values. Since

$$dz = \frac{\partial \phi}{\partial x} dx + \frac{d\phi}{\partial y} dy,$$

and since either x or y may be varied alone, the first necessary condition is that the coefficients $\frac{\partial \phi}{\partial x}$, $\frac{\partial \phi}{\partial y}$ change signs separately; otherwise it would be possible to find a mode (or direction) of variation in which dz does not change sign; for instance, if $\frac{\partial \phi}{\partial x}$ does not change sign, then dz preserves its sign when dy is zero and x increases through a.

Hence the critical values are those at which

$$\frac{\partial \phi}{\partial x} = 0, \quad \frac{\partial \phi}{\partial y} = 0,$$

or at which $\frac{\partial \phi}{\partial x}$, $\frac{\partial \phi}{\partial y}$ become infinite.

To determine whether these values of x, y will give a maximum or minimum value to z, it is usually impracticable to test the signs of $\frac{\partial \phi}{\partial x}$, $\frac{\partial \phi}{\partial y}$ for all neighboring values of x, y. It is consequently necessary to proceed to the higher derivatives. Usually, those values which make $\frac{\partial \phi}{\partial x}$, $\frac{\partial \phi}{\partial y}$ infinite, will also make successive derivatives infinite; hence such values will be excluded from the present mode of investigation.

As an example of a function which has a minimum, and yet has no partial derivatives, consider

$$z = (x^2 + y^2)^{\frac{1}{2}}.$$

When $x=0$ and $y=0$, then $z=0$; but for every other value of x and of y, z must be positive; hence $z=0$ is a minimum; but

$$\frac{\partial z}{\partial x} = \frac{0}{0}, \quad \frac{\partial z}{\partial y} = \frac{0}{0} \text{ at } x=0, \; y=0.$$

First expand the function $\phi(a+h, b+k)$ in the vicinity of (a, b) by Taylor's theorem; thus

$$\phi(a+h, b+k) - \phi(a, b) = h\frac{\partial \phi}{\partial x} + k\frac{\partial \phi}{\partial y}$$

$$+ \frac{1}{2!}\left\{h^2\frac{\partial^2 \phi}{\partial x^2} + 2hk\frac{\partial^2 \phi}{\partial x \partial y} + k^2\frac{\partial^2 \phi}{\partial y^2}\right\}$$

$$+ \text{ higher powers of } h, k;$$

but $\frac{\partial \phi}{\partial x} = 0$, $\frac{\partial \phi}{\partial y} = 0$; hence

$$\phi(a+h, b+k) - \phi(a, b) = \frac{1}{2!}\left\{h^2\frac{\partial^2 \phi}{\partial x^2} + 2hk\frac{\partial^2 \phi}{\partial x \partial y} + k^2\frac{\partial^2 \phi}{\partial y^2}\right\} + \cdots$$

Criteria. To distinguish between a maximum and a minimum, at both of which $\frac{\partial \phi}{\partial x} = 0$, $\frac{\partial \phi}{\partial y} = 0$, it is usually sufficient to consider the sign of the expression involving terms of the second degree in h, k; for h, k can generally be made so small that this expression numerically exceeds the sum of all the subsequent terms; hence its sign will determine the sign of $\phi(a+h, b+k) - \phi(a, b)$.

When $x=a$, $y=b$, let $\frac{\partial^2 \phi}{\partial x^2} = A$, $\frac{\partial^2 \phi}{\partial x \partial y} = B$, $\frac{\partial^2 \phi}{\partial y^2} = C$, then the quadratic expression can be written in either of the forms

$$\tfrac{1}{2}(Ah^2 + 2Bhk + Ck^2) = \frac{1}{2A}(A^2h^2 + 2ABhk + ACk^2)$$

$$= \frac{1}{2A}[(Ah + Bk)^2 + (AC - B^2)k^2].$$

The first term of the numerator of the last form is always positive or zero; the second term has the same sign as

$AC - B^2$. If the latter expression is positive, the numerator is positive for all values of h and k; but if it is negative, the sign of the fraction will depend upon the values of h and k, and hence there can be no maximum nor minimum; for instance, the numerator is positive when $k = 0$, and negative when h and k are so taken that $Ah + Bk = 0$.

The second indispensable condition for a maximum or minimum is, therefore,

$$B^2 < AC. \qquad (1)$$

This being satisfied, the numerator is positive, and hence the sign of the fraction is finally determined by the sign of the denominator A. If A is positive, $\phi(a, b)$ is a minimum; if A is negative, $\phi(a, b)$ is a maximum.

It follows from the condition (1) that, since B^2 is positive, A and C must have the same sign.

The whole process may be summarized as follows: to determine whether $\phi(x, y)$ has either a maximum or a minimum, equate its first partial derivatives to zero, and solve the resulting equations $\dfrac{\partial \phi}{\partial x} = 0$, $\dfrac{\partial \phi}{\partial y} = 0$, for x, y. Substitute these critical values in the three second derivatives $\dfrac{\partial^2 \phi}{\partial x^2}$, $\dfrac{\partial^2 \phi}{\partial x\, \partial y}$, $\dfrac{\partial^2 \phi}{\partial y^2}$; then if $\dfrac{\partial^2 \phi}{\partial x^2}$, $\dfrac{\partial^2 \phi}{\partial y^2}$ have the same sign, and

$$\left(\dfrac{\partial^2 \phi}{\partial x\, \partial y}\right)^2 < \dfrac{\partial^2 \phi}{\partial x^2} \cdot \dfrac{\partial^2 \phi}{\partial y^2},$$

there is a maximum when the common sign of $\dfrac{\partial^2 \phi}{\partial x^2}$ and $\dfrac{\partial^2 \phi}{\partial y^2}$ is negative, and a minimum when it is positive.

It is instructive to examine the form of the representative surface in the vicinity of the critical point, especially when some of the conditions for a complete maximum or minimum are not satisfied. The geometric meaning of all the conditions (except the one regarding the sign of B^2-AC) is immediately evident by considering the conditions that the ordinate may have a turning value in each of the vertical sections parallel to the coördinate planes. The deportment of the ordinate in the intermediate sections depends on the sign of B^2-AC, the discriminant of the quadratic expression in h, k, as will be illustrated in the examples.

Special cases can arise in which $A = 0$, $B = 0$, $C = 0$, or when $B^2 - AC = 0$. It is then necessary to consider the higher degree terms. Instead, however, of finding general test formulas for such cases, it is better to work special examples independently. The higher degree terms can in many other cases be made to give useful information regarding the deportment of the function in the vicinity of the critical value, especially in cases of incomplete maxima or minima.

EXERCISES

1. Find the maximum and minimum values of
$$\phi(x, y) = 3\,axy - x^3 - y^3.$$
Here $\quad \dfrac{\partial \phi}{\partial x} = 3\,ay - 3\,x^2; \quad \dfrac{\partial \phi}{\partial y} = 3\,ax - 3\,y^2.$

The critical values are therefore $x = a$, $y = a$; $x = 0$, $y = 0$.

$$\frac{\partial^2 \phi}{\partial x^2} = -6\,x; \quad \frac{\partial^2 \phi}{\partial x\,\partial y} = 3\,a; \quad \frac{\partial^2 \phi}{\partial y^2} = -6\,y.$$

At $x = 0$, $y = 0$, $A = 0$, $B = 3\,a$, $C = 0$, hence $(0, 0)$ is neither a maximum nor a minimum.

At $x = a$, $y = a$,
$$A = -6\,a, \quad B = 3\,a, \quad C = -6\,a.$$
In this case both $\dfrac{\partial^2 \phi}{\partial x^2}$ and $\dfrac{\partial^2 \phi}{\partial y^2}$ are negative, and $B^2 < AC$, hence $\phi(a, a)$ has a maximum value a^3.

2. Exhibit graphically the deportment of the function
$$z = 1 - 4\,x^2 + 21\,xy - 5\,y^2 + x^3 + y^3$$
in the vicinity of the critical point $(0, 0)$.

It is here unnecessary to find the derivatives, as the function is already expanded in the vicinity of the point (0, 0), the letters x and y taking the place of the increments h and k. The absence of the first degree terms shows that the point (0, 0) is a critical point. As the discriminant of the second degree terms, $21^2 - 4\cdot 4\cdot 5$, is positive, the quadratic expression has real factors, and can therefore be made to change its sign for different ratios of y to x; hence there is no complete maximum nor minimum.

To distinguish the sections that have a maximum ordinate at this point, from those that have a minimum ordinate, write the equation in the form

$$z = 1 - 5\left(y - \frac{x}{5}\right)(y - 4x) + x^3 + y^3,$$

which shows that the second degree expression is zero when the ratio of y to x is either $\frac{1}{5}$ or 4; positive when this ratio lies between $\frac{1}{5}$ and 4; and negative for all other values of the ratio. Hence, all vertical sections within the acute angle between the directions $\frac{y}{x} = \frac{1}{5}$ and $\frac{y}{x} = 4$ have a minimum ordinate at (0, 0); and all vertical sections within the obtuse angle have a maximum ordinate at this point. In the first transition direction $\frac{y}{x} = \frac{1}{5}$,

$$z = 1 + x^3 + \frac{x^3}{125} = 1 + \frac{126}{125}x^3;$$

hence the increment of the ordinate is positive when x is positive, and negative when x is negative; and there is neither a maximum nor mini-

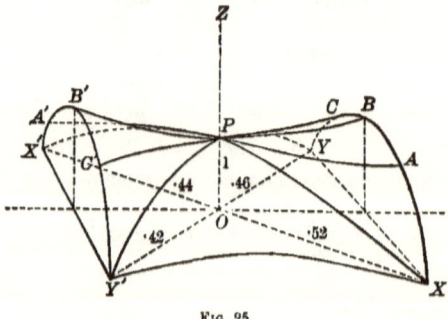

Fig. 25.

mum, but an inflexion, in the transition section. Similarly for the other transition direction. The two horizontal inflexional tangents in these vertical sections are also tangents to the contour of the section made by the horizontal tangent plane through P (Fig. 25).

Some idea of the form of the cubic surface at the critical point P is given in the figure. It shows the vertical sections XPX', YPY' in the coördinate planes, in both of which OP is a maximum; the transition sections APA', CPC', the contours of which bend upwards in the first quarter, and downwards in the third quarter; and an intermediate section BPB', in which OP is a minimum.

If the third degree terms were absent, the transition contours APA', CPC' would be straight lines, the surface would be a hyperbolic paraboloid, and $XYX'Y'$ would be a parallelogram.

3. Examine the deportment of the function

$$z = -70 + 38x - 60y - 10x^2 + 12xy - 15y^2 + 2x^3 - y^3$$

in the vicinity of the critical point $(1, -2)$.

Differentiation and substitution give

$$\frac{\partial z}{\partial x} = 0, \quad \frac{\partial z}{\partial y} = 0, \quad \frac{\partial^2 z}{\partial x^2} = -8, \quad \frac{\partial^2 z}{\partial x \partial y} = 12, \quad \frac{\partial^2 z}{\partial y^2} = -18;$$

$$\frac{\partial^3 z}{\partial x^3} = 12, \quad \frac{\partial^3 z}{\partial x^2 \partial y} = 0, \quad \frac{\partial^3 z}{\partial x \partial y^2} = 0, \quad \frac{\partial^3 z}{\partial y^3} = -6.$$

Hence the expansion of the function in the vicinity of the point $(1, -2)$ is

$$\phi(1+h, -2+k) = \phi(1, -2) - 4h^2 + 12hk - 9k^2 + 2h^3 - k^3.$$

This is one of the exceptional cases referred to above, in which the discriminant $B^2 - AC$ vanishes, and the terms of the second degree form a complete square. Thus,

$$\phi(1+h, -2+k) - \phi(1, -2) = -(2h - 3k)^2 + 2h^3 - k^3,$$

hence the increment of the function is negative for all small values of h and k, unless when $k = \dfrac{2h}{3}$; and thus the ordinate $\phi(1, -2) = 4$ is a maximum in every vertical section but one. In this section the increment of the function is $\Delta\phi = 2h^3 - (\tfrac{2}{3}h)^3 = \tfrac{46}{27}h^3$, hence the contour of the section bends upwards in the first quarter and downwards in the third quarter.

In Fig. 26, XPX' and YPY' are the contours of the sections parallel to the coördinate planes, and APA' is the contour of the vertical section in the intermediate direction $k = \tfrac{2}{3}h$. This may be regarded as a limiting case in which the two transition directions coincide. The horizontal tangent plane at P cuts the surface in a curve which has a cusp at that point; the cuspidal tangent coinciding with the inflexional tangent to the vertical section just mentioned.

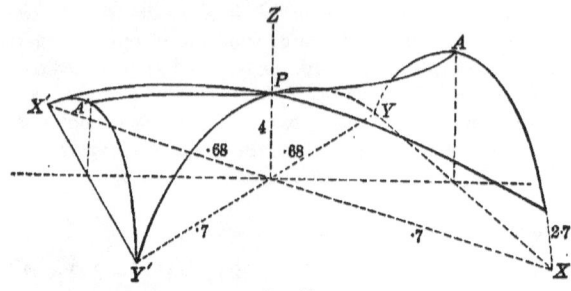

Fig. 26.

4. Find the transition directions in exercise 1 for the critical point $(0, 0)$, and show the form of the surface in the vicinity of the point.

5. Examine the function $z = x^2 - 6xy^2 + cy^4$ at the point $(0, 0)$. Show that if $c > 9$ there is a minimum; and if $c \not> 9$, neither maximum nor minimum. Draw graph.

6. Show that $xe^{y+z\sin y}$ has neither a maximum nor a minimum.

7. Divide a into three parts such that their continued product may be a maximum.

8. Find the minimum surface of a rectangular parallelopiped whose volume is a^3.

9. What value of x, y will make $\dfrac{1 + x^2 + y^2}{1 - ax - by}$ a maximum or a minimum?

10. Find the values of x and y that make $\sin x + \sin y + \cos(x + y)$ a maximum or a minimum.

11. Find the maximum of $(a - x)(a - y)(x + y - a)$.

12. The electric time constant of a cylindric coil of wire is

$$u = \frac{mxyz}{ax + by + cz},$$

where x is the mean radius, y is the difference between the internal and external radii, z is the axial length, and m, a, b, c are known constants. The volume of the coil is $nxyz = g$. Find the values of x, y, z to make u a minimum if the volume of the coil is fixed. (Perry's Calculus.)

108. Conditional maxima and minima. Maxima and minima of implicit functions.

In certain problems the maximum or minimum values of a function of two variables are desired, when the mode of variation of x and y is restricted by an imposed condition.

Let the function be $z = f(x, y)$, and let the assigned condition be $\phi(x, y) = 0$; then it is required to find the maximum or minimum values passed through by the function z, when x and y vary consistently with the relation $\phi(x, y) = 0$.

This problem may also be stated in the following geometrical form: A point moves on the surface $z = f(x, y)$ in the curve of intersection made by the cylindrical surface $\phi(x, y) = 0$; find the maximum and minimum values of its height above the horizontal coördinate plane.

Since the variables x and y always satisfy $\phi(x, y) = 0$, hence their rates of change are connected by the relation

$$d\phi = \frac{\partial \phi}{\partial x} dx + \frac{\partial \phi}{\partial y} dy = 0, \tag{1}$$

but, since z is at a turning value, its rate of change vanishes, hence

$$dz = \frac{\partial f}{\partial x} dx + \frac{\partial f}{\partial y} dy = 0; \tag{2}$$

therefore, by elimination of dx and dy,

$$\frac{\partial \phi}{\partial x} \frac{\partial f}{\partial y} - \frac{\partial \phi}{\partial y} \frac{\partial f}{\partial x} = 0. \tag{3}$$

This equation, together with $\phi(x, y) = 0$, will determine the critical values of x and y.

The value of the function z, corresponding to a critical value, will be a maximum or minimum according as d^2z is negative or positive; but

$$d^2z = \frac{\partial^2 f}{\partial x^2} dx^2 + 2 \frac{\partial^2 f}{\partial x \, \partial y} dx \, dy + \frac{\partial^2 f}{\partial y^2} dy^2 + \frac{\partial f}{\partial x} d^2x + \frac{\partial f}{\partial y} d^2y, \tag{4}$$

$$d^2\phi = \frac{\partial^2\phi}{\partial x^2}dx^2 + 2\frac{\partial^2\phi}{\partial x \partial y}dx\,dy + \frac{\partial^2\phi}{\partial y^2}dy^2 + \frac{\partial\phi}{\partial x}d^2x + \frac{\partial\phi}{\partial y}d^2y = 0; \quad (5)$$

to eliminate d^2x, d^2y, multiply (4) by $\frac{\partial\phi}{\partial y}$ and (5) by $\frac{\partial f}{\partial y}$, subtract, and take account of (3); then

$$\phi_2 d^2z = (\phi_2 f_{11} - f_2 \phi_{11})dx^2 + 2(\phi_2 f_{12} - f_1 \phi_{12})dx\,dy$$
$$+ (\phi_2 f_{22} - f_2 \phi_{22})dy^2, \quad (6)$$

in which the subscripts 1, 2, indicate differentiation with regard to x, y, respectively. The sign of the right hand member of (6) is not changed by dividing by dx^2, and then replacing $\dfrac{dy}{dx}$ by $-\dfrac{\phi_1}{\phi_2}$, from (1); hence the sign of d^2z at a critical point is the same as the sign of

$$\frac{1}{\phi_2}[(\phi_2 f_{11} - f_2 \phi_{11})\phi_2^2 - 2(\phi_2 f_{12} - f_1 \phi_{12})\phi_1 \phi_2$$
$$+ (\phi_2 f_{22} - f_2 \phi_{22})\phi_1^2];$$

which is therefore sufficient to discriminate a maximum from a minimum.

Ex. 1. If $z = x^3 + y^3 - 3axy$, and if x and y vary subject to the condition $x^2 + y^2 = 8a^2$; show that z passes through a minimum when $x = 2a$, $y = 2a$.

Here, $f_1 = 3(x^2 - ay)$, $f_2 = 3(y^2 - ax)$, $f_{11} = 6x$, $f_{12} = -3a$, $f_{22} = 6y$,
$\phi_1 = 2x$, $\phi_2 = 2y$, $\phi_{11} = 2$, $\phi_{12} = 0$, $\phi_{22} = 2$.

The critical values are found from $\phi_1 f_2 - \phi_2 f_1 = 0$ and $x^2 + y^2 = 8a^2$; and one pair is easily found to be $x = 2a$, $y = 2a$. At this critical point

$f_1 = 6a^2$, $f_2 = 6a^2$, $f_{11} = 12a$, $f_{12} = -3a$, $f_{22} = 12a$,
$\phi_1 = 4a$, $\phi_2 = 4a$, $\phi_{11} = 2$, $\phi_{12} = 0$, $\phi_{22} = 2$;

and the sign of the discriminating expression above is found to be positive, showing that z is a minimum.

Ex. 2. Show that the maximum and minimum of the function $x^2 + y^2$, subject to the condition $ax^2 + 2hxy + by^2 = 1$, are given by the roots of the quadratic equation

$$\left(a - \frac{1}{r^2}\right)\left(b - \frac{1}{r^2}\right) = 0;$$

hence show how to find the axes of the conic defined by the above equation of condition.

Ex. 3. Find the minimum value of x^2+y^2, subject to the condition
$$\frac{x}{a}+\frac{y}{b}=1.$$

NOTE. When the equation $\phi(x, y) = 0$ can be solved for one of the variables, the method of Art. 82 can also be used.

IMPLICIT FUNCTIONS

Let y be defined as a function of the single variable x by the implicit relation $f(x, y) = 0$; it is required to find at what values of x the function y passes through a maximum or a minimum.

By successive differentiation, leaving the independent variable at first arbitrary,

$$df = \frac{\partial f}{\partial x}dx + \frac{\partial f}{\partial y}dy = 0, \qquad (1)$$

$$d^2f = \frac{\partial^2 f}{\partial x^2}dx^2 + 2\frac{\partial^2 f}{\partial x\,\partial y}dx\,dy + \frac{\partial^2 f}{\partial y^2}dy^2 + \frac{\partial f}{\partial x}d^2x + \frac{\partial f}{\partial y}d^2y = 0. \quad (2)$$

From (1) $\qquad \dfrac{dy}{dx} = -\dfrac{\dfrac{\partial f}{\partial x}}{\dfrac{\partial f}{\partial y}};$

hence the values of x, y at which $\dfrac{dy}{dx}$ changes sign satisfy one of the equations

$$\frac{\partial f}{\partial x}=0, \quad \frac{\partial f}{\partial y}=0.$$

Thus the first set of critical values of x, with the corresponding values of y, are to be found from the simultaneous equations,

$$f(x, y) = 0, \ \frac{\partial f}{\partial x} = 0,$$

and the second set of critical values from

$$f(x, y) = 0, \frac{\partial f}{\partial y} = 0.$$

These two sets may or may not have values of x, y in common.

Those of the first set that do not belong to the second set make $\frac{\partial f}{\partial x} = 0$, $\frac{\partial f}{\partial y} \neq 0$, and hence make $\frac{dy}{dx} = 0$.

To test whether $\frac{dy}{dx}$ changes its sign, in passing through zero, the method of Art. 82 is available. Taking x as the independent variable in (2) and putting $\frac{dy}{dx} = 0$, $\frac{\partial f}{\partial x} = 0$, it gives

$$\frac{d^2y}{dx^2} = -\frac{\frac{\partial^2 f}{\partial x^2}}{\frac{\partial f}{\partial y}}.$$

Hence for the critical values under consideration, y is a maximum or a minimum according as $\frac{\partial^2 f}{\partial x^2}$, $\frac{\partial f}{\partial y}$ have the same or opposite signs.

Those of the second set of critical values that do not belong to the first set make $\frac{\partial f}{\partial y} = 0$, $\frac{\partial f}{\partial x} \neq 0$, and hence make $\frac{dy}{dx}$ infinite.

To find whether $\frac{dy}{dx}$ changes its sign, the second derivative is not available, since it and all subsequent derivatives are infinite; but methods of trial may be resorted to, in which assistance can often be derived from the graphical representation of the function.

The critical values that are common to the first and the second set make $\frac{\partial f}{\partial x} = 0$, $\frac{\partial f}{\partial y} = 0$, and hence render $\frac{dy}{dx}$ inde-

108.]   MAXIMA AND MINIMA IN TWO VARIABLES   195

terminate in form. When numerically evaluated it is either zero, infinite, or finite. In the last case $\dfrac{dy}{dx}$ cannot change its sign and there is no turning value of y. In the first two cases the question whether $\dfrac{dy}{dx}$ changes its sign as x passes through the critical values, and y changes correspondingly, is to be decided by trial.

Ex. 4.   Given $(x^2 + y^2)^2 - 2y(x^2 + y^2) - x^2 = 0$;
find the turning values of y, and the corresponding values of x.

Since
$$\frac{dy}{dx} = -\frac{2x(x^2+y^2) - 2xy - x}{2y(x^2+y^2) - x^2 - 3y^2}, \qquad (1)$$

the first set of critical values are found from
$$(x^2 + y^2)^2 - 2y(x^2 + y^2) - x^2 = 0, \qquad (2)$$
$$x[2(x^2 + y^2) - 2y - 1] = 0. \qquad (3)$$

Equation (3) is satisfied by $x = 0$, which, substituted in (2), gives $y = 0$, or $y = 2$. Equation (3) is also satisfied when
$$2(x^2 + y^2) - 2y - 1 = 0,$$
i.e., when $x^2 = -y^2 + y + \tfrac{1}{2}$, which substituted in (2) gives $y = -\tfrac{1}{4}$, whence $x = \pm .43 \cdots$. Thus the first set of critical values of (x, y) is composed of the four pairs:

$$(0, 0), \quad (0, 2), \quad (.43, -.25), \quad (-.43, -.25).$$

The second set is found from (2) and the equation
$$2y(x^2 + y^2) - x^2 - 3y^2 = 0, \qquad (4)$$
which, on eliminating x^2, gives $y = .75$ or 0, whence $x = \pm 1.3$ or 0.

Thus the second set of critical values of x, y is composed of
$$(0, 0), \quad (1.3, .75), \quad (-1.3, .75);$$
the values $(0, 0)$ being common to both sets.

To test the remaining critical values of the first set, use the second derivative
$$\frac{d^2y}{dx^2} = -\frac{\dfrac{\partial^2 f}{\partial x^2}}{\dfrac{\partial f}{\partial y}} = -\frac{6x^2 + 4y^2 - 2y - 1}{2y(x^2+y^2) - x^2 - 3y^2},$$

which, for $(0, 2)$ is negative, and for $(.43, -.25)$, $(-.43, -.25)$ positive; hence, when x passes through 0, the function y passes through a maximum value 2, and when x passes through $-.43$, $.43$, y passes through a minimum value $-.25$. It is to be observed that in the latter case the function has three other values (or branches), real or imaginary, that do not pass through turning values when x passes through $\pm .43$.

To test the critical values $(0, 0)$, for which equation (1) becomes indeterminate, evaluate the function in the usual way, by replacing both numerator and denominator by their respective total x-derivatives. This gives

$$\frac{dy}{dx} = -\frac{(6x^2 + 2y^2 - 2y - 1) + (4xy - 2x)\frac{dy}{dx}}{(4xy - 2x) + (2x^2 + 6y^2 - 6y)\frac{dy}{dx}};$$

$$\therefore (2x^2 + 6y^2 - 6y)\left(\frac{dy}{dx}\right)^2 + (8xy - 4x)\left(\frac{dy}{dx}\right) + (6x^2 + 2y^2 - 2y - 1) = 0,$$

a quadratic equation in $\frac{dy}{dx}$. Now put $x = 0$, $y = 0$; the two roots of the equation become infinite, hence $\frac{dy}{dx} = \infty$. In the present case it is easy to find by trial whether $\frac{dy}{dx}$ changes sign; for in the vicinity of the values $(0, 0)$ equation (1) may be written in the approximate form

$$\frac{dy}{dx} = \frac{-x}{x^2 + 3y^2},$$

in which only the important terms are retained; hence $\frac{dy}{dx}$ changes sign from $+$ to $-$ as x increases through zero, and thus y passes through a maximum.

The values $(0, 0)$ could also be shown to give a maximum without the use of derivatives, by observing that in the vicinity of the values $(0, 0)$ equation (2) can be replaced by

$$x^2 + 2y^3 = 0.$$

When x is small, and either positive or negative, y must be negative; but when $x = 0$, then $y = 0$; hence $y = 0$ is a maximum value of the function.

It is not easy to test the other critical values at which $\frac{dy}{dx}$ becomes infinite without anticipating the methods of curve tracing. It will appear by the methods of Chapter XVIII that the graph of the function is as in the accompanying figure, and that the critical values last mentioned are neither maxima nor minima values for y.

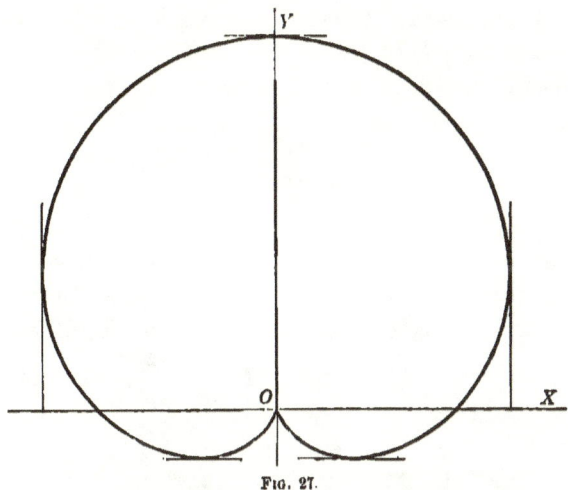

Fig. 27.

Ex. 5. Given $x^4 + 2ax^2y - ay^3 = 0$; find the maximum and minimum values of y, and of x.

Ex. 6. If $x^3 + y^3 - 3axy = 0$; find the maxima and minima of y.

Ex. 7. If $3a^2y^2 + xy^3 + 4ax^3 = 0$; find the turning values of x, y.

Ex. 8. Show that in the vicinity of a maximum or minimum value of $f(x, y)$, the increment $\Delta f(x, y)$ is an infinitesimal of an even order, when Δx and Δy are of the first order.

When is $\Delta f(x, y)$ of the third order?

CHAPTER XI

CHANGE OF THE VARIABLE

109. Interchange of dependent and independent variables. It has already been proved in Art. 21, as the direct consequence of the definition of a derivative, that if $y = \phi(x)$, then

$$\frac{dy}{dx} = \frac{1}{\frac{dx}{dy}}. \tag{1}$$

This process is known as changing the independent variable from x to y. The corresponding relation for the higher derivatives is less simple, and will now be developed.

To express $\frac{d^2y}{dx^2}$ in terms of $\frac{dx}{dy}$, $\frac{d^2x}{dy^2}$, differentiate (1) as to x,

$$\frac{d^2y}{dx^2} = \frac{d}{dx}\left[\frac{1}{\frac{dx}{dy}}\right] = \frac{d}{dy}\left[\frac{1}{\frac{dx}{dy}}\right] \cdot \frac{dy}{dx} = \frac{d}{dy}\left[\frac{1}{\frac{dx}{dy}}\right] \cdot \frac{1}{\frac{dx}{dy}},$$

but

$$\frac{d}{dy}\left[\frac{1}{\frac{dx}{dy}}\right] = -\frac{\frac{d^2x}{dy^2}}{\left(\frac{dx}{dy}\right)^2},$$

hence

$$\frac{d^2y}{dx^2} = -\frac{\frac{d^2x}{dy^2}}{\left(\frac{dx}{dy}\right)^3}. \tag{2}$$

In a similar manner,

$$\frac{d^3y}{dx^3} = -\frac{\dfrac{d^3x}{dy^3}\dfrac{dx}{dy} - 3\left(\dfrac{d^2x}{dy^2}\right)^2}{\left(\dfrac{dx}{dy}\right)^5}. \tag{3}$$

110. Change of the dependent variable. If y is a function of z, let it be required to express $\dfrac{dy}{dx}$, $\dfrac{d^2y}{dx^2}$, $\cdots$ in terms of $\dfrac{dz}{dx}$, $\dfrac{d^2z}{dx^2}$, $\cdots$.

Let $y = \phi(z)$, then

$$\frac{dy}{dx} = \frac{dy}{dz}\frac{dz}{dx} = \phi'(z)\frac{dz}{dx}.$$

$$\frac{d^2y}{dx^2} = \frac{d}{dz}\left(\frac{dy}{dx}\right) \cdot \frac{dz}{dx} = \frac{d}{dz}\left(\phi'(z)\frac{dz}{dx}\right) \cdot \frac{dz}{dx}$$

$$= \phi''(z)\left(\frac{dz}{dx}\right)^2 + \phi'(z)\frac{d}{dz}\left(\frac{dz}{dx}\right) \cdot \frac{dz}{dx};$$

but the second term can be expressed directly as $\phi'(z)\dfrac{d^2z}{dx^2}$, hence

$$\frac{d^2y}{dx^2} = \phi''(z)\left(\frac{dz}{dx}\right)^2 + \phi'(z)\frac{d^2z}{dx^2}. \tag{4}$$

The higher x-derivatives of y can be similarly expressed in terms of x-derivatives of z.

Ex. Show that (4) may be regarded as a special case of (6), Art. 102, in which one of the variables is replaced by a constant.

111. Change of the independent variable. Let y be a function of x, and let both x and y be functions of a new variable t. It is required to express $\dfrac{dy}{dx}$ in terms of $\dfrac{dy}{dt}$; and $\dfrac{d^2y}{dx^2}$ in terms of $\dfrac{dy}{dt}$ and $\dfrac{d^2y}{dt^2}$.

By Arts. 21, 51,

$$\frac{dy}{dx} = \frac{\frac{dy}{dt}}{\frac{dx}{dt}}, \qquad (1)$$

$$\frac{d^2y}{dx^2} = \frac{\frac{d^2y}{dt^2}\frac{dx}{dt} - \frac{d^2x}{dt^2}\frac{dy}{dt}}{\left(\frac{dx}{dt}\right)^3}. \qquad (2)$$

If x be given as an explicit function of t in the form $x = f(t)$, then $\frac{dx}{dt} = f'(t)$, $\frac{d^2x}{dt^2} = f''(t)$, and the last equation may be written

$$\frac{d^2y}{dx^2} = \frac{\frac{d^2y}{dt^2} - \frac{dy}{dt} \cdot \frac{f''(t)}{f'(t)}}{[f'(t)]^2}. \qquad (3)$$

In practical examples it is usually better to work by the methods here illustrated than to use the resulting formulas.

EXERCISES

1. Change the independent variable from x to z in the equation

$$x^2 \frac{d^2y}{dx^2} + x \frac{dy}{dx} + y = 0, \quad \text{when } x = e^z.$$

$$\frac{dy}{dx} = \frac{dy}{dz} e^{-z},$$

$$\frac{d^2y}{dx^2} = \frac{d^2y}{dz^2} e^{-2z} - \frac{dy}{dz} e^{-2z}.$$

Hence $x^2 \frac{d^2y}{dx^2} + x \frac{dy}{dx} + y = 0$ becomes $\frac{d^2y}{dz^2} + y = 0$.

2. Interchange the function and the variable in the equation

$$\frac{d^2y}{dx^2} + 2y\left(\frac{dy}{dx}\right)^2 = 0.$$

3. Interchange x and y in the equation

$$R = \frac{\left\{1 + \left(\frac{dy}{dx}\right)^2\right\}^{\frac{3}{2}}}{\frac{d^2y}{dx^2}}$$

4. Change the independent variable from x to y in the equation

$$3\left(\frac{d^2y}{dx^2}\right)^2 - \frac{dy}{dx}\frac{d^3y}{dx^3} - \frac{d^2y}{dx^2}\left(\frac{dy}{dx}\right)^2 = 0.$$

5. Change the dependent variable from y to z in the equation

$$\frac{d^2y}{dx^2} = 1 + \frac{2(1+y)}{1+y^2}\left(\frac{dy}{dx}\right)^2, \quad \text{when} \quad y = \tan z.$$

6. Change the independent variable from x to y in the equation

$$x^2\frac{d^2u}{dx^2} + x\frac{du}{dx} + u = 0, \quad \text{when} \quad y = \log x.$$

7. If y is a function of x, and x a function of the time t, express the y-acceleration in terms of the x-acceleration, and the x-velocity.

Since
$$\frac{dy}{dt} = \frac{dy}{dx}\frac{dx}{dt},$$

hence
$$\frac{d^2y}{dt^2} = \frac{dy}{dx}\frac{d^2x}{dt^2} + \frac{dx}{dt}\cdot\frac{d}{dt}\left(\frac{dy}{dx}\right),$$

but
$$\frac{d}{dt}\left(\frac{dy}{dx}\right) = \frac{d}{dx}\left(\frac{dy}{dx}\right)\frac{dx}{dt} = \frac{d^2y}{dx^2}\frac{dx}{dt},$$

hence
$$\frac{d^2y}{dt^2} = \frac{dy}{dx}\frac{d^2x}{dt^2} + \frac{d^2y}{dx^2}\left(\frac{dx}{dt}\right)^2.$$

In the abbreviated notation for t-derivatives,

$$d^2y = \frac{dy}{dx}d^2x + \frac{d^2y}{dx^2}(dx)^2.$$

Compare this result with (4), Art. 110, and with (6), Art. 102.

112. Change of two independent variables. Let $u = f(x, y)$ be a function of the two variables x, y which are themselves functions of two new variables w, z; it is required to express $\dfrac{\partial u}{\partial x}$, $\dfrac{\partial u}{\partial y}$ in terms of $\dfrac{\partial u}{\partial w}$, $\dfrac{\partial u}{\partial z}$.

I. The variables x, y explicit functions of w, z.

Let $u = f(x, y)$; $x = \phi_1(w, z)$; $y = \phi_2(w, z)$.

Since u is the function of w and z,

$$\frac{\partial u}{\partial w} = \frac{\partial f}{\partial x}\frac{\partial x}{\partial w} + \frac{\partial f}{\partial y}\frac{\partial y}{\partial w} \quad (z \text{ regarded as constant}).$$

The values of $\dfrac{\partial x}{\partial w}$, $\dfrac{\partial y}{\partial w}$ are to be found from $x = \phi_1$, $y = \phi_2$,

thus
$$\left.\begin{array}{l}\dfrac{\partial u}{\partial w} = \dfrac{\partial f}{\partial x}\dfrac{\partial \phi_1}{\partial w} + \dfrac{\partial f}{\partial y}\dfrac{\partial \phi_2}{\partial w},\\[1em]\dfrac{\partial u}{\partial z} = \dfrac{\partial f}{\partial x}\dfrac{\partial \phi_1}{\partial z} + \dfrac{\partial f}{\partial y}\dfrac{\partial \phi_2}{\partial z},\end{array}\right\} \quad (1)$$

Similarly,

(w regarded as constant).

In the expression for $\dfrac{\partial u}{\partial w}$, z is to be regarded as constant, and w as variable; and x, y as functions of w.

In the expression for $\dfrac{\partial u}{\partial z}$, w is to be regarded as constant, and z as variable; and x, y as functions of z.

If x, y be called the old variables, and z, w the new variables, then it appears from the above expressions that when the old variables are explicit functions of the new variables, the new derivatives $\dfrac{\partial u}{\partial w}$, $\dfrac{\partial u}{\partial z}$ are explicit functions of the old derivatives $\dfrac{\partial u}{\partial x}$, $\dfrac{\partial u}{\partial y}$. The last two equations may, when desired, be solved for $\dfrac{\partial f}{\partial x}$, $\dfrac{\partial f}{\partial y}$.

II. The variables w, z explicit functions of x, y.

Let $z = \psi_1(x, y)$; $w = \psi_2(x, y)$,

then
$$\frac{\partial u}{\partial x} = \frac{\partial u}{\partial w}\frac{\partial w}{\partial x} + \frac{\partial u}{\partial z}\frac{\partial z}{\partial x} \quad (y \text{ regarded as constant}),$$

$$\frac{\partial u}{\partial y} = \frac{\partial u}{\partial w}\frac{\partial w}{\partial y} + \frac{\partial u}{\partial z}\frac{\partial z}{\partial y} \quad (x \text{ regarded as constant}).$$

Substituting the values of $\frac{\partial w}{\partial x}$, $\frac{\partial w}{\partial y}$, $\frac{\partial z}{\partial x}$, $\frac{\partial z}{\partial y}$ from $z = \psi_1$, $w = \psi_2$, the last equations become

$$\left.\begin{aligned}\frac{\partial u}{\partial x} &= \frac{\partial u}{\partial w}\frac{\partial \psi_2}{\partial x} + \frac{\partial u}{\partial z}\frac{\partial \psi_1}{\partial x}, \\ \frac{\partial u}{\partial y} &= \frac{\partial u}{\partial w}\frac{\partial \psi_2}{\partial y} + \frac{\partial u}{\partial z}\frac{\partial \psi_1}{\partial y}.\end{aligned}\right\} \quad (2)$$

These equations may, when desired, be solved for $\frac{\partial u}{\partial z}$, $\frac{\partial u}{\partial w}$.

In this case the new variables are explicit functions of the old ones, while the old derivatives are explicit functions of the new ones.

Ex. Let $u \equiv x^2 - y^2$, $x = \rho \cos\theta$, $y = \rho \sin\theta$. Find $\frac{\partial u}{\partial \rho}$, $\frac{\partial u}{\partial \theta}$ by the method of I.

$$\begin{aligned}\frac{\partial u}{\partial \rho} &= \frac{\partial u}{\partial x}\frac{\partial x}{\partial \rho} + \frac{\partial u}{\partial y}\frac{\partial y}{\partial \rho} \quad (\theta \text{ regarded as constant}), \\ &= 2x\cos\theta - 2y\sin\theta, \\ &= 2\rho\cos^2\theta - 2\rho\sin^2\theta, \\ &= 2\rho\cos 2\theta,\end{aligned}$$

which agrees with the result of direct substitution.

Again, $\quad \begin{aligned}\frac{\partial u}{\partial \theta} &= \frac{\partial u}{\partial x}\frac{\partial x}{\partial \theta} + \frac{\partial u}{\partial y}\frac{\partial y}{\partial \theta} \quad (\rho \text{ regarded as constant}), \\ &= -2x\rho\sin\theta - 2y\rho\cos\theta, \\ &= -4\rho^2\cos\theta\sin\theta, \\ &= -2\rho^2\sin 2\theta,\end{aligned}$

which also agrees with the result of direct substitution.

Next suppose the new variables ρ, θ are expressed in terms of the old variables x, y in the form $\rho = \sqrt{x^2 + y^2}$, $\theta = \tan^{-1}\frac{y}{x}$; find $\frac{\partial u}{\partial \rho}$, $\frac{\partial u}{\partial \theta}$ by the method of II.

Here $\quad \dfrac{\partial \rho}{\partial x} = \dfrac{x}{\sqrt{x^2+y^2}} = \cos\phi; \quad \dfrac{\partial \theta}{\partial x} = \dfrac{-y}{x^2+y^2} = -\dfrac{\sin\theta}{\rho},$

$\quad\quad\quad \dfrac{\partial \rho}{\partial y} = \dfrac{y}{\sqrt{x^2+y^2}} = \sin\theta; \quad \dfrac{\partial \theta}{\partial y} = \dfrac{x}{x^2+y^2} = \dfrac{\cos\theta}{\rho},$

but $\quad \dfrac{\partial u}{\partial x} = 2x = \dfrac{\partial u}{\partial \rho}\dfrac{\partial \rho}{\partial x} + \dfrac{\partial u}{\partial \theta}\dfrac{\partial \theta}{\partial x} \quad$ (y regarded as constant),

hence $\quad 2\rho \cos\theta = \dfrac{\partial u}{\partial \rho}\cos\theta - \dfrac{\partial u}{\partial \theta}\cdot\dfrac{\sin\theta}{\rho};\quad$ (1)

also, $\quad \dfrac{\partial u}{\partial y} = -2y = \dfrac{\partial u}{\partial \rho}\dfrac{\partial \rho}{\partial y} + \dfrac{\partial u}{\partial \theta}\dfrac{\partial \theta}{\partial y} \quad$ (x regarded as constant),

hence $\quad -2\rho \sin\theta = \dfrac{\partial u}{\partial \rho}\sin\theta + \dfrac{\partial u}{\partial \theta}\dfrac{\cos\theta}{\rho}.\quad$ (2)

Now, solving (1) and (2) for $\dfrac{\partial u}{\partial \rho}$, $\dfrac{\partial u}{\partial \theta}$, it follows that

$$\dfrac{\partial u}{\partial \rho} = 2\rho \cos 2\theta, \quad \dfrac{\partial u}{\partial \theta} = -2\rho \sin 2\theta;$$

the same results as were obtained before.

III. **The relation between x, y and w, z defined by implicit equations.**

Let $\quad f_1(x, y, z, w) = 0, \quad f_2(x, y, z, w) = 0.$

In the first place, to find $\dfrac{\partial x}{\partial z}$, $\dfrac{\partial y}{\partial z}$ required in I (1), differentiate the two given equations partially with regard to z, then

$$\dfrac{\partial f_1}{\partial z} + \dfrac{\partial f_1}{\partial x}\dfrac{\partial x}{\partial z} + \dfrac{\partial f_1}{\partial y}\dfrac{\partial y}{\partial z} = 0,$$

(w regarded as constant)

$$\dfrac{\partial f_2}{\partial z} + \dfrac{\partial f_2}{\partial x}\dfrac{\partial x}{\partial z} + \dfrac{\partial f_2}{\partial y}\dfrac{\partial y}{\partial z} = 0;$$

solve the resulting equations for $\dfrac{\partial x}{\partial z}$, $\dfrac{\partial y}{\partial z}$, then substitute in (1) and proceed as before.

Similarly for $\dfrac{\partial x}{\partial w}$, $\dfrac{\partial y}{\partial w}$.

113. Change of three independent variables. The student will not have much difficulty in extending the theory to functions of three or more variables.

Let $u = f(x, y, z)$; and let x, y, z be functions of three new variables u, v, w, connected by the equations

$$x = \phi_1(u, v, w), \quad y = \phi_2(u, v, w), \quad z = \phi_3(u, v, w). \qquad (1)$$

It is required to express $\dfrac{\partial f}{\partial x}$ in terms of u, v, w.

$$\frac{\partial f}{\partial u} = \frac{\partial f}{\partial x}\frac{\partial x}{\partial u} + \frac{\partial f}{\partial y}\frac{\partial y}{\partial u} + \frac{\partial f}{\partial z}\frac{\partial z}{\partial u} \quad (v, w \text{ regarded as constants}),$$

$$\frac{\partial f}{\partial v} = \frac{\partial f}{\partial x}\frac{\partial x}{\partial v} + \frac{\partial f}{\partial y}\frac{\partial y}{\partial v} + \frac{\partial f}{\partial z}\frac{\partial z}{\partial v} \quad (u, w \text{ regarded as constants}),$$

$$\frac{\partial f}{\partial w} = \frac{\partial f}{\partial x}\frac{\partial x}{\partial w} + \frac{\partial f}{\partial y}\frac{\partial y}{\partial w} + \frac{\partial f}{\partial z}\frac{\partial z}{\partial w} \quad (u, v \text{ regarded as constants}).$$

From (1), $\dfrac{\partial x}{\partial u}, \dfrac{\partial x}{\partial v}, \dfrac{\partial x}{\partial w}, \dfrac{\partial y}{\partial u}, \cdots$ can be found; their values are to be substituted in the equations for $\dfrac{\partial f}{\partial u}, \cdots$, and the resulting equations solved for $\dfrac{\partial f}{\partial x}, \dfrac{\partial f}{\partial y}, \cdots$.

Similarly for the case in which u, v, w are explicit functions of x, y, z.

114. Application to higher derivatives. The second and higher derivatives can be obtained in the same way. As the general formulas become too complicated to be of much use, it is better to work out special examples independently.

Ex. Express $\dfrac{\partial^2 u}{\partial x^2} + \dfrac{\partial^2 u}{\partial y^2}$ in terms of ρ, θ, given

$$x = \rho \cos \theta, \quad y = \rho \sin \theta. \qquad (1)$$

The general formula is

$$\frac{\partial u}{\partial x} = \frac{\partial u}{\partial \rho}\frac{\partial \rho}{\partial x} + \frac{\partial u}{\partial \theta}\frac{\partial \theta}{\partial x}; \qquad (2)$$

in which $\dfrac{\partial \rho}{\partial x}, \dfrac{\partial \theta}{\partial x}$ are to be obtained from (1), by differentiating and solving.

Thus
$$1 = \frac{\partial \rho}{\partial x}\cos\theta - \rho\sin\theta\frac{\partial \theta}{\partial x},$$
$$0 = \frac{\partial \rho}{\partial x}\sin\theta + \rho\cos\theta\frac{\partial \theta}{\partial x},$$
(y regarded as constant);

hence
$$\frac{\partial \rho}{\partial x} = \cos\theta, \quad \frac{\partial \theta}{\partial x} = -\frac{\sin\theta}{\rho}. \tag{3}$$

Similarly $\frac{\partial \rho}{\partial y}, \frac{\partial \theta}{\partial y}$ can be obtained from (1):

$$0 = \frac{\partial \rho}{\partial y}\cos\theta - \rho\sin\theta\frac{\partial \theta}{\partial y},$$
$$1 = \frac{\partial \rho}{\partial y}\sin\theta + \rho\cos\theta\frac{\partial \theta}{\partial y},$$
(x regarded as constant);

hence
$$\frac{\partial \rho}{\partial y} = \sin\theta, \quad \frac{\partial \theta}{\partial y} = \frac{\cos\theta}{\rho}. \tag{4}$$

Substitution from (3) in (2) gives
$$\frac{\partial u}{\partial x} = \frac{\partial u}{\partial \rho}\cos\theta - \frac{\partial u}{\partial \theta}\frac{\sin\theta}{\rho}. \tag{5}$$

A repetition of this process gives
$$\frac{\partial^2 u}{\partial x^2} = \frac{\partial^2 u}{\partial \rho^2}\cos^2\theta - \frac{\partial^2 u}{\partial \theta\,\partial \rho}\frac{\sin\theta\cos\theta}{\rho} + \frac{\partial u}{\partial \rho}\frac{\sin^2\theta}{\rho} - \frac{\partial^2 u}{\partial \theta\,\partial \rho}\frac{\cos\theta\sin\theta}{\rho}$$
$$+ \frac{\partial^2 u}{\partial \theta^2}\frac{\sin^2\theta}{\rho^2} + \frac{\partial u}{\partial \theta}\frac{\cos\theta\sin\theta}{\rho^2} + \frac{\partial u}{\partial \theta}\frac{\sin\theta\cos\theta}{\rho^2}. \tag{6}$$

The expression, similar to (2), for $\frac{\partial u}{\partial y}$, combined with (4), leads to

$$\frac{\partial u}{\partial y} = \frac{\partial u}{\partial \rho}\sin\theta + \frac{\partial u}{\partial \theta}\frac{\cos\theta}{\rho}, \tag{7}$$

and when this step is repeated, there results,
$$\frac{\partial^2 u}{\partial y^2} = \frac{\partial^2 u}{\partial \rho^2}\sin^2\theta + \frac{\partial^2 u}{\partial \theta\,\partial \rho}\frac{\sin\theta\cos\theta}{\rho} + \frac{\partial u}{\partial \rho}\frac{\cos^2\theta}{\rho} + \frac{\partial^2 u}{\partial \theta\,\partial \rho}\frac{\cos\theta\sin\theta}{\rho}$$
$$+ \frac{\partial^2 u}{\partial \theta^2}\frac{\cos^2\theta}{\rho^2} - \frac{\partial u}{\partial \theta}\frac{\cos\theta\sin\theta}{\rho^2} - \frac{\partial u}{\partial \theta}\frac{\sin\theta\cos\theta}{\rho^2}; \tag{8}$$

and the addition of (6) and (8) gives the required identity
$$\frac{\partial^2 u}{\partial x^2} + \frac{\partial^2 u}{\partial y^2} = \frac{\partial^2 u}{\partial \rho^2} + \frac{1}{\rho}\frac{\partial u}{\partial \rho} + \frac{1}{\rho^2}\frac{\partial^2 u}{\partial \theta^2}.$$

EXERCISES

1. Given $x = \rho \cos \theta$, $y = \rho \sin \theta$, y being a function of x, show that

$$\frac{d^2 y}{dx^2} = \frac{\rho^2 + 2\left(\dfrac{d\rho}{d\theta}\right)^2 - \rho \dfrac{d^2\rho}{d\theta^2}}{\left(\cos\theta \dfrac{d\rho}{d\theta} - \rho \sin\theta\right)^3}.$$

2. Given $x = a(1 - \cos t)$, $y = a(nt + \sin t)$; prove that

$$\frac{d^2 y}{dx^2} = -\frac{n \cos t + 1}{a \sin^3 t}.$$

3. If $\xi = x \cos a - y \sin a$, $\eta = x \sin a + y \cos a$, prove that

$$\frac{\partial^2 u}{\partial x^2} + \frac{\partial^2 u}{\partial y^2} = \frac{\partial^2 u}{\partial \xi^2} + \frac{\partial^2 u}{\partial \eta^2}.$$

4. Given $x = \rho \cos \theta$, $y = \rho \sin \theta$, show that

$$x \frac{\partial u}{\partial x} - y \frac{\partial u}{\partial y} = \frac{\partial u}{\partial \theta}; \quad x \frac{\partial u}{\partial x} + y \frac{\partial u}{\partial y} = \rho \frac{\partial u}{\partial \rho}.$$

5. If $x = \rho \cos \theta$, $y = \rho \sin \theta$, show that the expression

$$\frac{\left\{1 + \left(\dfrac{dy}{dx}\right)^2\right\}^{\frac{3}{2}}}{\dfrac{d^2 y}{dx^2}} \quad \text{becomes} \quad \frac{\left\{\rho^2 + \left(\dfrac{d\rho}{d\theta}\right)^2\right\}^{\frac{3}{2}}}{\rho^2 + 2\left(\dfrac{d\rho}{d\theta}\right)^2 - \rho \dfrac{d^2\rho}{d\theta^2}}.$$

APPLICATIONS TO GEOMETRY

CHAPTER XII

TANGENTS AND NORMALS

115. It was shown in Art. 19 that if $f(x,y)=0$ be the equation of a plane curve, then $\dfrac{dy}{dx}$ measures the slope of the tangent to the curve at the point x, y. The slope at a particular point (x_1, y_1) will be denoted by $\dfrac{dy_1}{dx_1}$, meaning that x_1 is to be substituted for x, and y_1 for y in $\dfrac{dy}{dx}=-\dfrac{\frac{\partial f}{\partial x}}{\frac{\partial f}{\partial y}}$ after the differentiation has been performed.

116. **Equation of tangent and normal at a given point.** Since the tangent line goes through the given point (x_1, y_1) and has the slope $\dfrac{dy_1}{dx_1}$, its equation is

$$y-y_1=\frac{dy_1}{dx_1}(x-x_1). \tag{1}$$

The *normal* to the curve at the point (x_1, y_1) is the straight line through this point, perpendicular to the tangent.

Its equation is, since $\dfrac{dx}{dy} = \dfrac{1}{\dfrac{dy}{dx}}$ by Art. 21,

$$y - y_1 = -\dfrac{dx_1}{dy_1}(x - x_1),$$

i.e., $\quad (x - x_1) + \dfrac{dy_1}{dx_1}(y - y_1) = 0.$ \hfill (2)

117. Length of tangent, normal, subtangent, subnormal.
The portion of the tangent and normal intercepted between the point of tangency and the axis OX are called, respectively, the *tangent length* and the *normal length;* and their projections on OX are called the *subtangent* and the *subnormal*.

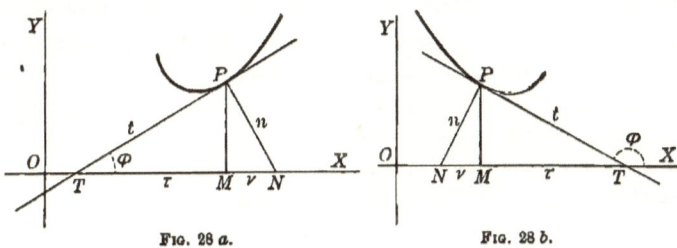

Fig. 28 a.　　　　　Fig. 28 b.

Thus, in Fig. 28, let the tangent and normal at P to the curve PC meet the axis OX in T and N, and let MP be the ordinate of P, then TP is the tangent length,

PN the normal length,
TM the subtangent,
MN the subnormal,

which will be denoted, respectively, by t, n, τ, ν.
Let the angle XTP be ϕ, then $\tan \phi = m$, say:

$$\cos \phi = \dfrac{1}{\sqrt{1 + m^2}}; \quad \sin \phi = \dfrac{m}{\sqrt{1 + m^2}};$$

$$t = \frac{y_1}{\sin \phi} = \frac{y_1 \sqrt{1 + m^2}}{m}; \quad n = \frac{y_1}{\cos \phi} = y_1 \sqrt{1 + m^2};$$

$$\tau = y_1 \cot \phi = y_1 \frac{dx_1}{dy_1} = \frac{y_1}{m}; \quad \nu = y_1 \tan \phi = y_1 \frac{dy_1}{dx_1} = my_1.$$

The subtangent is measured from the intersection of the tangent to the foot of the ordinate; it is therefore positive when the foot of the ordinate is to the right of the intersection of tangent. The subnormal is measured from the foot of the ordinate to the intersection of normal, and is positive when the normal cuts OX to the right of the foot of the ordinate. Both are therefore positive or negative, according as ϕ is acute or obtuse.

The expressions for τ, ν may also be obtained by finding from equations (1), (2), Art. 116, the intercepts made by the tangent and normal on the axis OX. The intercept of the tangent subtracted from x_1 gives τ, and x_1 subtracted from the intercept of the normal gives ν.

EXERCISES

1. In the curve $y(x-1)(x-2) = x-3$, show that the tangent is parallel to the axis of x at the points for which $x = 3 \pm \sqrt{2}$.

2. Write down the equations of the tangents and normals to the curve $y = \frac{ax^2}{a^2 + x^2}$ at the points for which $y = \frac{a}{4}$.

3. Find the equations of the tangents and normals at the point (x_1, y_1) on each of the following curves:

(a) $x^2 + y^2 = c^2$,   (c) $xy(x+y) = a^3$,
(b) $xy = k^2$,   (d) $e^y = \sin x$.

4. Prove that $\frac{x}{a} + \frac{y}{b} = 1$ touches the curve $y = be^{-\frac{x}{a}}$ at the point in which the latter crosses the axis of y.

5. Find the points on the curve
$$y = (x-1)(x-2)(x-3)$$
at which the tangent is parallel to the axis of x.

6. Find the intercepts made upon the axes by the tangent at (x_1, y_1) to the curve $\sqrt{x} + \sqrt{y} = \sqrt{a}$, and show that their sum is constant.

7. In the curve $x^2 y^2 = a^3(x + y)$, the tangent at the origin is inclined at an angle of $135°$ to the axis of x.

8. In the curve $x^{\frac{2}{3}} + y^{\frac{2}{3}} = a^{\frac{2}{3}}$, find the length of the perpendicular from the origin on the tangent at (x_1, y_1); and the length of that part of the tangent which is intercepted between the two axes. (A. G., p. 323.)

9. Show that all the curves represented by the equation

$$\left(\frac{x}{a}\right)^n + \left(\frac{y}{b}\right)^n = 2,$$

when different values are given to n, touch each other at the point (a, b).

10. Show that all the points of the curve

$$y^2 = 4a\left(x + a\sin\frac{x}{a}\right),$$

at which the tangent is parallel to the axis OX lie on a certain parabola.

11. Prove that the parabola $y^2 = 4ax$ has a constant subnormal.

12. Prove that the circle $x^2 + y^2 = a^2$ has a constant normal.

13. Show that in the tractrix, the length of the tangent is constant; the equation of the tractrix being

$$x = \sqrt{c^2 - y^2} + \frac{c}{2}\log\frac{c - \sqrt{c^2 - y^2}}{c + \sqrt{c^2 - y^2}}.$$

14. Show that the exponential curve $y = ae^{\frac{x}{c}}$ has a constant subtangent.

15. At what angle does the circle $x^2 + y^2 = 8ax$ intersect the cissoid $y^2 = \dfrac{x^3}{2a - x}$? (A. G., p. 309.)

16. Find the subtangent of the cissoid $y^2 = \dfrac{x^3}{2a - x}$.

17. Find the normal length of the catenary $y = \dfrac{a}{2}(e^{\frac{x}{a}} + e^{-\frac{x}{a}})$.

18. Show that the only Cartesian (x, y) curve in which the ratio of the subtangent to the subnormal is constant is a straight line.

19. Show that the equation of the tangent to the curve $f(x, y) = 0$ at the point (x_1, y_1) may be written

$$(x - x_1)\frac{\partial f}{\partial x_1} + (y - y_1)\frac{\partial f}{y_1} = 0.$$

20. Prove that the equation of the tangent to the curve

$$x^3 - 3axy + y^3 = 0$$

may be written $\quad x_1^2 x - ax_1 y - axy_1 + y_1^2 y = 0.$

POLAR COÖRDINATES

118. When the equation of a curve is expressed in polar coördinates, the vectorial angle θ is usually regarded as the independent variable. To determine the direction of the curve at any point, it is most convenient to express the angle between the tangent and the radius vector to the point of tangency.

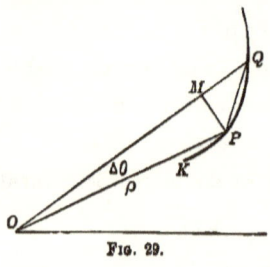

Fig. 29.

Let P, Q be two points on the curve (Fig. 29). Join P, Q with the pole O, and drop a perpendicular PM from P on OQ. Let ρ, θ be the coördinates of P; $\rho + \Delta\rho$, $\theta + \Delta\theta$ those of Q; then the angle $POQ = \Delta\theta$; $PM = \rho \sin \Delta\theta$; and $MQ = OQ - OM = \rho + \Delta\rho - \rho \cos \Delta\theta$;

hence $\quad \tan MQP = \dfrac{\rho \sin \Delta\theta}{\rho + \Delta\rho - \rho \cos \Delta\theta}.$

When Q moves to coincidence with P, the angle MQP approaches as a limit the angle between the radius vector and the tangent line at the point P. This angle will be designated by ψ.

Thus $\quad \tan \psi = \lim\limits_{\Delta\theta \doteq 0} \dfrac{\rho \sin \Delta\theta}{\rho + \Delta\rho - \rho \cos \Delta\theta},$

but $\quad \rho(1 - \cos \Delta\theta) = 2\rho \sin^2 \tfrac{1}{2} \Delta\theta,$

hence $\quad \tan \psi = \lim\limits_{\Delta\theta \doteq 0} \dfrac{\dfrac{\rho \sin \Delta\theta}{\Delta\theta}}{\rho \sin \tfrac{1}{2} \Delta\theta \cdot \dfrac{\sin \tfrac{1}{2}\Delta\theta}{\tfrac{1}{2}\Delta\theta} + \dfrac{\Delta\rho}{\Delta\theta}};$

but
$$\lim_{\Delta\theta \doteq 0} \frac{\sin \Delta\theta}{\Delta\theta} = 1, \quad \lim_{\Delta\theta \doteq 0} \frac{\sin \tfrac{1}{2} \Delta\theta}{\tfrac{1}{2} \Delta\theta} = 1.$$

Therefore
$$\tan \psi = \frac{\rho}{\dfrac{d\rho}{d\theta}} = \rho \frac{d\theta}{d\rho}.$$

Examples on dynamical interpretation.

Ex. 1. A point describes a circle of radius ρ; prove that at any instant the arc velocity is ρ times the angle velocity;

i.e.,
$$\frac{ds}{dt} = \rho \frac{d\theta}{dt}.$$

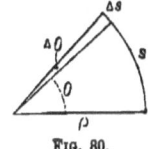

FIG. 80.

Ex. 2. When a point describes any curve, prove that at any instant the velocity $\dfrac{ds}{dt}$ has a radius component $\dfrac{d\rho}{dt}$ and a circle component $\rho\dfrac{d\theta}{dt}$, and hence that

$$\cos \psi = \frac{d\rho}{ds}, \; \sin \psi = \rho\frac{d\theta}{ds}, \; \tan \psi = \rho\frac{d\theta}{d\rho}.$$

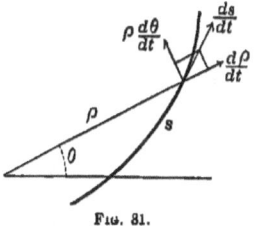

FIG. 81.

119. Relation between $\dfrac{dy}{dx}$ and $\dfrac{\rho d\theta}{d\rho}$. If the initial line be taken as the axis of x, the tangent line at P makes an angle ϕ with this line by Art. 117. Hence

$$\theta + \psi = \phi;$$

i.e., $\theta + \tan^{-1}\left(\dfrac{\rho d\theta}{d\rho}\right) = \tan^{-1}\left(\dfrac{dy}{dx}\right).$

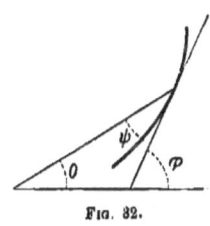

FIG. 82.

120. Length of tangent, normal, polar subtangent, and polar subnormal. The portions of the tangent and normal intercepted between the point of tangency P and the line through the pole perpendicular to the radius vector OP to the point of tangency, are called the *polar tangent length* and the *polar* *normal length;* and their projections on this perpendicular are called the *polar subtangent* and *polar subnormal*.

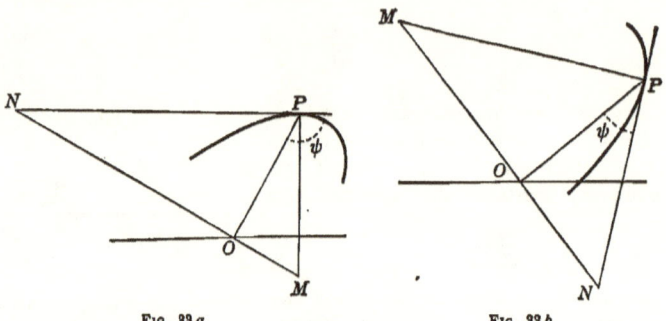

Fig. 33 a.   Fig. 33 b.

Thus, let the tangent and normal at P meet the perpendicular to OP in the points N, M. Then

PN is the polar tangent length,
PM is the polar normal length,
ON is the polar subtangent,
OM is the polar subnormal.

They are all seen to be independent of the position of the initial line. The lengths of these lines will now be considered.

Since $PN = OP \sec OPN = \rho \sec \psi = \rho \sqrt{\rho^2 \left(\dfrac{d\theta}{d\rho}\right)^2 + 1}$

$= \rho \dfrac{d\theta}{d\rho} \sqrt{\rho^2 + \left(\dfrac{d\rho}{d\theta}\right)^2}$,

hence   polar tangent length $= \rho \dfrac{d\theta}{d\rho} \sqrt{\rho^2 + \left(\dfrac{d\rho}{d\theta}\right)^2}$.

Again, $ON = OP \tan OPN = \rho \tan \psi = \rho^2 \dfrac{d\theta}{d\rho}$,

hence   polar subtangent $= \rho^2 \dfrac{d\theta}{d\rho}$.

$$PM = OP \csc OPN = \rho \csc \psi = \sqrt{\rho^2 + \left(\frac{d\rho}{d\theta}\right)^2},$$

hence polar normal length $= \sqrt{\rho^2 + \left(\frac{d\rho}{d\theta}\right)^2}.$

$$OM = OP \cot OPN = \frac{\rho d\rho}{\rho d\theta},$$

hence polar subnormal $= \frac{d\rho}{d\theta}.$

The signs of the polar tangent length and polar normal length are ambiguous on account of the radical. The direction of the subtangent is determined by the sign of $\rho^2 \frac{d\theta}{d\rho}$: when $\frac{d\theta}{d\rho}$ is positive, the distance ON should be measured to the right, and when negative, to the left of an observer placed at O and looking along OP; for when θ increases with ρ, $\frac{d\theta}{d\rho}$ is positive (Art. 20), and ψ is an acute angle (as in Fig. 33 b); when θ decreases as ρ increases, $\frac{d\theta}{d\rho}$ is negative, and ψ is obtuse (Fig. 33 a).

EXERCISES

1. Show that the polar subtangent is constant in the curve $\rho\theta = a$.

2. Show that in the curve $\rho = a \cdot e^{\theta \cot \alpha}$, the tangent makes a constant angle α with the radius vector. For this reason, this curve is called the equiangular spiral. (A. G., p. 330.)

3. For the same curve as in Ex. 2, find the polar subtangent and polar subnormal.

4. Find the angle of intersection of the curves
$$\rho = a(1 + \cos\theta), \quad \rho = b(1 - \cos\theta).$$

5. In the circle $\rho = a \sin\theta$, find ψ and ϕ.

6. In the curve $\rho = a\theta$, show that $\tan\psi = \theta$, and that the polar subnormal is constant. (A. G., p. 325.)

7. In the parabola $\rho = a \sec^2\frac{\theta}{2}$, show that $\phi + \psi = \pi$.

CHAPTER XIII

DERIVATIVE OF AN ARC, AREA, VOLUME AND SURFACE OF REVOLUTION

121. Derivative of an arc. The length s of the arc AP of a given curve $y = f(x)$, measured from a fixed point A to any point P, is a function of the abscissa x of the latter point, and may be expressed by a relation of the form $s = \phi(x)$.

The determination of the function ϕ when the form of f is known, is an important and sometimes difficult problem in the Integral Calculus. The first step in its solution is to determine the form of the derivative function $\dfrac{ds}{dx} = \phi'(x)$, which is easily done by the methods of the Differential Calculus.

Let PQ be two points on the curve (Fig 34); let x, y be the coördinates of P; $x + \Delta x$, $y + \Delta y$ those of Q; s the length of the arc AP; $s + \Delta s$ that of the arc AQ. Draw the ordinates MP, NQ; and draw PR parallel to MN; then $PR = \Delta x$, $RQ = \Delta y$; arc $PQ = \Delta s$. Hence

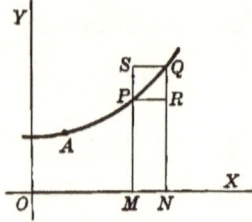

Fig. 34.

$$\text{Chord } PQ = \sqrt{(\Delta x)^2 + (\Delta y)^2},$$

$$\frac{PQ}{\Delta x} = \sqrt{1 + \left(\frac{\Delta y}{\Delta x}\right)^2}.$$

Therefore $\dfrac{\Delta s}{\Delta x} = \dfrac{\Delta s}{PQ} \cdot \dfrac{PQ}{\Delta x}, = \dfrac{\Delta s}{PQ}\sqrt{1+\left(\dfrac{\Delta y}{\Delta x}\right)^2}.$

Taking the limit of both members as $\Delta x \doteq 0$ and putting $\lim\limits_{\Delta x \doteq 0} \dfrac{\Delta s}{PQ} = 1$, by Art. 13, Th. 4, and Art. 10, Th. 10, Cor., it follows that

$$\dfrac{ds}{dx} = \sqrt{1+\left(\dfrac{dy}{dx}\right)^2}. \tag{1}$$

Similarly
$$\dfrac{ds}{dy} = \sqrt{1+\left(\dfrac{dx}{dy}\right)^2}, \tag{2}$$

and
$$\left(\dfrac{ds}{dt}\right)^2 = \left(\dfrac{dx}{dt}\right)^2 + \left(\dfrac{dy}{dt}\right)^2, \text{ Art. 89,} \tag{3}$$

$$ds^2 = dx^2 + dy^2. \tag{4}$$

122. Trigonometric meaning of $\dfrac{ds}{dx}, \dfrac{ds}{dy}$.

Since $\dfrac{\Delta x}{\Delta s} = \dfrac{\Delta x}{PQ} \cdot \dfrac{PQ}{\Delta s} = \cos RPQ \cdot \dfrac{PQ}{\Delta s},$

it follows, by taking the limit, as $\Delta x \doteq 0$, that

$$\dfrac{dx}{ds} = \cos \phi,$$

wherein ϕ, being the limit of the angle RPQ, is the angle which the tangent drawn at the point (x, y) makes with the x-axis.

Similarly, $\dfrac{dy}{ds} = \sin \phi$; whence $\dfrac{ds}{dx} = \sec \phi$; $\dfrac{ds}{dy} = \csc \phi.$

Using the idea of a rate or differential, all these relations may be conveniently exhibited by Fig. 35.

These results may also be derived from equations (1), (2) of Art. 121, by putting $\dfrac{dy}{dx} = \tan \phi.$

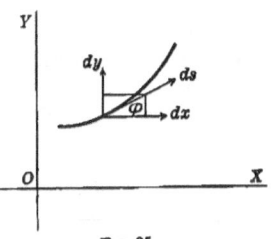

FIG. 35.

123. Derivative of the volume of a solid of revolution. Let the curve APQ revolve about the x-axis, and thus generate a surface of revolution; let V be the volume included between this surface, the fixed initial plane face generated by the ordinate AB, and the terminal face generated by any ordinate MP.

Let ΔV be the volume generated by the area $PMNQ$; then ΔV lies between the volumes of the cylinders generated by the rectangles $PMNR$ and $SMNQ$; that is,

$$\pi y^2 \Delta x < \Delta V < \pi(y + \Delta y)^2 \Delta x.$$

Dividing by Δx and taking limits,

$$\frac{dV}{dx} = \pi y^2.$$

124. Derivative of a surface of revolution. Let S be the area of the surface generated by the arc AP (Fig. 36); and ΔS that by the arc PQ, whose length is Δs.

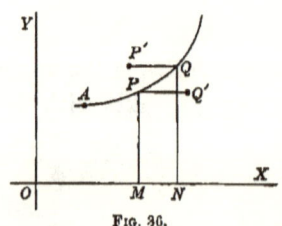

Fig. 36.

Draw PQ', QP' parallel to OX and equal in length to the arc PQ; then it may be assumed as an axiom that the area generated by PQ lies between the areas generated by PQ' and $P'Q$; i.e.,

$$2\pi y \Delta s < \Delta S < 2\pi(y + \Delta y)\Delta s.$$

Dividing by Δs and passing to the limit,

$$\frac{dS}{ds} = 2\pi y, \qquad (1)$$

$$\frac{dS}{dx} = \frac{dS}{ds} \cdot \frac{ds}{dx} = 2\pi y \sqrt{1 + \left(\frac{dy}{dx}\right)^2}. \qquad (2)$$

125. Derivative of arc in polar coördinates.

Let ρ, θ be the coördinates of P; $\rho + \Delta\rho$, $\theta + \Delta\theta$ those of Q; s the length of the arc KP; Δs that of arc PQ. Let PM be perpendicular to OQ; then

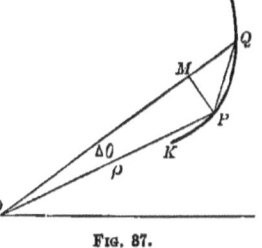

Fig. 87.

$$PM = \rho \sin \Delta\theta,$$

$$MQ = OQ - OM = \rho + \Delta\rho - \rho \cos \Delta\theta$$
$$= \rho(1 - \cos \Delta\theta) + \Delta\rho$$
$$= 2\rho \sin^2 \tfrac{1}{2}\Delta\theta + \Delta\rho.$$

Hence $\quad PQ^2 = (\rho \sin \Delta\theta)^2 + (2\rho \sin^2 \tfrac{1}{2}\Delta\theta + \Delta\rho)^2,$

$$\left(\frac{PQ}{\Delta\theta}\right)^2 = \rho^2 \left(\frac{\sin \Delta\theta}{\Delta\theta}\right)^2 + \left(\rho \sin \tfrac{1}{2}\Delta\theta \cdot \frac{\sin \tfrac{1}{2}\Delta\theta}{\tfrac{1}{2}\Delta\theta} + \frac{\Delta\rho}{\Delta\theta}\right)^2.$$

Replacing the first member by $\left(\dfrac{PQ}{\Delta s} \cdot \dfrac{\Delta s}{\Delta\theta}\right)^2$, passing to the limit when $\Delta\theta \doteq 0$, and putting $\lim \dfrac{PQ}{\Delta s} = 1$, $\lim \dfrac{\sin \Delta\theta}{\Delta\theta} = 1$, $\lim \dfrac{\sin \tfrac{1}{2}\Delta\theta}{\tfrac{1}{2}\Delta\theta} = 1$, it follows that

$$\left(\frac{ds}{d\theta}\right)^2 = \rho^2 + \left(\frac{d\rho}{d\theta}\right)^2;$$

i.e.,
$$\frac{ds}{d\theta} = \sqrt{\rho^2 + \left(\frac{d\rho}{d\theta}\right)^2}.$$

In the rate or differential notation this relation may be conveniently written

$$ds^2 = d\rho^2 + \rho^2 d\theta^2,$$

and its dynamic interpretation is shown in the figure of Art. 118 (Fig. 31).

126. Derivative of area in polar coördinates. Let A be the area of OKP measured from a fixed radius vector OK to any other radius vector OP; let ΔA be the area of OPQ. Draw arcs PM, QN, with O as a center; then the area POQ lies between the areas of the sectors OPM and ONQ; i.e.,

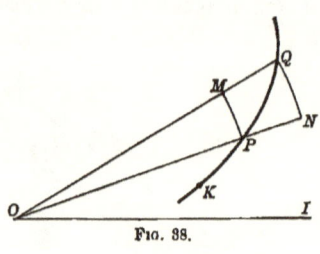

Fig. 38.

$$\tfrac{1}{2}\rho^2 \Delta\theta < \Delta A < \tfrac{1}{2}(\rho + \Delta\rho)^2 \Delta\theta.$$

Dividing by $\Delta\theta$ and passing to the limit, when $\Delta\theta \doteq 0$, it follows that

$$\frac{dA}{d\theta} = \tfrac{1}{2}\rho^2.$$

For the derivative of the area of a curve in x, y coördinates, see Art. 17. The result is $\dfrac{dA}{dx} = y$.

EXERCISES

1. Given $\dfrac{x^2}{a^2} + \dfrac{y^2}{b^2} = 1$; find $\dfrac{ds}{dx}$, $\dfrac{dA}{dx}$, $\dfrac{dS}{dx}$, $\dfrac{dy}{dx}$.

2. Similarly for the parabola $y^2 = 4ax$.

3. In the curve $e^y(e^x - 1) = e^x + 1$, show that $\dfrac{ds}{dx} = \dfrac{e^{2x}+1}{e^{2x}-1}$

4. If ϕ be the eccentric angle of the ellipse $\dfrac{x^2}{a^2} + \dfrac{y^2}{b^2} = 1$, prove that $\dfrac{ds}{d\phi} = a\sqrt{1 - e^2 \cos^2 \phi}$, e being the eccentricity.

[$dx = -a \sin \phi\, d\phi$, $dy = b \cos \phi\, d\phi$, $ds^2 = (a^2 \sin^2 \phi + b^2 \cos^2 \phi)d\phi^2$, etc.]

5. Given $\rho = a \cos \theta$; find $\dfrac{ds}{d\theta}$, $\dfrac{dA}{d\theta}$.

6. In $\rho^2 = a^2 \cos 2\theta$, show that $\dfrac{ds}{d\theta} = \dfrac{a^2}{\rho}$.

7. Given $\rho = a(1 + \cos \theta)$, prove $\dfrac{ds}{dt} = \sqrt{2a\rho}$.

CHAPTER XIV

ASYMPTOTES

127. When a curve has a branch extending to infinity, the tangents drawn at successive points of this branch may tend to coincide with a definite fixed line as in the familiar case of the hyperbola; or, on the other hand, the successive tangents may move further and further out of the field as in the parabola. These two kinds of infinite branches may be called *hyperbolic* and *parabolic*.

The character of each of the infinite branches of a curve can always be determined when the equation of the curve is known.

128. Definition of a rectilinear asymptote. If the tangents at successive points of a curve approach a fixed straight line as a limiting position when the point of contact moves further and further along any infinite branch of the given curve, then the fixed line is called an *asymptote* of the curve.

This definition may be stated more briefly but less precisely as follows: An asymptote to a curve is a tangent whose point of contact is at infinity, but which is not itself entirely at infinity.

DETERMINATION OF ASYMPTOTES

129. Method of limiting intercepts. The equation of the tangent at any point (x_1, y_1) being

$$y - y_1 = \frac{dy_1}{dx_1}(x - x_1),$$

the intercepts made by this line on the coördinate axes are

$$\left.\begin{array}{l} y_0 = y_1 - x_1 \dfrac{dy_1}{dx_1}, \\ x_0 = x_1 - y_1 \dfrac{dy_1}{dx_1}. \end{array}\right\} \quad (1)$$

Suppose the curve has a branch on which $x \doteq \infty$ and $y \doteq \infty$; then from (1) the limits can be found to which the intercepts x_0, y_0 approach as the coördinates x_1, y_1 of the point of contact tend to become infinite. If these limits be denoted by a, b, the equation of the corresponding asymptote is

$$\frac{x}{a} + \frac{y}{b} = 1.$$

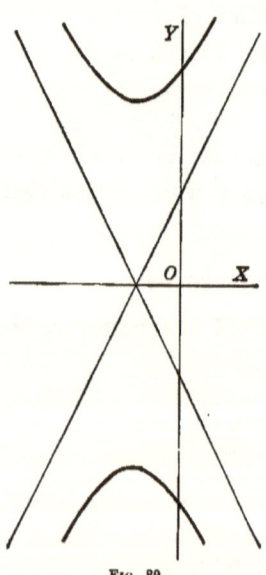

Fig. 89.

Ex. 1. Find the asymptotes of the curve

$$y^2 = 4x^2 + 2x + 6.$$

Since $\quad \dfrac{dy}{dx} = \dfrac{4x+1}{y}$,

hence $x_0 = x - y\dfrac{dx}{dy} = \dfrac{4x^2 + x - y^2}{4x+1}$

$= \dfrac{-x-6}{4x+1}$, and this $\doteq -\dfrac{1}{4}$

when $\quad x \doteq \infty$.

$$y_0 = y - x\frac{dy}{dx} = \frac{y^2 - 4x^2 - x}{y}$$

$$= \frac{x+6}{\sqrt{4x^2 + 2x + 6}}.$$

To evaluate this expression, square both terms, and then apply the rule of Art. 73. The value of the square is $\frac{1}{4}$; thus, $y_0 = \pm \frac{1}{2}$.

Hence the asymptotes are

$$\frac{x}{-\frac{1}{4}} + \frac{y}{\pm \frac{1}{2}} = 1,$$

i.e., $y = 2x + \frac{1}{2}$, $y = -2x - \frac{1}{2}$.

Ex. 2. Find the equations of the asymptotes of the curve
$$x^2 + 3xy + 2y^2 + 3x - 2y + 1 = 0.$$

Here
$$\frac{dy}{dx} = -\frac{2x + 3y + 3}{3x + 4y - 2};$$

hence substituting in (1), and omitting the subscripts throughout the right-hand member,
$$y_0 = \frac{2(x^2 + 3xy + 2y^2) + 3x - 2y}{3x + 4y - 2}.$$

Replacing $x^2 + 3xy + 2y^2$ by $-3x + 2y + 1$ from the given equation, this becomes

$$y_0 = -\frac{-3y + 2y - 2}{3x + 4y - 2} = \frac{-3 + 2\left(\frac{y}{x}\right) - \frac{2}{x}}{3 + 4\left(\frac{y}{x}\right) - \frac{2}{x}}.$$

Next, to find the limit of $\frac{y}{x}$ as $y \doteq \infty$, $x \doteq \infty$, observe that the terms $3x, 2y, 1$ are infinities of a lower order (1 is an infinite of order 0) than x^2, xy, y^2; hence, for large values of x and y, the terms of the second degree would have most effect in fixing the form of the curve; and in the limit, when $x \doteq \infty$ and $y \doteq \infty$, the smaller terms can be neglected. Then the equation becomes

$$x^2 + 3xy + 2y^2 = 0,$$
$$(x + 2y)(x + y) = 0,$$
$$\left(\frac{y}{x} + \frac{1}{2}\right)\left(\frac{y}{x} + 1\right) = 0.$$

Hence, on one branch $\frac{y}{x} \doteq -\frac{1}{2}$, and on the other, $\frac{y}{x} \doteq -1$.

Using these limiting values for $\frac{y}{x}$ in the values of y_0,

$$y_0 \doteq \frac{-3 + 2(-\frac{1}{2})}{3 + 4(-\frac{1}{2})} \doteq -4, \text{ and } y_0 \doteq \frac{-3 + 2(-1)}{3 + 4(-1)} \doteq 5,$$

on the respective branches.

Similarly for the x-intercept, after reduction,
$$x_0 = \frac{-3x + 2y - 2}{2x + 3y + 3}$$
$$= \frac{-3 + 2\left(\frac{y}{x}\right) - \frac{2}{x}}{2 + 3\left(\frac{y}{x}\right) + \frac{3}{x}},$$
$$\doteq 5, \text{ when } \frac{y}{x} \doteq -1; \text{ and } \doteq -8, \text{ when } \frac{y}{x} \doteq -\frac{1}{2}.$$

The equations of the asymptotes are therefore

$$\frac{x}{5} + \frac{y}{5} = 1, \text{ and } \frac{x}{-8} + \frac{y}{-4} = 1,$$

i.e,
$$x + y = 5, \quad x + 2y + 8 = 0.$$

Except in special cases this method is usually too complicated to be of practical use in determining the equations of the asymptotes of a given curve. There are three other principal methods, of which at least one will always suffice to determine the asymptotes of curves whose equations involve only algebraic functions. These may be called the methods of inspection, of substitution, and of expansion.

130. Method of inspection. Infinite ordinates, asymptotes parallel to axes.

When an algebraic equation in two coördinates x and y is rationalized, cleared of fractions, and arranged according to powers of one of the coördinates, say y, it takes the form

$$ay^n + (bx + c)y^{n-1} + (dx^2 + ex + f)y^{n-2} + \cdots + u_{n-1}y + u_n = 0,$$

in which u_n is a polynomial of the degree n in terms of the other coördinate x.

When any value is given to x, the equation gives n values to y.

Let it be required to find for what value of x the corresponding ordinate y has an infinite value.

Suppose at first that the term in y^n is present; in other words, that the coefficient a is not zero. Then when any finite value is given to x, all of the n values of y are finite, and there are thus no infinite ordinates for finite values of the abscissa.

Next suppose that a is zero, and b, c not zero. In this case one value of y is infinite for every finite value of x, and thus one branch of the curve lies entirely at infinity. It is shown in projective geometry that this branch always has the form of a straight line. In this work no account will be taken of such branches, and the wording of the theorems will in no case refer to them.

There is one particular value of x that gives one additional infinite value to y, namely, the value $x = -\dfrac{c}{b}$; for this makes $bx + c$ (the coefficient of the highest power of y) zero, and hence from the theory of equations one corresponding value of y must be infinite; and this value is finite when $x \neq -\dfrac{c}{b}$.

The equation of the infinite ordinate is $bx + c = 0$.

Again, if not only a, but also b and c, are zero, there are two values of x that make y infinite; namely, those values of x that make $dx^2 + ex + f = 0$, and the equations of the infinite ordinates are found by factoring this last equation; and so on.

Similarly, by arranging the equation of the curve according to powers of x, it is easy to find what values of y give an infinite value to x.

Ex. 3. In the curve
$$2x^3 + x^2y + xy^2 = x^2 - y^2 - 5,$$
find the equation of the infinite ordinate, and determine the finite point in which this line meets the curve.

This is a cubic equation in which the coefficient of y^3 is zero.

Arranged in powers of y it is
$$y^2(x+1) + yx^2 + (2x^3 - x^2 + 5) = 0.$$

When $x = -1$, the equation for y becomes
$$0 \cdot y^2 + y + 2 = 0,$$
the two roots of which are $y = \infty$, $y = -2$; hence the equation of the infinite ordinate is $x + 1 = 0$. The infinite ordinate meets the curve again in the finite point $(-1, -2)$.

Since the term in x^3 is present, there are no infinite values of x for finite values of y.

Ex. 4. In the curve

$$x^2y + 5xy^2 + 2x^2 = 3x^2y + 6,$$

find what values of x make y infinite and what values of y make x infinite.

131. Infinite ordinates are asymptotes. Applying to the general equation of the last article the method of Art. 119, the slope of the tangent at (x, y) is $\dfrac{dy}{dx}$

$$= -\frac{by^{n-1} + (2dx+e)y^{n-2} + \cdots}{nay^{n-1} + (n-1)(bx+c)y^{n-2} + (n-2)(dx^2+ex+f)y^{n-3} + \cdots}.$$

Now, the first condition that y may become infinite for a finite value of x, is $a = 0$; but when a is zero, x finite, and y infinite, the numerator is an infinite of higher order than the denominator, hence $\dfrac{dy}{dx} \doteq \infty$, when $x \doteq -\dfrac{c}{b}$ and $y \doteq \infty$. Therefore the inclination of the tangent approaches nearer and nearer to 90°, and the tangent approaches to coincidence with the ordinate through the point $x = -\dfrac{c}{b}$; and thus this line is an asymptote parallel to the y-axis.

Similarly, if the value $y = k$ gives an infinite value to x, then the line $y = k$ is an asymptote parallel to the x-axis.

Thus, to determine all the asymptotes parallel to the y-axis, equate to zero the coefficient of the highest power of y, if it be not a constant. If this equation be of the first degree, it represents an asymptote parallel to the y-axis. If it be of higher degree, it may be resolved into first degree equations, each of which represents such an asymptote.

Similarly, to determine all the asymptotes parallel to the x-axis, equate to zero the coefficient of the highest power of x, if it be not a constant.

Ex. 5. In the curve $a^2x = y(x-a)^2$, the line $y = 0$ is an asymptote coincident with the x-axis, and the line $x = a$ is an asymptote parallel to the y-axis.

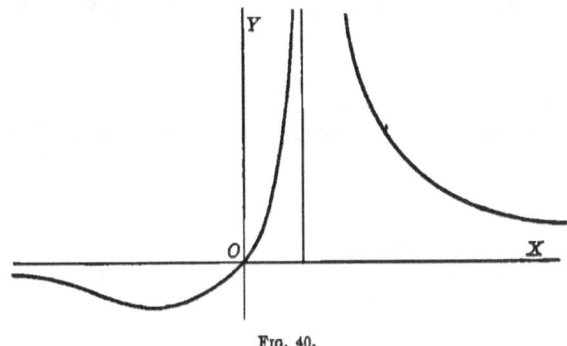

Fig. 40.

Ex. 6. Find the asymptotes of the curve $x^2(y-a) + xy^2 = a^3$.

132. Method of substitution. Oblique asymptotes. The asymptotes that are not parallel to either axis can be found by the method of substitution, which is applicable to all algebraic curves, and is of especial value when the equation is given in the implicit form

$$f(x, y) = 0. \qquad (1)$$

Consider the straight line

$$y = mx + b, \qquad (2)$$

and let it be required to determine m and b so that this line shall be an asymptote to the curve $f(x, y) = 0$.

Since an asymptote is the limiting position of a line that meets the curve in two points that tend to coincide at infinity, then, by making (1) and (2) simultaneous, the resulting equation in x,

$$f(x, mx + b) = 0,$$

is to have two of its roots infinite. This requires that the coefficients of the two highest powers of x shall vanish.

These coefficients, equated to zero, furnish two equations, from which the required values of m and b can be determined; and these values, substituted in (2), will give the equation of an asymptote.

Ex. 7. Find the asymptotes to the curve $y^3 = x^2(2a - x)$.

In the first place, there are evidently no asymptotes parallel to either of the coördinate axes. To determine the oblique asymptotes, make the equation of the curve simultaneous with $y = mx + b$, and eliminate y, then
$$(mx + b)^3 = x^2(2a - x),$$
or, arranged in powers of x,
$$(1 + m^3)x^3 + (3m^2b - 2a)x^2 + 3b^2mx + b^3 = 0.$$
Let $\quad m^3 + 1 = 0 \quad$ and $\quad 3m^2b - 2a = 0$,

then $\quad m = -1, \quad b = \dfrac{2a}{3};$

hence $\quad y = -x + \dfrac{2a}{3}$

is the equation of an asymptote.

The third intersection of this line with the given cubic is found from the equation $3mb^2x + b^3 = 0$.

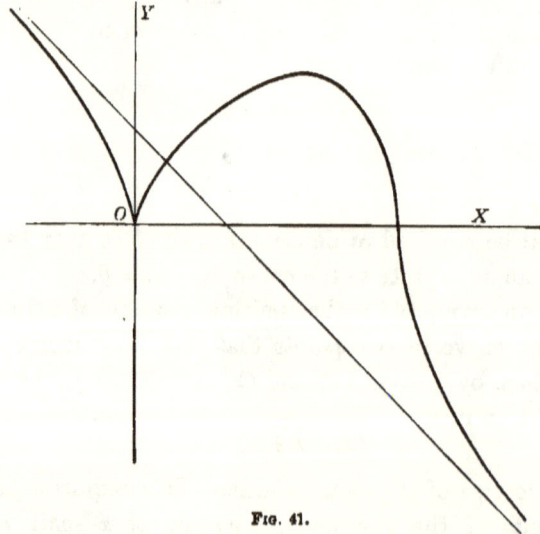

Fig. 41.

whence $$x = \frac{2a}{9}.$$

This is the only oblique asymptote, as the other roots of the equation for m are imaginary.

Ex. 8. Find the asymptotes to the curve $y(a^2 + x^2) = a^2(a - x)$.

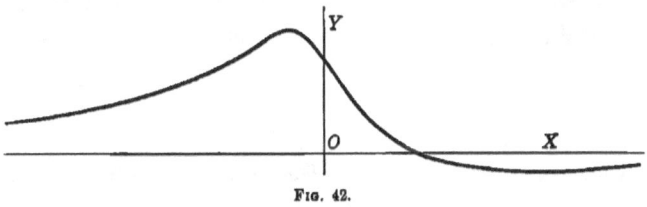

Fig. 42.

Here the line $y = 0$ is a horizontal asymptote by Art. 130. To find the oblique asymptotes, put $y = mx + b$,
then $$(mx + b)(a^2 + x^2) = a^2(a - x);$$
i.e., $$mx^3 + a^2bx^2 + (ma^2 + a^2)x + (a^2b - a^3) = 0,$$
hence $m = 0$, $b = 0$, for an asymptote.

Thus the only asymptote is the line $y = 0$, already found.

133. Number of asymptotes. The illustrations of the last article show that if all the terms be present in the general equation of an nth degree curve, then the equation for determining m is of the nth degree, and there are accordingly n values of m, real or imaginary. The equation for finding b is usually of the first degree, but for certain curves, when y has been replaced by $mx + b$, one or more values of m, say m_1, may cause the coefficient of x^n and x^{n-1} both to vanish, irrespective of b. In such cases any line whose equation is of the form $y = m_1x + c$ will satisfy the definition of an asymptote, independent of c; but by equating the coefficient of x^{n-2} to zero, two values of b can be found such that the resulting lines have three points at infinity in common with the curve. These two lines are parallel; and it will be seen that in each case in which this happens the equation defining m has a double root, so that the total number of asymptotes is not increased. Hence the total number of asymptotes, real and imaginary, is in general equal to the degree of the equation of the curve.

It is to be observed, however, that in special cases (*i.e.*, for certain special values of the given coefficients) two or more of these lines may coincide, and moreover that some of these n "tangents at infinity" may be situated entirely at infinity and thus be improperly called asymptotes.

Since the imaginary values of m occur in pairs, it is evident that a curve of odd degree has an odd number of real asymptotes; and that a curve of even degree has either no real asymptotes or an even number. Thus, a cubic curve has either one real asymptote or three; a conic has either two real asymptotes or none.

134. Method of expansion. Explicit functions. Although the two foregoing methods are in all cases sufficient to find the asymptotes of algebraic curves; yet in certain special cases the oblique asymptotes are most conveniently found by the method of expansion in descending powers. It is based on the following principle : a straight line will be an asymptote to a curve when the difference between the ordinates of the curve and of the line, corresponding to a common abscissa, approaches zero as a limit as the abscissa becomes larger and larger.

It will appear from the process of applying this principle that a line answering the condition just stated will also satisfy the original definition of an asymptote.

Suppose that the equation of the given curve can be solved for y in the form of a descending series of powers of x, beginning with the first power, and let the equation then be

$$y = a_0 x + a_1 + \frac{a_2}{x} + \frac{a_3}{x^2} + \cdots. \tag{1}$$

The line whose equation is

$$y = a_0 x + a_1 \tag{2}$$

is an asymptote to the curve represented by (1); for the difference between the ordinate of the curve and line, corresponding to the same abscissa x, is

$$\frac{a_2}{x} + \frac{a_3}{x^2} + \cdots,$$

which approaches zero when $x \doteq \infty$.

It is also evident that the line (2) satisfies the original definition of an asymptote; for, from (1), the slope of the tangent at the point whose abscissa is x, is

$$\frac{dy}{dx} = a_0 - \frac{a_2}{x^2} \cdots,$$

and the intercept made by the tangent on the y-axis is

$$y - x\frac{dy}{dx} = a_1 + \frac{2a_2}{x} + \cdots,$$

hence when $x \doteq \infty$, the slope approaches the limit a_0, and the intercept $\doteq a_1$; thus the equation of the asymptote is

$$y = a_0 x + a_1.$$

Ex. 9. Find the asymptotes of the curve

$$y^2 = \frac{x^3}{x-1}.$$

The line $x = 1$ is an asymptote parallel to the y-axis.

To obtain the oblique asymptotes, write the equation in the form

$$y = \frac{x^3}{x\left(1 - \frac{1}{x}\right)} = \frac{x^2}{1 - \frac{1}{x}} = x^2 \left(1 - \frac{1}{x}\right)^{-1},$$

$$y = \pm x\left(1 - \frac{1}{x}\right)^{-\frac{1}{2}} = \pm x\left(1 + \frac{1}{2}\frac{1}{x} + \frac{3}{8}\frac{1}{x^2} + \frac{5}{16}\frac{1}{x^3} + \cdots\right),$$

$$y = \pm\left(x + \frac{1}{2} + \frac{3}{8}\frac{1}{x} + \frac{5}{16}\frac{1}{x^2} + \cdots\right).$$

Hence the two oblique asymptotes are

$$y = \pm (x + \tfrac{1}{2}).$$

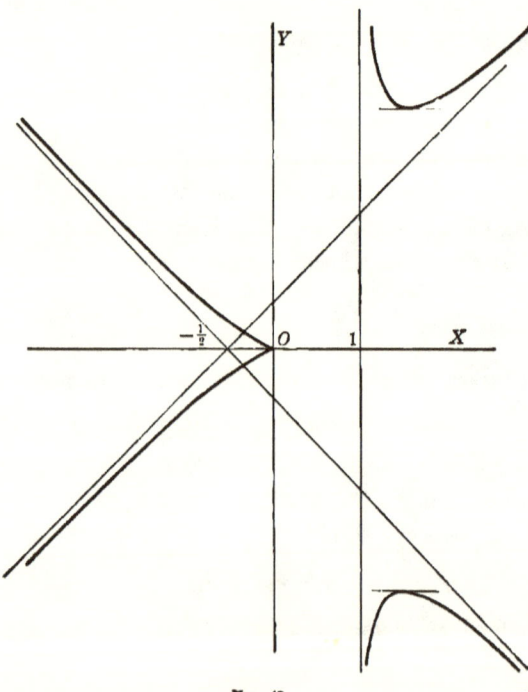

Fig. 43.

The sign of the term $\dfrac{3}{8}\dfrac{1}{x}$ shows that when $x \doteq +\infty$, the curve is above the first asymptote, and below the second, as in figure; and that when $x \doteq -\infty$, the curve is below the first asymptote, and above the second.

Ex. 10. Find the asymptotes of the curve

$$y^2 = \frac{(x-1)(2-x)^2}{x-3}.$$

Here
$$y^2 = \frac{x^2\left(1-\frac{1}{x}\right)\left(1-\frac{2}{x}\right)^2}{\left(1-\frac{3}{x}\right)},$$

$$y = \pm x\left(1-\frac{1}{x}\right)^{\frac{1}{2}}\left(1-\frac{2}{x}\right)\left(1-\frac{3}{x}\right)^{-\frac{1}{2}}$$

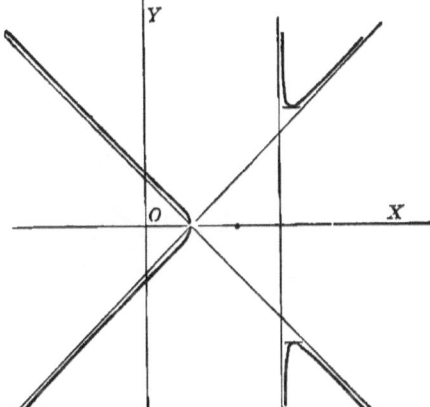

Fig. 44.

$$= \pm x\left(1-\frac{1}{2x}-\frac{1}{8x^2}\cdots\right)\left(1-\frac{2}{x}\right)\left(1+\frac{3}{2x}+\frac{27}{8x^2}\cdots\right)$$

$$= \pm x\left(1-\frac{1}{x}+\frac{1}{2x^2}\right) = \pm\left(x-1+\frac{1}{2x}\cdots\right).$$

Hence the oblique asymptotes are

$$y = \pm(x-1).$$

The same method may be applied to cases in which x is an explicit function of y.

Ex. 11. Find the asymptotes of

$$x^2 y^2 = (y+2)^2(y^2+1).$$

Here
$$x^2 = y^2\left(1+\frac{2}{y}\right)^2\left(1+\frac{1}{y^2}\right),$$

$$x = \pm y\left(1+\frac{2}{y}\right)\left(1+\frac{1}{y^2}\right)^{\frac{1}{2}}$$

$$= \pm y\left(1+\frac{2}{y}\right)\left(1+\frac{1}{2y^2}+\cdots\right),$$

$$x = \pm\left(y+2+\frac{1}{2y}\cdots\right).$$

Hence the asymptotes are $x = \pm(y+2)$. The next term $\dfrac{1}{2y}$ shows that when $y \doteq +\infty$, the curve is to the right of the first asymptote, and to

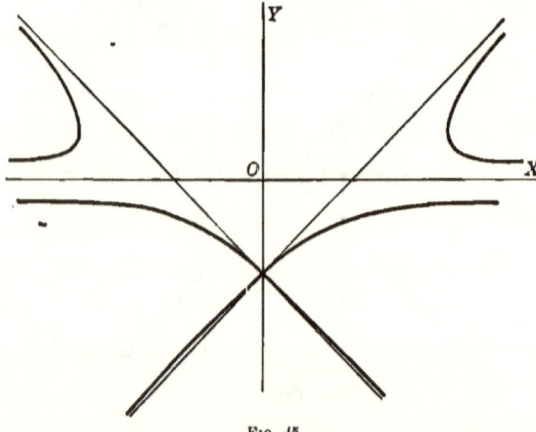

Fig. 45.

the left of the second; and *vice versa* when $y \doteq -\infty$. The form of the equation shows that the curve has a horizontal asymptote $y = 0$.

135. Method of expansion. Implicit functions. It was shown in Art. 132 that the direction of each oblique asymptote is determined by equating each factor of the terms of highest degree, in the equation of the curve, separately to zero. The subsequent procedure will be shown by an example.

Ex. 1. Determine the asymptotes to the curve

$$y^4 - x^4 - 2ax^2y - b^2x = 0,$$

and the manner in which the corresponding branch of the curve approaches each.

The terms of highest degree are $y^4 - x^4$, and this expression has but two real linear factors, hence the curve cannot have more than two real asymptotes; and these are parallel to the lines $y \pm x = 0$. To find the asymptote parallel to $y - x = 0$, arrange the equation of the curve thus:

$$y - x = \frac{2ax^2y + b^2x}{(x^2 + y^2)(y + x)}$$

$$= \frac{2a\dfrac{y}{x} + \dfrac{b^2}{x^2}}{\left(1 + \dfrac{y^2}{x^2}\right)\left(\dfrac{y}{x} + 1\right)} \qquad (1)$$

When y, x becomes infinite, $\dfrac{y}{x} \doteq 1$; hence

$$\lim_{x \doteq \infty} \frac{2a\dfrac{y}{x} + \dfrac{b^2}{x^2}}{\left(1 + \dfrac{y^2}{x^2}\right)\left(\dfrac{y}{x} + 1\right)} = \frac{2a}{4} = \frac{a}{2}, \qquad (2)$$

and the equation of the asymptote is

$$y = x + \frac{a}{2}. \qquad (3)$$

To obtain the next term in the equation of the curve, use (3) as a first approximation, which gives

$$\frac{y}{x} = 1 + \frac{a}{2x}, \text{ correct as far as the order } \frac{1}{x},$$

$$\frac{y^2}{x^2} = \left(1 + \frac{a}{2x}\right)^2 = 1 + \frac{a}{x}, \text{ to the same order}; \qquad (4)$$

$$1 + \frac{y}{x} = 2 + \frac{a}{2x} = 2\left(1 + \frac{a}{4x}\right),$$

$$1 + \frac{y^2}{x^2} = 2 + \frac{a}{x} = 2\left(1 + \frac{a}{2x}\right).$$

These values substituted in (1) give as a second approximation

$$y - x = \frac{1}{2}a\left(1 + \frac{a}{2x}\right)\left(1 + \frac{a}{2x}\right)^{-1}\left(1 + \frac{a}{4x}\right)^{-1}$$

$$= \frac{1}{2}a\left(1 + \frac{a}{2x}\right)\left(1 - \frac{a}{2x}\right)\left(1 - \frac{a}{4x}\right) = \frac{1}{2}a\left(1 + \frac{a}{2x} - \frac{a}{2x} - \frac{a}{4x}\right). \quad (5)$$

Hence the curve approaches the lower side of the asymptote on the right, and the upper side on the left.

Similarly the equation of the branch approaching the direction $y + x = 0$ will be found to have the successive approximations

$$y = -x + \frac{a}{2}, \quad y = -x + \frac{a}{2} + \frac{1}{8}\frac{a^2}{x} \cdots,$$

and thus on the right the curve approaches the upper side of the asymptote, and on the left, the lower side.

If the term in $\frac{1}{x}$ should happen to disappear from the result, a third approximation may be obtained by keeping the terms of order $\frac{1}{x^2}$ in the equations that correspond to (1), (4), (5), (6).

Ex. 2. $y^3 - x^2y + 2y^2 + 4y + x = 0$.

Ex. 3. $x^3 + 2x^2y - xy^2 - 2y^3 + 4y^2 + 2xy + y = 1$.

Ex. 4. $y^3 = x^3 + a^2x$.

136. Curvilinear asymptotes. When two curves are so situated that the difference between their ordinates corresponding to the same abscissa approaches zero as a limit when the common abscissa is made larger and larger, then each curve is said to be an asymptote of the other. This definition will also apply if the words "ordinate" and "abscissa" be interchanged.

E.g., suppose that the equation of a given curve can be brought to the form

$$y = ax^2 + bx + c + \frac{d}{x} + \frac{e}{x^2} + \frac{f}{x^3} + \cdots,$$

then it follows from the definition that the curve

$$y = ax^2 + bx + c$$

is a second degree asymptote to the given curve; and
$$y = ax^2 + bx + c + \frac{d}{x},$$
i.e., $\qquad xy = ax^3 + bx^2 + cx + d$

is a third degree asymptote, and so on.

Ex. Find the second and third degree asymptotes to the curves of examples 8–11, Arts. 132–134.

137. Examples of asymptotes of transcendental curves.

1. Consider the curve
$$y = \log x.$$
Here, when
$$x \doteq 0,$$
$$y \doteq -\infty,$$
and $\qquad \dfrac{dy}{dx} \doteq \infty;$

hence the line $x = 0$ is an asymptote, by Art. 131.

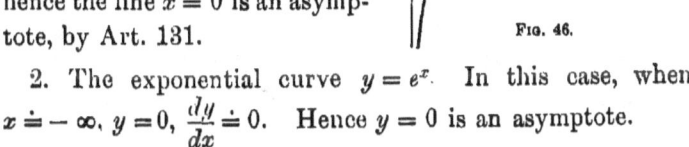

Fig. 46.

2. The exponential curve $y = e^x$. In this case, when $x \doteq -\infty$, $y = 0$, $\dfrac{dy}{dx} \doteq 0$. Hence $y = 0$ is an asymptote.

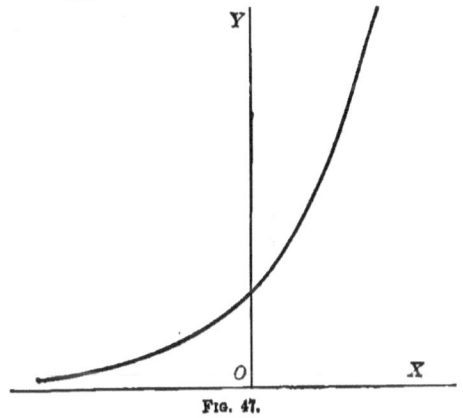

Fig. 47.

238   DIFFERENTIAL CALCULUS   [CH. XIV.

3. Find the asymptotes to the curve $1 + y = e^{\frac{1}{x}}$.

When x approaches zero from the positive side, $y \doteq +\infty$, and $\frac{dy}{dx} \doteq +\infty$; but when x approaches zero from the negative side, $x \doteq 0$, and $\frac{dy}{dx} \doteq 0$. Hence the line $x = 0$ is an

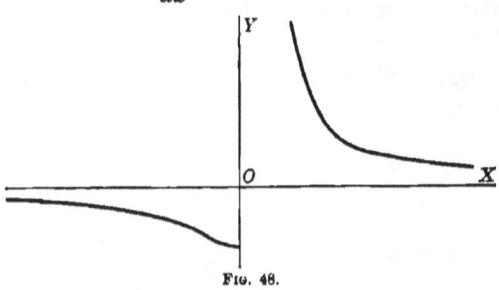

Fig. 48.

asymptote at $y = +\infty$ on the positive side of the y-axis. Again, when $x \doteq \pm\infty$, $y \doteq 0$; hence the line $y = 0$ is an asymptote both at $x = +\infty$ and $-\infty$.

4. The probability curve,

$$y = e^{-x^2}.$$

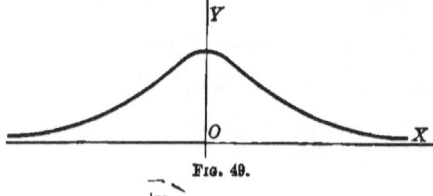

Fig. 49.

5. The curve

$$y^2 = \frac{x^2 - 1}{e^{2x}}.$$

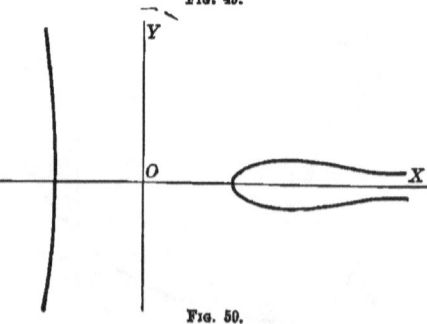

Fig. 50.

ASYMPTOTES

EXERCISES

Find the asymptotes of the following curves:

1. $(x + a)y^2 = (y + b)x^2$.
2. $x^2y^2 + ax(x+y)^2 - 2a^2y^2 - a^4 = 0$.
3. $x^4y^4 - (x^2 - y^2)^2 + y^2 - 1 = 0$.
4. $(x^2 - y^2)^2 - 4y^2 + y = 0$.
5. $x^2(x - y)^2 - a^2(x^2 + y^2) = 0$.
6. $y = \dfrac{x^3}{x^2 + 3a^2}$.
7. $y^2 = \dfrac{x^3}{2a - x}$.
8. $(x - 2a)y^2 = x^3 - a^3$.
9. $y^3 = x^2(2a - x)$.
10. $y(a^2 + x^2) = a^2(a - x)$.
11. $xy^2 + yx^2 = a^3$.
12. $(x^2 + a^2)x^2 = (a^2 - x^2)y^2$.
13. $x^2y^2 = x^3 + x + y$.
14. $x^2y^2 = (a + y)^2(b^2 - y^2)$.
15. $y(x - y)^3 = y(x - y) + 2$.

138. Asymptotes in polar coördinates. When a curve defined by an equation in polar coördinates has an asymptote, this line must be parallel to the radius vector to the point at infinity on the curve.

In Fig. 51, consider the curve $KP'P$, having the asymptote PT. The radius vector to the point at infinity must be parallel to the asymptote, for these two lines must inter-

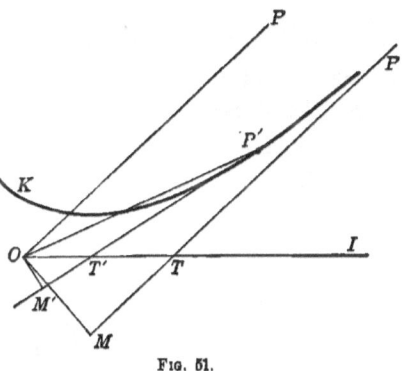

Fig. 51.

sect at infinity; and, moreover, the asymptote, according to the definition in Art. 128, must pass within a finite distance of this radius vector.

The polar subtangent OM, being by definition perpendicular to the radius vector OP, will, when P passes to infinity, become a common perpendicular to the radius vector OP

240   DIFFERENTIAL CALCULUS   [Ch. XIV.

and to the asymptote MP; hence the measure of the common perpendicular is

$$\lim_{\rho \doteq \infty} \left(\rho^2 \frac{d\theta}{d\rho} \right).$$

139. Determination of asymptotes to polar curves. To determine whether a given curve has asymptotes, first find for what values of θ, the vector ρ becomes infinite; then substitute each of these values of θ in the expression for the polar subtangent. If the result of any such substitution is finite, there is a corresponding asymptote.

To construct the asymptote, look along the direction of the infinite radius vector from the pole, and turn through a right angle, to the right if $\lim_{\rho \doteq \infty} \rho^2 \frac{d\theta}{d\rho}$ be positive, and to the left if it be negative (Art. 120). Measure a distance from the pole in this perpendicular direction equal to $\lim_{\rho \doteq \infty} \rho^2 \frac{d\theta}{d\rho}$, and through its extremity draw a line parallel to the infinite radius vector; this line will be the required asymptote.

Circular asymptotes. In some cases it may happen that when θ is made larger and larger without limit, the value of ρ may approach a definite limit a; thus $\lim_{\theta \doteq \infty} \rho = a$. The circle whose equation is $\rho = a$ is then called an *asymptotic circle*.

E.g. The curve $\rho = \dfrac{\theta + \sin \theta}{\theta + \cos \theta}$ has an asymptotic circle $\rho = 1$; the curve being exterior to the circle from the middle of the first quarter to the middle of the third quarter, and interior for the remainder of the circle; it approaches nearer to the circle with every revolution of θ.

Ex. 1. Find the rectilinear asymptotes to the curve $\rho = \dfrac{a\theta}{\sin \theta}$.

When $\theta = 0$, $\rho = a$; but when $\theta = n\pi$, n being any positive or negative integer, ρ becomes infinite.

Since
$$\frac{d\rho}{d\theta} = \frac{a(\sin \theta - \theta \cos \theta)}{\sin^2 \theta},$$

hence
$$\rho^2 \frac{d\theta}{d\rho} = \frac{a\theta^2}{\sin \theta - \theta \cos \theta}.$$

When $\theta = n\pi$, this expression becomes $an\pi$ or $-an\pi$, according as n is odd or even, and may therefore be written in the form $(-1)^{n-1}an\pi$.

There are thus an infinite number of asymptotes, all parallel to the initial line, and situated at intervals $a\pi$ from each other.

When n is positive, the asymptotes are above the initial line; when n is negative, they are below it. There are no circular asymptotes.

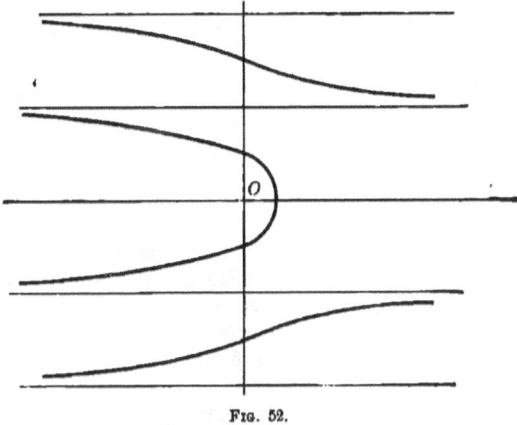

Fig. 52.

In many problems it shortens the work to substitute $\dfrac{1}{u}$ for ρ in the equation of the curve, and then to find what values of θ will make u vanish. The expression $\rho^2 \dfrac{d\theta}{d\rho}$ for the length of the polar subtangent then becomes $-\dfrac{d\theta}{du}$; and hence $\lim\limits_{u \doteq 0} \left(-\dfrac{d\theta}{du}\right)$, taken for any of the values of θ just found, measures the distance of the corresponding asymptote from the pole.

Ex. 2. Find the asymptotes to the curve

$$\rho \sin 4\theta = a \sin 3\theta.$$

Put $\rho = \dfrac{1}{u}$, then $\quad \sin 4\theta = au \sin 3\theta,$

and $u = 0$, when $\theta = \pm\frac{\pi}{4},\ \pm\frac{\pi}{2},\ \pm\frac{3\pi}{4},\ \pm\frac{5\pi}{4}\ \ldots$.

By differentiation, $4\cos 4\theta = a\dfrac{du}{d\theta}\sin 3\theta + 3au\cos 3\theta$,

$$-\frac{du}{d\theta} = \frac{-4\cos 4\theta + 3au\cos 3\theta}{a\sin 3\theta}.$$

This expression becomes $\dfrac{4\sqrt{2}}{a}$ when $\theta = \dfrac{\pi}{4}$; hence the distance to the corresponding asymptote is $\dfrac{a}{4\sqrt{2}}$. To construct the asymptote, look from the pole along the direction of 45°, measure a distance $\dfrac{a}{4\sqrt{2}}$ units to the right, perpendicular to this radius vector; then draw a line through the end of the perpendicular, parallel to the infinite radius vector (Fig 53). The student should determine the number and position of the remaining asymptotes.

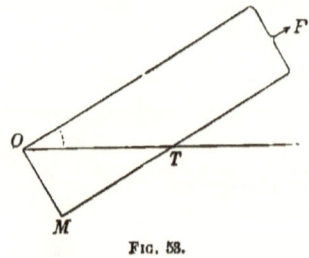

Fig. 53.

EXERCISES

Find and draw the asymptotes to the following curves:

1. The reciprocal spiral $\rho\theta = a$.
2. $\rho\cos\theta = a\cos 2\theta$.
3. $\rho = b\sec a\theta$.
4. $\rho\cos 2\theta = a\sin 3\theta$.
5. $\rho(e^\theta - 1) = a(e^\theta + 1)$.
6. Show that the curve $\rho = \dfrac{a}{1 - \cos\theta}$ has no asymptote.
7. Show that the initial line is an asymptote to two branches of the curve $\rho^2\sin\theta = a^2\cos 2\theta$.
8. Find the rectilinear and circular asymptotes of the curve

$$\rho = \frac{a\theta^2}{\theta^2 - 1}.$$

9. Which of the curves in 1-7 have circular asymptotes?

CHAPTER XV

DIRECTION OF BENDING. POINTS OF INFLEXION

140. Concavity upward and downward. A curve is said to be *concave downward* in the vicinity of a point P when, for a finite distance on each side of P, the curve is situated

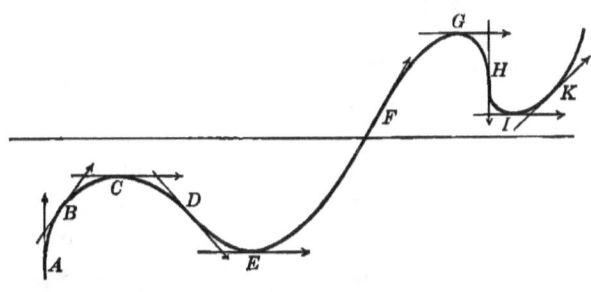

Fig. 54.

below the tangent drawn at that point, as in the arcs AD, FH. It is *concave upward* when the curve lies above the tangent, as in the arcs DF, HK.

It is evident, by drawing successive tangents to the curve, as in the figure, that if the point of contact advances to the right, the tangent swings in the positive direction of rotation when the concavity is upward, and in the negative direction when the concavity is downward. Hence upward concavity may be called a positive bending of the curve, and downward concavity, negative bending.

A point at which the direction of bending changes continuously from positive to negative, as at F, is called a *point of inflexion*, and the tangent at such a point is called a *stationary tangent*.

The points of the curve that are situated just before and just after the point of inflexion are thus on opposite sides of the stationary tangent, and hence the tangent crosses the curve, as at D, F, H.

141. Algebraic test for positive and negative bending. Let the inclination of the tangent line, measured from the right-hand end of the x-axis toward the forward (right-hand) end of the tangent, be denoted as usual by ϕ, then ϕ is an increasing or decreasing function of the abscissa according as the bending is positive or negative; for instance, in the arc AD, the angle ϕ diminishes from $+\frac{\pi}{2}$ through zero to $-\frac{\pi}{4}$; in the arc DF, ϕ increases from $-\frac{\pi}{4}$ through zero to $\frac{\pi}{3}$; in the arc FH, ϕ decreases from $+\frac{\pi}{3}$ through zero to $-\frac{\pi}{2}$; and in the arc HK, ϕ increases from $-\frac{\pi}{2}$ through zero to $+\frac{\pi}{4}$.

At a point of inflexion ϕ has evidently a turning value which is a maximum or minimum, according as the concavity changes from upward to downward, or conversely.

Thus in Fig. 54, ϕ is a maximum at F, and a minimum at D and at H.

Instead of recording the variation of the inclination ϕ, it is generally convenient to consider the variation of the slope $\tan \phi$, which is easily expressed as a function of x by the equation

$$\tan \phi = \frac{dy}{dx}.$$

Since $\tan \phi$ is always an increasing function of ϕ, it follows that, according as the concavity is upward or downward, the slope function $\frac{dy}{dx}$ is an increasing or a decreasing function of x, and hence that its x-derivative is positive or negative.

Thus the bending of the curve is in the positive or negative direction of rotation, according as the function $\frac{d^2y}{dx^2}$ is positive or negative.

At a point of inflexion the slope $\frac{dy}{dx}$ is a maximum or minimum; and its derivative $\frac{d^2y}{dx^2}$ changes sign from positive to negative or from negative to positive. This latter condition is evidently both necessary and sufficient in order that the point (x, y) may be a point of inflexion on the given curve.

Hence, the coördinates of the points of inflexion on the curve

$$y = f(x)$$

may be found by solving the equations

$$f''(x) = 0, \quad f''(x) = \infty,$$

and then testing whether $f''(x)$ changes its sign as x passes through the critical values thus obtained. To any critical value a that satisfies the test, corresponds the point of inflexion $(a, f(a))$.

Ex. 1. For the curve

$$y = (x^2 -)1^2,$$

find the points of inflexion, and show the mode of variation of the slope and of the ordinate.

Here
$$\frac{dy}{dx} = 4 x (x^2 - 1),$$

$$\frac{d^2y}{dx^2} = 4 (3 x^2 - 1),$$

hence the critical values for inflexions are $x = -\frac{1}{\sqrt{3}} = -.58$ approximately; and $x = +.58$. It will be seen that as x increases through $-.58$, the second derivative changes sign from positive to negative, hence there is an inflexion at which the concavity changes from upward to downward. Similarly, at $x = +.58$ the concavity changes from downward to upward. The following numerical table will help to show the mode of variation of the ordinate and of the slope, and the direction of bending.

x	y	$\dfrac{dy}{dx}$	$\dfrac{d^2y}{dx^2}$
$-\infty$	$+\infty$	$-\infty$	$+$
-2	$+25$	-24	$+$
-1	0	0	$+$
$-.58$	$+.44$	$+1.5$	0
0	1	0	$-$
$+.58$	$+.44$	-1.5	0
1	0	0	$+$
$+\infty$	$+\infty$	$+\infty$	$+$

As x increases from $-\infty$ to $-.58$, the bending is positive, and the slope continually increases from $-\infty$ through zero to a maximum value, 1.5, which is the slope of the stationary tangent drawn at the point $(-.58, .44)$.

As x continues to increase from $-.58$ to $+.58$, the bending is negative, and the slope decreases from $+1.5$ through zero to a minimum value, -1.5, which is the slope of the stationary tangent drawn at the point $(+.58, .44)$.

Finally, as x increases from $+.58$ to $+\infty$, the bending is positive, and the slope increases from the value -1.5 through zero to $+\infty$.

The values $x = -1, 0, +1$, at which the slope passes through zero, correspond to turning values of the ordinate.

Ex. 2. Examine for inflexions the curve
$$x + 4 = (y - 2)^3.$$

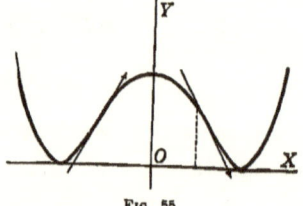

Fig. 55.

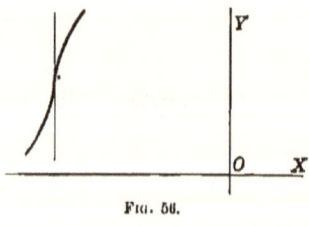

Fig. 56.

In this case
$$y = 2 + (x + 4)^{\frac{1}{3}},$$
$$\frac{dy}{dx} = \frac{1}{3}(x+4)^{-\frac{2}{3}},$$
$$\frac{d^2y}{dx^2} = -\frac{2}{9}(x+4)^{-\frac{5}{3}}.$$

Hence, at the point $(-4, 2)$, $\dfrac{dy}{dx}$ and $\dfrac{d^2y}{dx^2}$ are infinite. When $x < -4$, $\dfrac{d^2y}{dx^2}$ is positive, and when $x > -4$, $\dfrac{d^2y}{d^2x}$ is negative.

Thus there is a point of inflexion at $(-4, 2)$, at which the slope is infinite, and the bending changes from the positive to the negative direction.

Ex. 3. Consider the curve $y = x^4$.

$$\frac{dy}{dx} = 4\,x^3, \quad \frac{d^2y}{dx^2} = 12\,x^2.$$

At $(0, 0)$, $\frac{d^2y}{dx^2}$ is zero, but the curve has no inflexion, for $\frac{d^2y}{dx^2}$ never changes sign (Fig. 57).

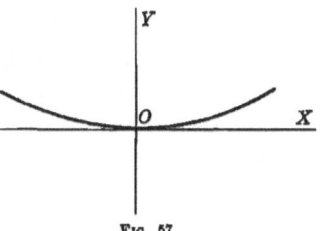

FIG. 57.

142 Analytical proof of the test for the direction of bending. Let the equation of a curve be $y = f(x)$, and let P, (x_1, y_1), be a point upon it; then the equation of the tangent at P is

$$y - y_1 = f'(x_1)(x - x_1).$$

When x changes from x_1 to $x_1 + h$, let the ordinate of the tangent change from y_1 to y', and that of the curve from y_1 to y''; then it is proposed to determine the sign of the difference of ordinates $y'' - y'$ corresponding to the same abscissa $x_1 + h$.

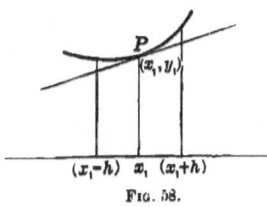

FIG. 58.

By Taylor's theorem,

$$y'' = f(x_1 + h) = f(x_1) + hf'(x_1) + \frac{h^2}{1 \cdot 2} f''(x_1 + \theta h);$$

and from the above equation of the tangent,

$$y' - y_1 = f'(x_1) \cdot (x_1 + h - x_1),$$

hence $\quad y' = y_1 + hf'(x_1) = f(x_1) + hf'(x_1),$

and it follows that

$$y'' - y' = \frac{h^2}{2} f''(x_1 + \theta h).$$

248   DIFFERENTIAL CALCULUS   [Ch. XV.

As h is made smaller and smaller, $f''(x_1 + \theta h)$ will have the same sign as $f''(x_1)$; but the factor h^2 is always positive, hence when $f''(x_1)$ is positive, $y'' - y'$ is positive, and thus the curve is above the tangent, at both sides of the point of contact, that is, the concavity is upward. Similarly when $f''(x_1)$ is negative, the concavity is downward.

This agrees with the former result.

143. Concavity and convexity towards the axis. A curve is said to be convex or concave toward a line, in the vicinity of a given point on the curve, according as the tangent at the point does or does not lie between the curve and the line, for a finite distance on each side of the point of contact.

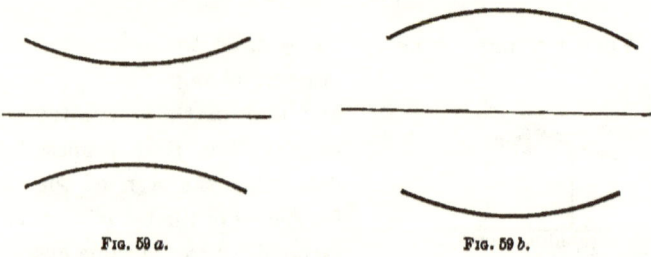

Fig. 59 a.   Fig. 59 b.

First, let the curve be convex toward the x-axis, as in the left-hand figure; then if y is positive, the bending is positive and $\dfrac{d^2y}{dx^2}$ is positive; but if y is negative, the bending is negative and $\dfrac{d^2y}{dx^2}$ is negative. Thus in either case the product $y\dfrac{d^2y}{dx^2}$ is positive.

Next, let the curve be concave toward the x-axis, as in the right-hand figure; then if y is positive, the bending is negative and $\dfrac{d^2y}{dx^2}$ is negative; but if y is negative, the bending is positive and $\dfrac{d^2y}{dx^2}$ is positive. Thus in either case the product $y\dfrac{d^2y}{dx^2}$ is negative. Hence:

In the vicinity of a given point (x, y) the curve is convex or concave to the x-axis, according as the product $y\dfrac{d^2y}{dx^2}$ is positive or negative.

EXERCISES

1. Show that the curve $y = \dfrac{x^3}{a^2 + x^2}$ has a point of inflexion at the origin, and also when $x = \pm a\sqrt{3}$.

2. In the curve $y(a^4 - b^4) = x(x-a)^4 - xb^4$, there is a point of inflexion at $x = \dfrac{2a}{5}$. Examine the points at which $x = a$.

3. Find the points of inflexion of the curve
$$\{y - 2\sqrt[3]{a^2x}\}^2 = 4ax.$$

4. Show that the curve $y(x^2 + a^2) = a^2(a - x)$ has three points of inflexion on the same straight line.

5. Find the points of inflexion on the curve $y^2(x-1) = x^2$.

6. Show that the curve $6x(1-x)y = 1 + 3x$ has one point of inflexion, and three asymptotes.

7. Show why a conic section cannot have a point of inflexion.

8. Draw the part of the curve $a^2y = \dfrac{x^3}{3} - ax^2 + 2a^3$ near its point of inflexion.

144. Concavity and convexity; polar coördinates. A curve referred to polar coördinates is said to be concave or convex to the pole, at a given point on the curve, according as the curve in the neighborhood of that point does or does not lie between the tangent and the pole.

Let p be the perpendicular from the pole to the tangent at the point (ρ, θ). Then when the curve is concave to the pole, p evidently increases with ρ, as in the arc AB, and diminishes with ρ, as in the arc BC (Fig. 60 a); hence $\dfrac{dp}{d\rho}$ is positive (Art. 19).

Again, when the curve is convex to the pole, p increases when ρ diminishes, as in the arc DE (Fig. 60 b), and p diminishes when ρ increases, as in the arc EF; hence $\dfrac{dp}{d\rho}$ is negative.

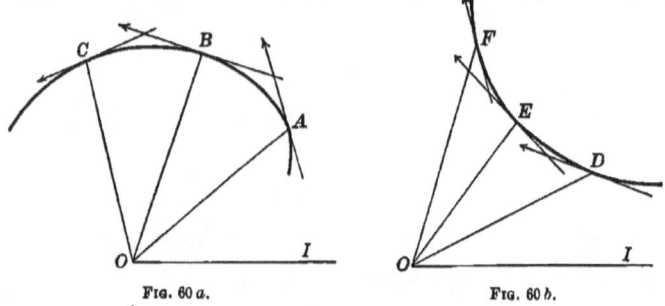

Fig. 60 a.   Fig. 60 b.

Thus the curve is concave or convex to the pole at the point (ρ, θ), according as $\dfrac{dp}{d\rho}$ is positive or negative.

To express this condition in terms of θ-derivatives of ρ, use the equation $\qquad p = \rho \sin \psi,$

$$\frac{1}{p^2} = \frac{1}{\rho^2}\csc^2\psi = \frac{1}{\rho^2}(1+\cot^2\psi) = \frac{1}{\rho^2}\left[1+\frac{1}{\rho^2}\left(\frac{dp}{d\rho}\right)^2\right], \quad (1)$$

because $\tan \psi = \rho \dfrac{d\theta}{d\rho}$, by Art. 107.

This may be simplified by putting $\dfrac{1}{\rho} = u$, $\rho = \dfrac{1}{u}$, whence $\dfrac{d\rho}{d\theta} = -\dfrac{1}{u^2}\cdot\dfrac{du}{d\theta}$, and equation (1) becomes

$$\frac{1}{p^2} = u^2 + \left(\frac{du}{d\theta}\right)^2. \quad (2)$$

Differentiation as to u gives

$$-\frac{2}{p^3}\cdot\frac{dp}{du} = 2u + 2\frac{du}{d\theta}\cdot\frac{d^2u}{d\theta^2}\cdot\frac{d\theta}{du},$$

$$\frac{dp}{du} = -p^3\left(u + \frac{d^2u}{d\theta^2}\right). \quad (3)$$

but
$$\frac{dp}{d\rho} = \frac{dp}{du} \cdot \frac{du}{d\rho} = -\frac{dp}{du} \cdot \frac{1}{\rho^2} = -\frac{dp}{du} \cdot u^2,$$

hence, from (3),
$$\frac{dp}{d\rho} = p^3 u^2 \left(u + \frac{d^2 u}{d\theta^2} \right).$$

Since p is always taken positively, hence

The curve is concave or convex to the pole at the point (ρ, θ), *according as* $u + \dfrac{d^2 u}{d\theta^2}$ *is positive or negative.*

EXERCISES

Trace the following curves near their points of inflexion:

1. $\rho = \dfrac{a 2 \theta}{2 \theta - 1}$ (find its asymptotes). 2. $\rho = \dfrac{a\theta^2}{\theta^2 - 1}$. 3. $\rho = b\theta^2$.

4. In the curve defined by the two equations
$$x = a(1 - \cos\phi), \quad y = a(n\phi + \sin\phi),$$
show that there is an inflexion at the point where $\cos\phi = -\dfrac{1}{n}$.

5. Locate the inflexions on the curve $\rho = \dfrac{a\theta}{\sin\theta}$. (See Fig. 52.)

6. Find the coördinates of the inflexion in Fig. 40.

7. In Fig. 41, show that the inflexional tangent is vertical.

8. Show that there are three real inflexions in Fig. 42.

9. How many inflexions are there in Figs. 44, 45?

10. In the logarithmic curve, the curvature is always negative; and in the exponential curve it is always positive. (Figs. 46, 47.)

11. Locate the points of inflexion in Figs. 48, 49, 50.

CHAPTER XVI

CONTACT AND CURVATURE

145. Order of contact. The points of intersection of the two curves
$$y = \phi(x), \quad y = \psi(x)$$
are found by making the two equations simultaneous; that is, by finding those values of x for which
$$\phi(x) = \psi(x).$$
Suppose $x = a$ is one value that satisfies this equation, then the point $x = a$, $y = \phi(a) = \psi(a)$ is common to the curves.

If, moreover, the two curves have the same tangent at this point, they are said to touch each other, or to have *contact* of the first order with each other. The values of y and of $\dfrac{dy}{dx}$ are thus the same for both curves at the point in question, and this requires that
$$\phi(a) = \psi(a),$$
$$\phi'(a) = \psi'(a).$$

If, in addition, the values of $\dfrac{d^2y}{dx^2}$ be the same for each curve at the point, then
$$\phi''(a) = \psi''(a),$$
and the curves are said to have a contact of the second order with each other at the point for which $x = a$.

If $\phi(a) = \psi(a)$, and all the derivatives up to the nth order be equal to each other, the curves are said to have contact of the nth order. This is seen to require $n+1$ conditions; hence if the equation of the curve $y = \phi(x)$ be given, and if the equation of a second curve be written in the form $y = \psi(x)$, in which $\psi(x)$ proceeds in powers of x with undetermined coefficients, then $n + 1$ of these coefficients could be determined by requiring the second curve to have contact of the nth order with the given curve at a given point.

146. Number of conditions implied by contact. A straight line has two arbitrary constants, which can be determined by two conditions; thus, a straight line can be drawn which touches a given curve at any specified point.

In general no line can be drawn having contact of an order higher than the first with a given curve; but there are certain points at which this can be done. For instance, if the equation of a line be written $y = mx + b$, then

$$\frac{dy}{dx} = m, \quad \frac{d^2y}{dx^2} = 0;$$

hence, through any arbitrary point $x = a$ on a given curve $y = \phi(x)$, a line can be drawn which has contact of the first order with the curve, but which has not in general contact of the second order; for the two conditions for first order contact are

$$ma + b = \phi(a),$$
$$m = \phi'(a),$$

which are just sufficient to determine m and b; and the additional condition for second-order contact is $0 = \phi''(a)$, which is satisfied only when the point $x = a$ is a point of inflexion on the given curve $y = \phi(x)$. Thus the tangent at a point of inflexion on a curve has contact of the second order with the curve.

The equation of a circle has three independent constants. It is therefore possible to determine a circle having contact of the second order with a given curve at any assigned point.

The equation of a parabola has four constants, hence a parabola can be found which has contact of the third order with the given curve at any point.

The general equation of a central conic has five independent constants, hence a conic can be found which has contact of the fourth order with a given curve at any given point.

As in the case of the tangent line, special points may be found for which these curves have contact of higher order.

147. Contact of odd and of even order.

THEOREM. At a point where two curves have contact of an odd order they do not cross each other; but they do cross where they have contact of an even order.

For, let the curves $y = \phi(x)$, $y = \psi(x)$ have contact of the nth order at the point whose abscissa is a; and let y_1, y_2 be the ordinates of these curves at the point whose abscissa is $a + h$; then

$$y_1 = \phi(a + h), \quad y_2 = \psi(a + h),$$

and by Taylor's theorem

$$y_1 = \phi(a) + \phi'(a) \cdot h + \frac{\phi''(a)}{2!} \cdot h^2 + \cdots$$
$$+ \frac{\phi^n(a)}{n!} \cdot h^n + \frac{h^{n+1}}{(n+1)!} \phi^{n+1}(a + \theta h);$$

$$y_2 = \psi(a) + \psi'(a) \cdot h + \frac{\psi''(a)}{2!} \cdot h^2 + \cdots$$
$$+ \frac{\psi^n(a)}{n!} \cdot h^n + \frac{h^{n+1}}{(n+1)!} \cdot \psi^{n+1}(a + \theta_1 h).$$

Since by hypothesis the two curves have contact of the nth order at the point whose abscissa is a,

hence $\phi(a) = \psi(a)$, $\phi'(a) = \psi'(a)$, ..., $\phi^n(a) = \psi^n(a)$,

and $y_1 - y_2 = \dfrac{h^{n+1}}{(n+1)!} [\phi^{n+1}(a + \theta h) - \psi^{n+1}(a + \theta_1 h)];$

but this expression, when h is sufficiently diminished, has the same sign as

$$h^{n+1}[\phi^{n+1}(a) - \psi^{n+1}(a)];$$

hence, if n be odd, $y_1 - y_2$ does not change sign when h is changed into $-h$, and thus the two curves do not cross each other at the common point. On the other hand, if n be even, $y_1 - y_2$ changes sign with h; and therefore when the contact is of even order the curves cross each other at their common point.

For example, the tangent line usually lies entirely on one side of the curve, but at a point of inflexion the tangent crosses the curve.

Again, the circle of second-order contact crosses the curve except at the special points, noted later, in which the circle has contact of the third order.

148. Circle of curvature. The circle that has contact of the closest (*i.e.*, second) order with a given curve at a specified point is called the *osculating circle* or circle of curvature of the curve at the given point. The radius of this circle is called the radius of curvature, and its center is called the center of curvature at the assigned point.

149. Length of radius of curvature; coördinates of center of curvature.

Let the equation of a circle be

$$(X - \alpha)^2 + (Y - \beta)^2 = R^2, \qquad (1)$$

in which R is the radius, and α, β are the coördinates of the center, the current coördinates being denoted by X, Y, to distinguish them from the coördinates of a point on the given curve.

It is required to determine R, α, β, such that this circle may have contact of the second order with the given curve at the point (x, y).

From (1), by successive differentiation,

$$(X-\alpha)+(Y-\beta)\frac{dY}{dX}=0, \\ 1+\left(\frac{dY}{dX}\right)^2+(Y-\beta)\frac{d^2Y}{dX^2}=0. \qquad (2)$$

If the circle (1) has contact of the second order at the point (x, y) with the given curve, then the common abscissa $x = X$ makes

$$Y = y, \\ \frac{dY}{dX}=\frac{dy}{dx}, \quad \frac{d^2Y}{dX^2}=\frac{d^2y}{dx^2}; \qquad (3)$$

hence, from (2),
$$(x-\alpha)+(y-\beta)\frac{dy}{dx}=0, \\ 1+\left(\frac{dy}{dx}\right)^2+(y-\beta)\frac{d^2y}{dx^2}=0, \qquad (4)$$

whence

$$y-\beta = -\frac{1+\left(\frac{dy}{dx}\right)^2}{\frac{d^2y}{dx^2}}, \quad x-\alpha = \frac{\frac{dy}{dx}\left[1+\left(\frac{dy}{dx}\right)^2\right]}{\frac{d^2y}{dx^2}}, \qquad (5)$$

and finally, by substitution in (1),

$$R = \frac{\left\{1+\left(\frac{dy}{dx}\right)^2\right\}^{\frac{3}{2}}}{\frac{d^2y}{dx^2}}. \qquad (6)$$

If, for shortness, m, n be written for $\dfrac{dy}{dx}$, $\dfrac{d^2y}{dx^2}$, then the coördinates of the center and the radius of the circle of curvature are given by the equations

$$x - \alpha = \frac{m(1+m^2)}{n}; \quad y - \beta = -\frac{1+m^2}{n}; \quad R = \frac{(1+m^2)^{\frac{3}{2}}}{n}.$$

150. Second method. The osculating circle is sometimes defined as the limiting position of a circle passing through three points on the curve when two of these points move towards the third as a limit.

It is proposed to find the equation of this circle, and thus to show that the two definitions lead to the same result.

Let $x - h$, x, $x + h$ be the abscissas of three points on the curve, and $y - k$, y, $y + k'$ the corresponding ordinates, in which k' is not in general equal to k.

Let these three points lie on the circle whose equation is

$$(x - \alpha)^2 + (y - \beta)^2 = R^2, \tag{1}$$

then
$$(x - h - \alpha)^2 + (y - k - \beta)^2 = R^2,$$
$$(x + h - \alpha)^2 + (y + k' - \beta)^2 = R^2.$$

Subtracting the second and third from the first,

$$\left.\begin{array}{l} 2h(x-\alpha) - h^2 + 2k(y-\beta) - k^2 = 0, \\ -2h(x-\alpha) - h^2 - 2k'(y-\beta) - k'^2 = 0, \end{array}\right\} \tag{2}$$

whence by adding, and solving for $y - \beta$,

$$y - \beta = \frac{2h^2 + k^2 + k'^2}{2(k - k')}. \tag{3}$$

To find the limit of this fraction as $h \doteq 0$, let $y = \phi(x)$ be the equation of the given curve, then

$$y - k = \phi(x - h), \quad y + k' = \phi(x + h),$$

whence, by Taylor's theorem,

$$y - k = \phi(x) - h\phi'(x) + \frac{h^2}{2!}\phi''(x - \theta_1 h),$$

$$y + k' = \phi(x) + h\phi'(x) + \frac{h^2}{2!}\phi''(x + \theta h),$$

and
$$k = h\phi'(x) - \frac{h^2}{2!}\phi''(x - \theta_1 h),$$

$$k' = h\phi'(x) + \frac{h^2}{2!}\phi''(x + \theta h);$$

hence, when $h \doteq 0$,

$$\frac{k}{h} \doteq \phi'(x), \quad \frac{k'}{h} \doteq \phi'(x), \quad \frac{k' - k}{h^2} \doteq \phi''(x). \tag{4}$$

Equation (3) may now be written

$$y - \beta = -\frac{1 + \frac{1}{2}\left(\frac{k}{h}\right)^2 - \frac{1}{2}\left(\frac{k'}{h}\right)^2}{\frac{k' - k}{h^2}},$$

therefore, by (4),

$$y - \beta = -\frac{1 + [\phi'(x)]^2}{\phi''(x)} = -\frac{1 + \left(\dfrac{dy}{dx}\right)^2}{\dfrac{d^2y}{dx^2}}.$$

To find $(x - \alpha)$, divide the first of equations (2) by $2h$ and pass to the limit, then

$$x - \alpha \doteq -\frac{k}{h}(y - \beta)$$

$$\doteq \frac{\phi'(x)\{1 + [\phi'(x)]^2\}}{\phi''(x)} = \frac{\dfrac{dy}{dx}\left\{1 + \left(\dfrac{dy}{dx}\right)^2\right\}}{\dfrac{d^2y}{dx^2}}.$$

Thus the coördinates (α, β) of the center of the osculating circle at the point (x, y) are the same by either definition. The value of R is then found as before.

151. Direction of radius of curvature. Since the given curve and its osculating circle at a point P have the same value of $\frac{dy}{dx}$ at that point, it follows that they have the same tangent and normal at P, and hence that the radius of curvature coincides with the normal. Again, since the curve and its osculating circle have the same value of $\frac{d^2y}{dx^2}$ at P, it follows from Art. 141, that they have the same direction of bending at that point, and hence that the center of curvature lies on the concave side of the given curve (Fig. 61).

This could also be seen from the fact (Art. 150) that the osculating circle is the limiting position of a circle passing through three points on the curve when these points move into coincidence.

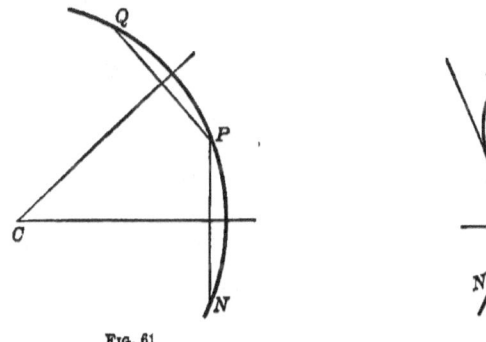

FIG. 61.   FIG. 62.

The radius of curvature is usually regarded as positive or negative according as the bending of the curve is positive or negative (Art. 141), that is, according as the value of $\frac{d^2y}{dx^2}$ is positive or negative; hence, in the expression for R, the radical in the numerator is always to be given the positive sign. The sign of R changes as the point P passes through a point of inflexion on the given curve (Fig. 62). It is evident from the figure that in this case R passes through an infinite value; for the circle through the points N, P, Q approaches coincidence with the inflexion tangent when N and Q approach coincidence with P; and thus the center of this circle at the same time passes to infinity.

152. Other forms for R.

I. Expression for R, when x and y are functions of an independent variable t.

By Arts 21, 51,

$$\frac{dy}{dx} = \frac{\frac{dy}{dt}}{\frac{dx}{dt}}, \quad \frac{d^2y}{dx^2} = \frac{\left(\frac{d^2y}{dt^2} \cdot \frac{dx}{dt} - \frac{d^2x}{dt^2} \cdot \frac{dy}{dt}\right)}{\left(\frac{dx}{dt}\right)^3}.$$

Therefore the expression of Art. 149 becomes

$$R = \frac{\left\{\left(\frac{dx}{dt}\right)^2 - \left(\frac{dy}{dt}\right)^2\right\}^{\frac{3}{2}}}{\frac{d^2y}{dt^2} \cdot \frac{dx}{dt} - \frac{d^2x}{dt^2} \cdot \frac{dy}{dt}}.$$

II. Expression for R, when the curve is defined by an implicit equation.

Let $f(x, y) = 0$ be the equation of the curve; then when the value of $\frac{dy}{dx}$, $\frac{d^2y}{dx^2}$ are expressed in terms of

$$\frac{\partial f}{\partial x}, \quad \frac{\partial f}{\partial y}, \quad \frac{\partial^2 f}{\partial x^2}, \quad \frac{\partial^2 f}{\partial x \partial y}, \quad \frac{\partial^2 f}{\partial y^2},$$

as in Art. 88, the expression for R becomes

$$R = \frac{\left[\left(\frac{\partial f}{\partial x}\right)^2 + \left(\frac{\partial f}{\partial y}\right)^2\right]^{\frac{3}{2}}}{\left(\frac{\partial f}{\partial y}\right)^2 \frac{\partial^2 f}{\partial x^2} + 2 \frac{\partial f}{\partial x} \frac{\partial f}{\partial y} \frac{\partial^2 f}{\partial x \partial y} + \left(\frac{\partial f}{\partial x}\right)^2 \frac{\partial^2 f}{\partial y^2}}.$$

III. Expression for R in polar coördinates.

If the equation of the curve be given in the form $\rho = f(\theta)$, the expression for R may be found by transforming the equation of Art. 149, by means of the relations

$$x = \rho \cos \theta, \quad y = \rho \sin \theta.$$

The result is

$$R = \frac{\left[\rho^2 + \left(\dfrac{\partial \rho}{\partial \theta}\right)^2\right]^{\frac{3}{2}}}{\rho^2 - \rho \dfrac{\partial^2 \rho}{\partial \theta^2} + 2\left(\dfrac{\partial \rho}{\partial \theta}\right)^2}.$$

153. Total curvature of a given arc; average curvature. The total curvature of an arc PQ (Fig. 63) in which the bending is continuous and in one direction, is the angle through which the tangent swings as the point of contact moves from the initial point P to the terminal point Q; or, in other words, it is the angle between the tangents at P and Q, measured from the forward end of the former to that of the latter. Thus the

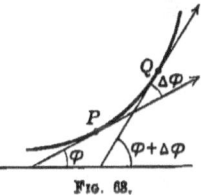

Fig. 63.

total curvature of a given arc is positive or negative according as the bending is in the positive or negative direction of rotation.

The total curvature of an arc divided by the length of the arc is called the *average* curvature of the arc, or the curvature for unit of length. Thus, if the length of the arc PQ be Δs centimeters, and if its total curvature be $\Delta \phi$ radians, then its average curvature is $\dfrac{\Delta \phi}{\Delta s}$ radians per centimeter.

154. Measure of curvature at a given point. The *measure of the curvature* of a given curve at a given point P is the limit which the average curvature of the arc PQ approaches when the point Q approaches coincidence with P.

Since the average curvature of the arc PQ is $\dfrac{\Delta\phi}{\Delta s}$, the measure of the curvature at the point P is

$$\kappa = \lim_{\Delta s \doteq 0} \frac{\Delta\phi}{\Delta s} = \frac{d\phi}{ds},$$

and may be regarded as the rate of deflection of the arc from the tangent estimated per unit of length; or again, as the ratio of the angular velocity of the tangent to the linear velocity of the point of contact.

To express κ in terms of x, y, and their derivatives. Since

$$\tan\phi = \frac{dy}{dx},$$

then
$$\phi = \tan^{-1}\frac{dy}{dx},$$

and
$$\frac{d\phi}{ds} = \frac{d}{ds}\left(\tan^{-1}\frac{dy}{dx}\right)$$

$$= \frac{d}{dx}\left(\tan^{-1}\frac{dy}{dx}\right) \cdot \frac{dx}{ds}$$

$$= \frac{\dfrac{d^2y}{dx^2}}{1+\left(\dfrac{dy}{dx}\right)^2} \cdot \frac{1}{\dfrac{ds}{dx}};$$

therefore
$$\kappa = \frac{d\phi}{ds}, = \frac{\dfrac{d^2y}{dx^2}}{\left\{1+\left(\dfrac{dy}{dx}\right)^2\right\}^{\frac{3}{2}}}. \qquad \text{[Art. 121}$$

155. Curvature of an arc of a circle. In the case of a circular arc the normals are radii;

hence $\quad \Delta s = r \cdot \Delta \phi, \quad \dfrac{\Delta \phi}{\Delta s} = \dfrac{1}{r},\quad$ (1)

thus $\quad\quad\quad \kappa = \dfrac{1}{r}.$

Thus the average curvature of all arcs of the same circle is constant and equal to $\dfrac{1}{r}$ radians per unit of length.

For example, in a circle of 2 feet radius the total curvature of an arc of 3 feet is $\frac{3}{2} = 1.5$ radians, and the average curvature is .5 radian per foot.

It also follows from (1) that in different circles, arcs of the same length have a total curvature inversely proportional to their radii.

Thus on a circumference of 1 meter radius, an arc of 5 decimeters has a total curvature of .5 radian, and an average curvature of .1 radian per decimeter; but on a circumference of half a meter radius, the same length of arc has a total curvature of 1 radian and an average curvature of .2 radian per decimeter.

156. Curvature of osculating circle. A curve and its osculating circle at P have the same measure of curvature at that point.

For, let κ, κ' be their respective measures of curvature at the point of contact (x, y); then from Art. 154,

$$\kappa = \dfrac{\dfrac{d^2 y}{dx^2}}{\left\{1 + \left(\dfrac{dy}{dx}\right)^2\right\}^{\frac{3}{2}}},$$

and from Art. 149,

$$\kappa' = \dfrac{1}{r} = \dfrac{\dfrac{d^2 y}{dx^2}}{\left\{1 + \left(\dfrac{dy}{dx}\right)^2\right\}^{\frac{3}{2}}}, \quad \text{hence } \kappa = \kappa'.$$

It is on account of this property that the osculating circle is called the circle of curvature. This is sometimes used as the defining property of the circle of curvature. The radius of curvature at P would then be defined as the radius of the circle, whose measure of curvature is the same as that of the given curve at the point P. Its value, as found from Art. 153 and Art. 155, accords with that given in Art. 149.

EXERCISES

Find the order of contact of the two curves
$$y = x^3, \quad y = 3x^2 - 3x + 1.$$

2. Find the order of contact of the parabola $y^2 = 4x$, and the straight line $3y = x + 9$.

3. Find the order of contact of
$$9y = x^3 - 3x^2 + 27 \quad \text{and} \quad 9y + 3x = 28.$$

4. Find the order of contact of
$$y = \log(x - 1) \quad \text{and} \quad x^2 - 6x + 2y + 8 = 0 \text{ at } (2, 0).$$

5. Show that the circle $\left(x - \dfrac{3a}{4}\right)^2 + \left(y - \dfrac{3a}{4}\right)^2 = \dfrac{a^2}{2}$ and the curve $\sqrt{x} + \sqrt{y} = \sqrt{a}$ have contact of the third order at the point $x = y = \dfrac{a}{4}$.

6. What must be the value of a in order that the parabola
$$y = x + 1 + a(x - 1)^2$$
may have contact of the second order with the hyperbola $xy = 3x - 1$?

7. Find the order of contact of the parabola
$$(x - 2a)^2 + (y - 2a)^2 = 2xy,$$
and the hyperbola $xy = a^2$.

EXERCISES ON CURVATURE

8. In the curve $y = x^4 - 4x^3 - 18x^2$, the radius of curvature at the origin is $\tfrac{1}{36}$.

9. Show that the two radii of curvature of the curve
$$y^2 = x^2 \cdot \dfrac{a + x}{a - x}$$
at the origin are $\pm a\sqrt{2}$; and that $R = \tfrac{1}{4}a$ at $(-a, 0)$.

Find the radius of curvature in each of the following curves:

10. The parabola $\qquad y^2 = 4ax$.

11. The ellipse $\qquad \dfrac{x^2}{a^2} = \dfrac{y^2}{b^2} = 1$.

12. The catenary $\qquad y = \dfrac{c}{2}(e^{\frac{x}{c}} + e^{-\frac{x}{c}})$.

13. The exponential curve $\quad y = ae^{\frac{x}{c}}$.

14. The parabola $\qquad \sqrt{x} + \sqrt{y} = 2\sqrt{a}$.

15. The hypocycloid $\qquad x^{\frac{2}{3}} + y^{\frac{2}{3}} = a^{\frac{2}{3}}$.

16. The curve $\qquad y = \log \sec x$. Catenary of uniform strength.

17. Derive the formula $\dfrac{1}{R^2} = \left(\dfrac{d^2x}{ds^2}\right)^2 + \left(\dfrac{d^2y}{ds^2}\right)^2$.

$$\left[\dfrac{dx}{ds} = \cos\phi;\ \dfrac{d^2x}{ds^2} = -\sin\phi \cdot \dfrac{d\phi}{ds} = -\dfrac{\sin\phi}{R};\ \text{etc.}\right]$$

157. Direct derivation of the expressions for κ and R in polar coördinates.

Using the notation of Art. 119,

$$\phi = \theta + \psi,$$

hence $\qquad \kappa = \dfrac{d\phi}{ds} = \dfrac{\dfrac{d\phi}{d\theta}}{\dfrac{ds}{d\theta}} = \dfrac{\left(1 + \dfrac{d\psi}{d\theta}\right)}{\dfrac{ds}{d\theta}} \qquad (1)$

$$= \dfrac{\left(1 + \dfrac{d\psi}{d\theta}\right)}{\left[\rho^2 + \left(\dfrac{d\rho}{d\theta}\right)^2\right]^{\frac{1}{2}}}. \qquad \text{[Art. 124}$$

But $\tan\psi = \rho\dfrac{d\theta}{d\rho}, \quad \psi = \tan^{-1}\left[\dfrac{\rho}{\dfrac{d\rho}{d\theta}}\right]$, therefore, by differentiating as to θ and reducing,

$$\frac{d\psi}{d\theta} = \frac{\left(\frac{d\rho}{d\theta}\right)^2 - \rho\frac{d^2\rho}{d\theta^2}}{\rho^2 + \left(\frac{d\rho}{d\theta}\right)^2},$$

which, substituted in (1), gives

$$\kappa = \frac{\rho^2 - \rho\frac{d^2\rho}{d\theta^2} + 2\left(\frac{d\rho}{d\theta}\right)^2}{\left[\rho^2 + \left(\frac{d\rho}{d\theta}\right)^2\right]^{\frac{3}{2}}};$$

and the relation $R = \frac{1}{\kappa}$ then reproduces the result obtained in Art. 152 by transformation of coördinates.

When $u = \frac{1}{\rho}$ is taken as dependent variable, the expression for κ assumes the simpler form

$$\kappa = \frac{u^3\left(u + \frac{d^2u}{d\theta^2}\right)}{\left[u^2 + \left(\frac{du}{d\theta}\right)^2\right]^{\frac{3}{2}}}.$$

Since at a point of inflexion κ vanishes and changes sign, hence the condition for a point of inflexion, expressed in polar coördinates, is that $u + \frac{d^2u}{d\theta^2}$ shall pass through zero and change its sign. See Art. 144.

EXERCISES

1. Show that the radius of curvature of the curve

$$\rho = a \sin n\theta \text{ at } (0, 0) \text{ is } \frac{na}{2}$$

2. Find the radius of curvature of $\rho^m = a^m \cos m\theta$.

Find the value of R in each of the following curves:

3. The circle $\rho = a \sin \theta$.

4. The lemniscate $\rho^2 = a^2 \cos 2\theta$.

5. The logarithmic spiral $\rho = e^{a\theta}$.

6. The trisectrix $\rho = 2a\cos\theta - a$.

7. The equilateral hyperbola $\rho^2 \cos 2\theta = a^2$.

8. For any curve prove the formula

$$R = \frac{\rho}{\sin\psi \cdot \left(1 + \frac{d\psi}{d\theta}\right)} \cdot \quad \left[\tan\psi = \frac{\rho d\theta}{d\rho}\cdot\right]$$

EVOLUTES AND INVOLUTES

158. Definition of an evolute. When the point P moves along the given curve, the center of curvature C describes another curve which is called the *evolute* of the first.

Let $f(x, y) = 0$ be the equation of the given curve, then the equation of the locus described by the point C is found by eliminating x and y from the three equations

$$f(x, y) = 0,$$

$$x - \alpha = \frac{\frac{dy}{dx}\left[1 + \left(\frac{dy}{dx}\right)^2\right]}{\frac{d^2y}{dx^2}},$$

$$y - \beta = -\frac{1 + \left(\frac{dy}{dx}\right)^2}{\frac{d^2y}{dx^2}},$$

and thus obtaining a relation between α, β, the coördinates of the center of curvature.

No general process of elimination can be given; the method to be adopted depends upon the form of the given equation $f(x, y) = 0$.

Ex. 1. Find the evolute of the parabola $y^2 = 4px$.

Since $y = 2p^{\frac{1}{2}}x^{\frac{1}{2}}$, $\dfrac{dy}{dx} = p^{\frac{1}{2}}x^{-\frac{1}{2}}$, $\dfrac{d^2y}{dx^2} = -\dfrac{1}{2}p^{\frac{1}{2}}x^{-\frac{3}{2}}$,

hence $x - \alpha = -p^{\frac{1}{2}}x^{-\frac{1}{2}}(1 + px^{-1})\,2\,p^{-\frac{1}{2}}x^{\frac{3}{2}} = -2(x+p)$,

and $y - \beta = (1 + px^{-1})\,2\,p^{-\frac{1}{2}}x^{\frac{3}{2}} = 2(p^{-\frac{1}{2}}x^{\frac{3}{2}} + p^{\frac{1}{2}}x^{\frac{1}{2}})$;

therefore $\alpha = 2p + 3x$, $\beta = -2p^{-\frac{1}{2}}x^{\frac{3}{2}}$,

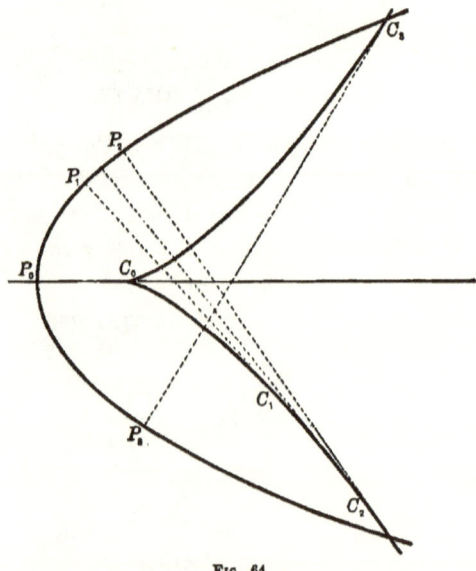

Fig. 64.

when, by eliminating x, $\tfrac{1}{27}(\alpha - 2p)^3 = \tfrac{1}{4}(p^{\frac{1}{2}}\beta)^2$,

i.e., $4(\alpha - 2p)^3 = 27\,p\beta^2$,

is the equation of the evolute of the parabola, in which α, β are current coördinates.

Ex. 2. Find the evolute of the ellipse

$$\frac{x^2}{a^2} + \frac{y^2}{b^2} = 1. \tag{1}$$

Here $\dfrac{x}{a^2} + \dfrac{y}{b^2} \cdot \dfrac{dy}{dx} = 0, \quad \dfrac{dy}{dx} = -\dfrac{b^2 x}{a^2 y},$

$\dfrac{d^2y}{dx^2} = -\dfrac{b^2}{a^2} \cdot \dfrac{y - x\dfrac{dy}{dx}}{y^2} = \dfrac{-b^2}{a^4 y^2}\left(y + \dfrac{b^2 x^2}{a^2 y}\right) = \dfrac{-b^2}{a^4 y^3}(a^2 y^2 + b^2 x^2) = \dfrac{-b^4}{a^2 y^3},$

whence

$y - \beta = \dfrac{(a^4 y^2 + \beta^4 x^2) y}{a^2 b^4} = \left(\dfrac{a^2 y^2}{b^4} + \dfrac{x^2}{a^2}\right) y = \left(\dfrac{a^2 y^2}{b^4} + 1 - \dfrac{y^2}{b^2}\right) y.$

Therefore $\qquad -\beta = \dfrac{a^2 - b^2}{b^4} y^3.$ \hfill (2)

Similarly, $\qquad a = \dfrac{a^2 - b^2}{a^4} x^3.$ \hfill (3)

Eliminating x, y between (1), (2), (3), the equation of the locus described by (α, β) is

$$(a\alpha)^{\frac{2}{3}} + (b\beta)^{\frac{2}{3}} = (a^2 - b^2)^{\frac{2}{3}}. \qquad \text{(Fig. 69.)}$$

159. Properties of the evolute. The evolute has two important properties that will now be established.

I. *The normal to the curve is tangent to the evolute.*
The relations connecting the coördinates (α, β) of the center of curvature with the coördinates (x, y) of the corresponding point on the curve are, by Art. 149,

$$x - \alpha + (y - \beta)\dfrac{dy}{dx} = 0, \qquad (1)$$

$$1 + \left(\dfrac{dy}{dx}\right)^2 + (y - \beta)\dfrac{d^2y}{dx^2} = 0. \qquad (2)$$

From these equations α, β may be considered functions of x; hence, by differentiating (1), regarding α, β, y as functions of x,

$$1 + \left(\dfrac{dy}{dx}\right)^2 + (y - \beta)\dfrac{d^2y}{dx^2} - \dfrac{d\alpha}{dx} - \dfrac{d\beta}{dx}\dfrac{dy}{dx} = 0. \qquad (3)$$

Subtracting (3) from (2) gives

$$\frac{d\alpha}{dx} + \frac{d\beta}{dx}\frac{dy}{dx} = 0, \qquad (4)$$

whence $\qquad \dfrac{d\beta}{d\alpha} = -\dfrac{dx}{dy};$

but $\dfrac{d\beta}{d\alpha}$ is the slope of the tangent to the evolute at (α, β); and $-\dfrac{dx}{dy}$ is the slope of the normal to the given curve at (x, y). Hence these lines have the same slope; but they pass through the same point (α, β), therefore they are coincident.

II. The difference between two radii of curvature of the given curve, touching the evolute at the points C_1, C_2 (Fig. 65), is equal to the arc $C_1 C_2$ of the evolute.

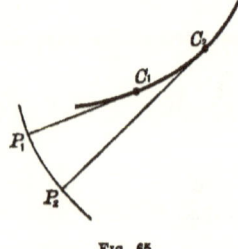

Fig. 65.

Since R is the distance between points (x, y), (α, β), hence

$$(x - \alpha)^2 + (y - \beta)^2 = R^2. \qquad (5)$$

When the point (x, y) moves along the given curve, the point (α, β) moves along the evolute, and thus α, β, R, y are all functions of x.

Differentiation of (5) as to x gives

$$(x - \alpha)\left(1 - \frac{d\alpha}{dx}\right) + (y - \beta)\left(\frac{dy}{dx} - \frac{d\beta}{dx}\right) = R\frac{dR}{dx}, \qquad (6)$$

hence, subtracting (6) from (1),

$$(x - \alpha)\frac{d\alpha}{dx} + (y - \beta)\frac{d\beta}{dx} = -R\frac{dR}{dx}. \qquad (7)$$

CONTACT AND CURVATURE

Again, from (1) and (4),

$$\frac{\frac{d\alpha}{dx}}{x-\alpha} = \frac{\frac{d\beta}{dy}}{y-\beta}. \tag{8}$$

Hence, each of these fractions is equal to

$$\frac{\sqrt{\left(\frac{d\alpha}{dx}\right)^2 + \left(\frac{d\beta}{dx}\right)^2}}{\sqrt{(x-\alpha)^2 + (y-\beta)^2}} = \pm \frac{\frac{d\sigma}{dx}}{R}, \tag{9}$$

in which σ is the arc of the evolute.

Next, multiplying numerator and denominator of the first member of (8) by $x - \alpha$, and those of the second member by $y - \beta$, and combining new numerators and denominators, it follows that each of the fractions in (8) is equal to

$$\frac{(x-\alpha)\frac{d\alpha}{dx} + (y-\beta)\frac{d\beta}{dx}}{(x-\alpha)^2 + (y-\beta)^2},$$

which equals $-\dfrac{RdR}{R^2}$, by (7) and (5).

Whence, by (9), $\qquad \dfrac{d\sigma}{dx} = \pm \dfrac{dR}{dx},$

that is, $\qquad \dfrac{d}{dx}(\sigma \pm R) = 0;$

therefore $\qquad \sigma \pm R = \text{constant}, \tag{10}$

wherein σ is measured from a fixed point A on the evolute.

Now, let C_1, C_2 be the centers of curvature for the points P_1, P_2 on the given curve; let $P_1 C_1 = R_1$, $P_2 C_2 = R_2$; and let the arcs AC_1, AC_2 be denoted by σ_1, σ_2; then

$$\sigma_1 \pm R_1 = \sigma_2 \pm R_2, \text{ by (10)};$$

that is, $\sigma_1 - \sigma_2 = \pm (R_2 - R_1)$;

hence, arc $C_1 C_2 = R_2 \sim R_1$. (11)

Thus, in figure 66,

$$P_1 C_1 + C_1 C_2 = P_2 C_2,$$
$$P_2 C_2 + C_2 C_3 = P_3 C_3, \text{ etc.}$$

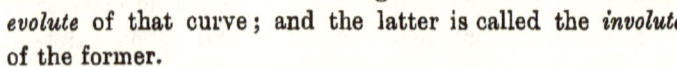

Fig. 66.

Hence, if a thread be wrapped around the evolute, and then be unwound, the free end of it can be made to trace out the original curve. From this property the locus of the center of curvature of a given curve is called the *evolute* of that curve; and the latter is called the *involute* of the former.

When the string is unwound, each point of it describes a different involute; hence to the same curve correspond an infinite number of involutes; but a curve has but one evolute.

Any two of these involutes intercept a constant distance on their common normal, and are called *parallel* curves.

EXERCISES

Find the coördinates of the center of curvature of the following curves:

1. The parabola $y^2 = 4ax$.
2. The semicubical parabola $x^3 = ay^2$.
3. The four-cusped hypocycloid $x^{\frac{2}{3}} + y^{\frac{2}{3}} = a^{\frac{2}{3}}$.
4. The catenary $y = \dfrac{c}{2}\left(e^{\frac{x}{c}} + e^{-\frac{x}{c}}\right)$.
5. The equation of the equilateral hyperbola being $xy = a^2$, prove that

$$\alpha + \beta = \frac{a}{2}\left(\frac{a}{x} + \frac{x}{a}\right)^2; \quad \alpha - \beta = \frac{a}{2}\left(\frac{a}{x} - \frac{x}{a}\right)^2;$$

and derive the equation of the evolute.

6. Show that the curvature of an ellipse is a minimum at the end of the minor axis, and that the osculating circle at this point has contact of the third order with the curve.

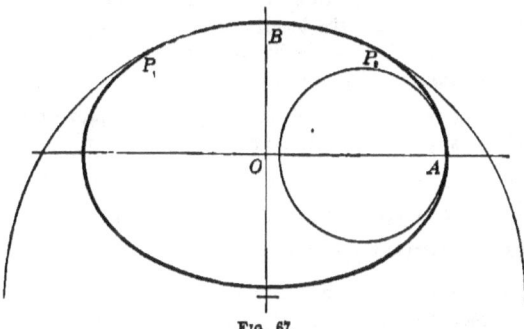

Fig. 67.

This circle of curvature must be entirely outside the ellipse (Fig. 67); for, consider two points P_1, P_2, one on each side of B, the end of the minor axis. At these points the curvature is greater than at B, hence these points must be farther from the tangent at B than the circle of curvature, which has everywhere the same curvature as at B.

7. Similarly, show that the curvature at A, the end of the major axis, is a maximum, and that the circle of curvature at A lies entirely within the ellipse (Fig. 67).

8. Show how to sketch the circle of curvature for points between A and B. The circle of curvature for points between A and B has three coincident points in common with the ellipse (Art. 137), hence the circle

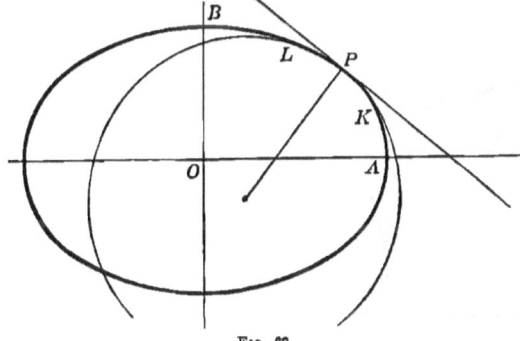

Fig. 68.

crosses the curve (Art. 134). Let K, P, L be three points on the arc, such that K is nearest A, L nearest B. The center of curvature for P lies on the normal to P, and on the concave side of the curve. The circle crosses at P, lying outside of the ellipse at K (on the side towards A), and inside the ellipse at L; for the bending of the ellipse increases from B to P and from P to K, while the bending (curvature) of the osculating circle remains constant (Fig. 68).

9. Two centers of curvature lie on every normal; prove geometrically that the normals to the curve are tangents to the evolute.

10. Show that the entire length of the evolute of the ellipse is $4\left(\dfrac{a^2}{b^2} - \dfrac{b^2}{a^2}\right)$. [From equation (11) above, take R_1, R_2 as the radii of curvature at the extremities of the major and minor axes.]

11. Show that in the parabola $y^2 = 4ax$ the length of the part of the evolute intercepted within the parabola is $4a(3\sqrt{3} - 1)$.

12. Show that in the parabola $\sqrt{x} + \sqrt{y} = \sqrt{a}$ the relation exists, $a + \beta = 3(x + y)$.

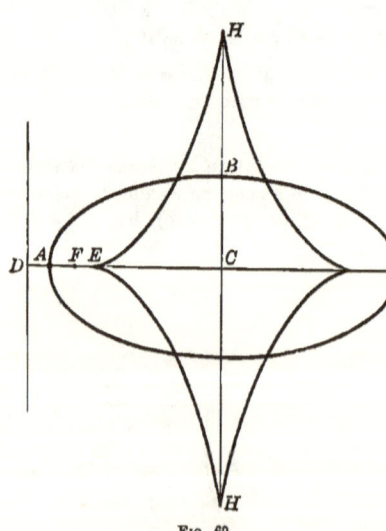

Fig. 69.

13. If E be the center of curvature at the vertex A (Fig. 69), prove that $CE = ae^2$, in which e is the eccentricity of the ellipse; and hence that CD, CA, CF, CE form a geometric series whose common ratio is e. Show also that DA, AF, FE form a similar series.

14. If H be the center of curvature at B, show that the point H is without or within the ellipse, according as $a >$ or $< b\sqrt{2}$, or according as $e^2 >$ or $< \frac{1}{2}$.

15. Show by inspection of the figure that four real normals can be drawn to the ellipse from any point within the evolute.

CHAPTER XVII

SINGULAR POINTS

160. Definition of a singular point. If the equation $f(x, y) = 0$ be represented by a curve, the derivative $\frac{dy}{dx}$, when it has a determinate value, expresses the slope of the tangent at the point (x, y). There may be certain points on the curve, however, at which the expression for the derivative assumes an illusory or indeterminate form; and, in consequence, any line whatever drawn through such a point may be regarded as a tangent at the point. Such values of x, y are called *singular values*, and the corresponding points on the curve are called *singular points*.

161. Determination of singular points of algebraic curves. When the equation of the curve is rationalized and cleared of fractions, let it take the form $f(x, y) = 0$.

This gives, by differentiation with regard to x, as in Art. 49,

$$\frac{\partial f}{\partial x} + \frac{\partial f}{\partial y}\frac{dy}{dx} = 0,$$

whence
$$\frac{dy}{dx} = -\frac{\frac{\partial f}{\partial x}}{\frac{\partial f}{\partial y}}. \qquad (1)$$

In order that $\frac{dy}{dx}$ may become illusory, it is therefore necessary that
$$\frac{\partial f}{\partial x} = 0, \ \frac{\partial f}{\partial y} = 0. \qquad (2)$$

Thus to determine whether a given curve $f(x, y) = 0$ has singular points, put $\dfrac{\partial f}{\partial x}$ and $\dfrac{\partial f}{\partial y}$ each equal to zero and solve these equations for x and y.

If any pair of values of x and y, so found, satisfy the equation $f(x, y) = 0$, the point thus determined is a singular point on the curve.

To determine the appearance of the curve in the vicinity of a singular point, (x_1, y_1) evaluate the indeterminate form

$$\frac{dy}{dx} = -\frac{\dfrac{\partial f}{\partial x}}{\dfrac{\partial f}{\partial y}} = \frac{0}{0},$$

by finding the limit approached continuously by the slope of the tangent when $x \doteq x_1,\ y \doteq y_1$,

thus
$$\frac{dy}{dx} = -\frac{\dfrac{d}{dx}\left(\dfrac{\partial f}{\partial x}\right)}{\dfrac{d}{dx}\left(\dfrac{\partial f}{\partial y}\right)}$$

$$= -\frac{\dfrac{\partial^2 f}{\partial x^2} + \dfrac{\partial^2 f}{\partial x\, \partial y}\dfrac{dy}{dx}}{\dfrac{\partial^2 f}{\partial x\, \partial y} + \dfrac{\partial^2 f}{\partial y^2}\dfrac{dy}{dx}}. \qquad \text{[Arts. 72, 101.}$$

This equation cleared of fractions gives, to determine the slope at (x_1, y_1), the quadratic

$$\frac{\partial^2 f}{\partial y^2}\left(\frac{dy}{dx}\right)^2 + 2\frac{\partial^2 f}{\partial x\, \partial y}\left(\frac{dy}{dx}\right) + \frac{\partial^2 f}{\partial x^2} = 0. \qquad (3)$$

This quadratic equation has in general two roots. The only exception is when simultaneously, at the point in question,

$$\frac{\partial^2 f}{\partial x^2} = 0, \quad \frac{\partial^2 f}{\partial x\, \partial y} = 0, \quad \frac{\partial^2 f}{\partial y^2} = 0, \qquad (4)$$

in which case $\dfrac{dy}{dx}$ is still indeterminate in form, and must be evaluated as before. The result of the next evaluation is a cubic in $\dfrac{dy}{dx}$, which gives three values of the slope, unless all the third partial derivatives vanish simultaneously at the point.

The geometric interpretation of the two roots of equation (3) will now be given, and similar principles will apply when the quadratic is replaced by an equation of higher degree.

The two roots of (3) are real and distinct, real and coincident, or imaginary, according as

$$H \equiv \left(\frac{\partial^2 f}{\partial x\, \partial y}\right)^2 - \frac{\partial^2 f}{\partial x^2} \frac{\partial^2 f}{\partial y^2}$$

is positive, zero, or negative. These three cases will be considered separately.

162. Multiple points. First let H be positive. Then at the point (x, y) for which $\dfrac{\partial f}{\partial x} = 0$, $\dfrac{\partial f}{\partial y} = 0$, there are two values of the slope, and hence two distinct singular tangents; thus the curve goes through the point in two directions, or, in other words, two branches of the curve cross at this point. Such a point is called a real double point of the curve, or simply a *node*. The conditions, then, to be satisfied at a node (x_1, y_1) are that

$$f(x_1, y_1) = 0, \quad \frac{\partial f}{\partial x_1} = 0, \quad \frac{\partial f}{\partial y_1} = 0,$$

and that $H(x_1, y_1)$ be positive.

Ex. Examine for singular points the curve
$$3x^2 - xy - 2y^2 + x^3 - 8y^3 = 0.$$

Here $\dfrac{\partial f}{\partial x} = 6x - y + 3x^2$, $\dfrac{\partial f}{\partial y} = -x - 4y - 2y^2$.

The values $x = 0$, $y = 0$ will satisfy these three equations, hence $(0, 0)$ is a singular point.

Since $\dfrac{\partial^2 f}{\partial x^2} = 6 + 6x = 6$ at $(0, 0)$,

$$\dfrac{\partial^2 f}{\partial x \partial y} = -1 = -1 \text{ at } (0, 0),$$

$$\dfrac{\partial^2 f}{\partial y^2} = -4 - 48y = -4 \text{ at } (0, 0),$$

hence the equation of the slope is, from (3),

$$-4\left(\dfrac{dy}{dx}\right)^2 - 2\left(\dfrac{dy}{dx}\right) + 6 = 0,$$

of which the roots are 1 and $-\frac{3}{2}$. Thus $(0, 0)$ is a double point at which the tangents have the slopes 1, $-\frac{3}{2}$.

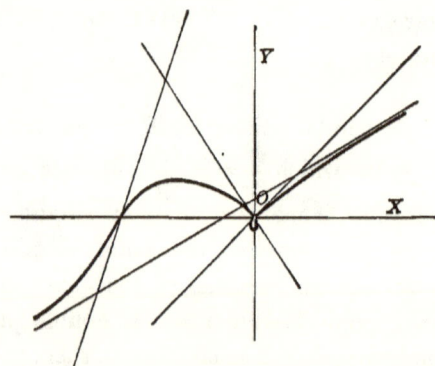

Fig. 70.

163. Cusps. Next let $H = 0$. The two tangents are then coincident, and there are two cases to consider. If the curve recedes from the tangent in both directions from the point of tangency, it is called a point of osculation; and two branches of the curve touch each other at this point. If both branches of the curve recede from the tangent in only one direction from the point of tangency, the point is called a *cusp*.

Here again there are two cases to be distinguished. If the branches recede from the point on opposite sides of the double tangent, the cusp is said to be of the first kind; if they recede on the same side, it is called a cusp of the second kind.

The method of investigation will be illustrated by a few examples.

Ex. 1. $\qquad f(x, y) = a^4 y^2 - a^2 x^4 + x^6 = 0.$

$$\frac{\partial f}{\partial x} = -4 a^2 x^3 + 6 x^5; \quad \frac{\partial f}{\partial y} = 2 a^4 y.$$

The point $(0, 0)$ will satisfy $f(x, y) = 0$, $\dfrac{\partial f}{\partial x} = 0$, $\dfrac{\partial f}{\partial y} = 0$; hence it is a singular point. Proceeding to the second derivatives,

$$\frac{\partial^2 f}{\partial x^2} = -12 a^2 x^2 + 30 x^4 = 0 \text{ at } (0, 0),$$

$$\frac{\partial^2 f}{\partial x \, \partial y} = 0,$$

$$\frac{\partial^2 f}{\partial y^2} = 2 a^4.$$

The two values of $\dfrac{\partial y}{\partial x}$ are therefore coincident, and each equal to zero. From the form of the equation, the curve is evidently symmetrical with regard to both axes; hence the point $(0, 0)$ is a point of osculation.

No part of the curve can be at a greater distance from the y-axis than $\pm a$, at which points $\dfrac{\partial y}{\partial x}$ is infinite. The maximum value of y corresponds to $x = \pm a \sqrt{\tfrac{2}{3}}$. Between $x = 0$, $x = a \sqrt{\tfrac{2}{3}}$ there is a point of inflexion (Fig. 71).

280   DIFFERENTIAL CALCULUS   [Ch. XVII.

Ex. 2. $f(x, y) = y^2 - x^3 = 0$;

$\dfrac{\partial f}{\partial x} = -3x^2, \quad \dfrac{\partial f}{\partial y} = 2y$.

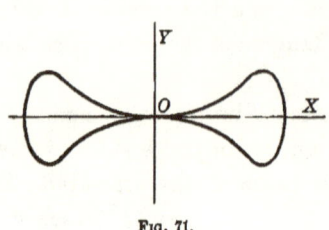

Fig. 71.

Hence the point (0, 0) is a singular point.

Again, $\dfrac{\partial^2 f}{\partial x^2} = -6x = 0$ at (0, 0);

$\dfrac{\partial^2 f}{\partial x \, \partial y} = 0; \quad \dfrac{\partial^2 f}{\partial y^2} = 2$.

Therefore the two roots of the quadratic equation defining $\dfrac{dx}{dy}$ are both equal to zero. Thus far, this case is exactly like the last one, but here no part of the curve lies to the left of the axis of y. On the right side, the curve is symmetric with regard to the x-axis. As x increases, y increases; there are no maxima nor minima, and no inflexions (Fig. 72).

Ex. 3. $f(x, y) = x^4 - 2ax^2y - axy^2 + a^2y^2 = 0$.

The point (0, 0) is a singular point, and the roots of the quadratic defining $\dfrac{dy}{dx}$ has two roots equal to zero.

Let a be positive. Solving the equation for y,

$$y = \dfrac{x^2}{a - x}\left(1 \pm \sqrt{\dfrac{x}{a}}\right).$$

When x is negative, y is imaginary; when $x = 0$, $y = 0$; when x is positive, but less than a, y has two positive values, therefore two branches

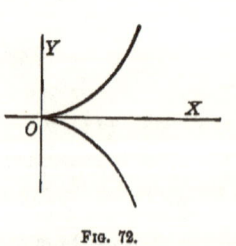
Fig. 72.

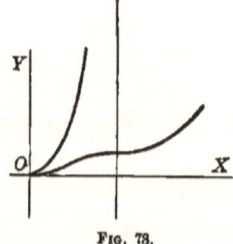
Fig. 73.

are above the x-axis. When $x = a$, one branch becomes infinite, having the asymptote $x = a$; the other branch has the ordinate $\tfrac{1}{2}a$. The origin is therefore a cusp of the second kind (Fig. 73).

164. Conjugate points. Lastly, let H be negative. In this case there are no real tangents; hence at the point in question, no points in the immediate vicinity of the given point satisfy the equation of the curve.

Such an isolated point is called a *conjugate point*.

Ex. $f(x, y) = ay^2 - x^3 + bx^2 = 0$.

Here $(0, 0)$ is a singular point of the locus, and

$$\frac{dy}{dx} = \pm \sqrt{\frac{-b}{a}},$$

both roots being imaginary if a and b have the same sign.

To show the form of the curve, solve the given equation for y,

then $\qquad y = \pm x\sqrt{\dfrac{x-b}{a}},$

hence, if a and b are positive, there are no real points on the curve between $x=0$ and $x=b$. Thus O is an isolated point (Fig. 74).

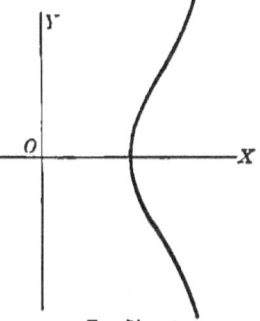

Fig. 74.

These are all the singularities that algebraic curves can have, though complicated combinations of them may appear. In all the foregoing examples, the singular point was $(0, 0)$; but for any other point, the same reasoning will apply.

Ex. $f(x, y) = x^2 + 3 y^3 - 13 y^2 - 4 x + 19y - 3 = 0$,

$$\frac{\partial f}{\partial x} = 2x - 4, \qquad \frac{\partial f}{\partial y} = 9 y^2 - 26 y + 19.$$

At the point $(2, 1)$, $f(2, 1) = 0$, $\dfrac{\partial f}{\partial x} = 0$, $\dfrac{\partial f}{\partial y} = 0$; hence $(2, 1)$ is a singular point.

Also $\dfrac{\partial^2 f}{\partial x^2} = 2$; $\dfrac{\partial^2 f}{\partial x \, \partial y} = 0$; $\dfrac{\partial^2 f}{\partial y^2} = 18y - 26, = -8$ at $(2, 1)$.

Hence $\dfrac{dy}{dx} = \pm 2$; and thus the equations of the two tangents at the node $(2, 1)$ are $y - 1 = 2(x - 2)$, $y - 1 = -2(x - 2)$.

When, at a singular point, H is negative, the point is necessarily a conjugate point, but the converse is not always true. A singular point may be a conjugate point when $H = 0$ (cf. Ex. 9).

Transcendental singularities. A curve whose equation involves a transcendental function may have a *stop-point*, at which the curve terminates abruptly (Fig. 48) or a *salient point* at which two branches of the curve meet and stop without having a common tangent. In the first case there is a discontinuity in the function; in the second, a discontinuity in the derivative.

They are usually discovered by inspection in tracing the curve.

EXERCISES

Find the multiple points, and the direction of the tangents at them, in the following curves:

1. $a^2y^2 = a^2x^2 - 4x$.
2. $x^4 - 2ay^3 - 3a^2y^2 - 2a^2x^2 + a^4 = 0$.
3. $(x^2 + y^2)^3 = 4a^2x^2y^2$.
4. $y^2 = x^4 + x^2$.

5. If $ay^2 = (x-a)^2(x-b)$, show that, when $x = a$, there is a conjugate point if a be less than b, a double point if a be greater than b, and a cusp if a be equal to b.

6. Show that the curve $y^3 = (x-a)^2(x-c)$ has a cusp of the first kind.

7. Draw the curve $x^3 + y^3 = x^2 + y^2$ in the vicinity of the origin.

8. Prove that the curve $x^4 - 2ax^2y - axy^2 + a^2y^2 = 0$ has a cusp of the second kind at the origin.

9. What change in the coefficient of x^2y in the last example will make the origin a conjugate point? Show that the tangents at this point are still real and coincident.

10. Trace the curve $x^4 + 2ax^2y - ay^3 = 0$ for points near the origin.

11. In the curve $y(1 + e^{\frac{1}{x}}) = x$, show that if $x \doteq 0$ from positive side, $\frac{y}{x} \doteq 0$; if from negative side, $\frac{y}{x} \doteq 1$; hence a discontinuity in slope, *i.e.*, a salient point.

CHAPTER XVIII

CURVE TRACING

165. Tracing a curve consists in finding its general form when its equation is given.

Three kinds of equations present themselves.

1. Cartesian equations :
   (*a*) algebraic ;
   (*b*) transcendental.
2. Polar equations.

There is no fixed method of procedure applicable to all cases. A few general suggestions for Cartesian equations will be given, and then some examples worked out in detail.

Find $\dfrac{dy}{dx}$; this will give the direction of the curve at any point, and will serve to locate maximum and minimum ordinates.

Examine for asymptotes, and construct them. Determine on which side of each asymptote the corresponding infinite branch is situated.

Find $\dfrac{d^2y}{dx^2}$; this will give the direction of bending at any point, and will determine the points of inflexion.

Examine algebraic curves for singular points, and determine whether they are nodes, cusps, or conjugate points.

If the minute configuration of a curve at any particular point is desired, it is often expeditious to transform the origin to that point, and then neglect the higher powers of x and y, as relatively unimportant. This principle will be used and discussed in some of the examples that follow.

166. Illustration. Ex. Trace the curve

$$f \equiv x^4 - y^4 + 6xy^2 = 0.$$

This curve goes through the origin; and it is symmetric with regard to the x-axis, for the equation is not changed when y is changed to $-y$; but it is not symmetric with regard to the y-axis.

Putting $x = 0$ gives $y^4 = 0$; and putting $y = 0$ gives $x^4 = 0$; hence the curve does not intersect either of the coördinate axes, except at the origin.

Since $\dfrac{\partial f}{\partial x} = 4x^3 + 6y^2$, $\dfrac{\partial f}{\partial y} = -4y^3 + 12xy$,

hence $\dfrac{dy}{dx} = -\dfrac{2x^3 + 3y^2}{(6x - 2y^2)y}$,

which becomes indeterminate only for $x = 0$, $y = 0$.

Thus the origin of a singular point of the curves. The second partial derivatives are

$$\dfrac{\partial^2 f}{\partial x^2} = 12x^2, \quad \dfrac{\partial^2 f}{\partial x \partial y} = 12y, \quad \dfrac{\partial^2 f}{\partial y^2} = 12x,$$

which all vanish at the origin; hence those of the third order must also be obtained:

$$\dfrac{\partial^3 f}{\partial x^3} = 24x; \quad \dfrac{\partial^3 f}{\partial x^2 \partial y} = 0; \quad \dfrac{\partial^3 f}{\partial x \partial y^2} = 12; \quad \dfrac{\partial^3 f}{\partial y^3} = 0.$$

The general equation determining $\dfrac{dy}{dx}$, derived similarly to that in Art. 149, is

$$\dfrac{\partial^3 f}{\partial y^3}\left(\dfrac{dy}{dx}\right)^3 + 3\dfrac{\partial^3 f}{\partial x \partial y^2}\left(\dfrac{dy}{dx}\right)^2 + 3\dfrac{\partial^2 f}{\partial x \partial y}\left(\dfrac{dy}{dx}\right) + \dfrac{\partial^3 f}{\partial x^3} = 0,$$

which becomes in this case

$$0\left(\frac{dy}{dx}\right)^3 + 36\left(\frac{dy}{dx}\right)^2 + 0\left(\frac{dy}{dx}\right) + 0 = 0.$$

Thus two values of $\frac{dy}{dx}$ are 0, and the third root is infinite; showing that the x-axis is tangent to two branches, and the y-axis to a third branch.

To obtain the form of the first branches in the vicinity of the origin it may be observed that since on these branches y is evidently an infinitesimal of a higher order than x, hence y^4 may be neglected in comparison with the other terms, and there results $x^3 = -6y^2$, as the equation of a curve approximately coinciding with the two branches in question near the origin.

This curve, and hence also the given curve, has obviously a cusp of the first kind lying to the left of the axis of y.

Similarly, in the case of the branch that is tangent to the y-axis, x^4 may be neglected, and the resulting curve is $y^2 = 6x$, which is a parabola situated on the right side of the y-axis.

Thus, the third branch is parabolic in form near the origin.

By solving for y,

$$y = \pm \sqrt{3x + \sqrt{9x^2 + x^4}},$$

in which only the positive sign is to be retained before the inner radical, as the negative sign would give imaginary values to y. Any line parallel to the y-axis will therefore meet the curve in only two points.

Again, regarding y as given, the resulting equation in x has one positive root between 0 and y because $f(0, y)$ is negative, and $f(y, y)$ is positive, and similarly one negative root numerically greater than y; the others being imaginary. Thus no branch of the curve crosses the lines $x = \pm y$, except at the origin.

To examine for asymptotes. Put $y = mx + b$, then

$$x^4 - (mx+b)^4 + 6x(mx+b)^2 = 0;$$

i.e., $(1-m^4)x^4 + (-4m^3b + 6m^2)x^3 + (-6m^2b^2 + 12mb)x^2$
$\qquad + (-4mb^3 + 6b^3)x = b^4.$

Let $\qquad 1 - m^4 = 0 \quad$ and $\quad -4m^3b + 6m^2 = 0,$

then $\qquad\qquad\qquad m = \pm 1, \qquad b = \pm \tfrac{3}{2},$

thus $\qquad\qquad\qquad y = x + \tfrac{3}{2}, \quad y = -x - \tfrac{3}{2}$

are the asymptotes. The other two asymptotes are imaginary.

To find the finite points in which this asymptote cuts the curve, put $m = 1$, $b = \tfrac{3}{2}$ in the above equation for x; it then becomes

$$0 \cdot x^4 + 0 \cdot x^3 + \tfrac{9}{2} x^2 + 0 \cdot x - \tfrac{81}{16} = 0,$$

of which the four roots are

$$\infty, \ \infty, \ +\tfrac{3}{4}\sqrt{2}, \ -\tfrac{3}{4}\sqrt{2};$$

hence the approximate values of the finite roots are ± 1.06.

The manner in which the infinite branches approach their asymptotes is best shown by the method of expansion, in which y is expressed in a series of descending powers of x.

Write the equation in the form

$$y^2 = 3x + \sqrt{9x^2 + x^4};$$

then $\qquad\qquad \dfrac{y^2}{x^2} = \dfrac{3}{x} + \left(1 + \dfrac{9}{x^2}\right)^{\frac{1}{2}}$

$$= \dfrac{3}{x} + 1 + \dfrac{9}{2x^2} - \cdots$$

$$\frac{y}{x} = \pm \left(1 + \frac{3}{x} + \frac{9}{2x^2} - \cdots \right)^{\frac{1}{2}}$$

$$= \pm \left[1 + \frac{1}{2}\left(\frac{3}{x} + \frac{9}{2x^2} \cdots\right) - \frac{1}{8}\left(\frac{3}{x} + \frac{9}{2x^2} \cdots\right)^2 \cdots \right]$$

$$= \pm \left[1 + \frac{3}{2x} + \frac{9}{4x^2} - \frac{9}{8x^2} \cdots \right],$$

$$\frac{y}{x} = \pm \left(1 + \frac{3}{2x} + \frac{9}{8x^2} \cdots \right).$$

Hence $\qquad y = \pm \left(x + \frac{3}{2} + \frac{9}{8x} \cdots \right).$

This verifies the equations of the asymptotes already found; and, moreover, the sign of the third term shows that the curve is above the first asymptote for large positive

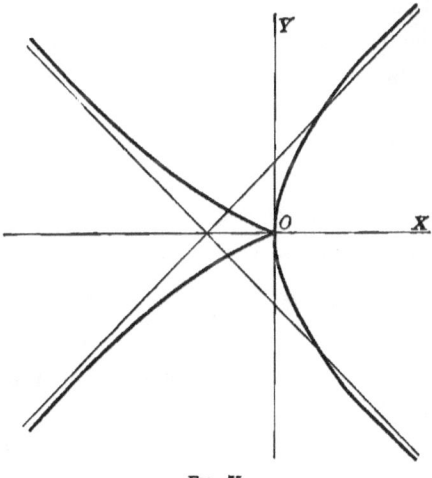

Fig. 75.

values of x, and below it for large negative values. On the other hand, the curve is below the second asymptote for large positive values of x, and above it for large negative values.

The form of the second derivative $\frac{d^2y}{dx^2}$ is too complicated to be of practical use in determining the direction of bending.

Since each infinite branch is convex to its asymptote for large values of x, hence on the upper right hand branch the concavity is ultimately upwards. Near the origin the concavity is downwards, hence there must be a point of inflexion on this branch, and also on the branch symmetrical to it.

On the left hand branches there are no points of inflexion, for if there were one on either branch there would be two on that branch, and it would then be possible to draw a line cutting the given fourth degree curve in more than four points.

167. Form of a curve near the origin. In the above example, in the vicinity of the origin, the curve approaches the form of an ordinary parabola on one side of the y-axis (which is the tangent at its vertex), and has a cusp of the first kind on the other side, the axis of x being the cuspidal tangent.

In the first case x was neglected in comparison with y, since $\lim_{x \doteq 0} \frac{x}{y} = 0$; while in the second case y is neglected in comparison with x, since $\lim_{y \doteq 0} \frac{y}{x} = 0$.

In many cases it is not so obvious which terms can be rejected, especially when the lowest terms in the expression are of high degree.

Before proceeding to the more difficult curves, a few elementary type forms will be given. The branches of every algebraic curve approximate to combinations of these forms in the vicinity of any assigned point as origin.

1. Trace the curve $y^2 = x$.

Here $\dfrac{dy}{dx} = \pm \dfrac{1}{2 x^{\frac{1}{2}}}$,

$\dfrac{d^2y}{dx^2} = \mp \dfrac{1}{4 x^{\frac{3}{2}}}$;

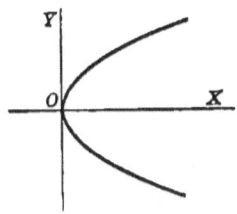

Fig. 76.

hence the slope is infinite at the origin, and diminishes to zero at infinity, showing that the curve becomes more and more horizontal; the bending is negative on the upper branch, and positive on the lower.

2. The curve $y = x^2$.

Here $\dfrac{dy}{dx} = 2x$,

$\dfrac{d^2y}{dx^2} = 2$,

hence the slope is zero at the origin, and becomes infinite at infinity, showing that the curve becomes more and more vertical; the bending is always positive.

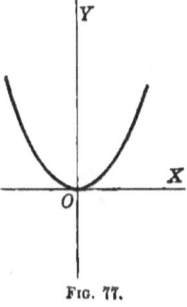

Fig. 77.

3. The curve $y = x^3$.

In this case $\dfrac{dy}{dx} = 3x^2$,

$\dfrac{d^2y}{dx^2} = 6x$,

hence the slope is zero at the origin, is elsewhere always positive, and becomes infinite at infinity. The bending changes sign where the curve passes through the origin, the x-axis being the inflexional tangent.

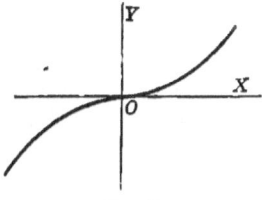

Fig. 78.

4. Show the form of the curve $y^2 = x^3$.

Here $y = \pm x^{\frac{3}{2}}$, $\dfrac{dy}{dx} = \pm \dfrac{3}{2} x^{\frac{1}{2}}$, $\dfrac{d^2y}{dx^2} = \pm \dfrac{3}{4} x^{-\frac{1}{2}}$.

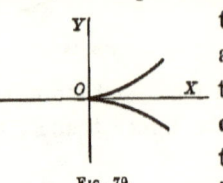

Fig. 79.

The curve is symmetrical with regard to the x-axis; and since the slope is zero at the origin, the axis of x is tangent to the upper and the lower branch. Since a negative value of x makes y imaginary, the curve does not extend to the left of the origin, hence there is a cusp of the first kind at this point. The slope increases numerically to infinity when x becomes infinite, and the bending is always positive on the upper branch, and negative on the lower. This curve is called the *semicubical* parabola because the ordinate is proportional to the square root of the cube of the abscissa.

In each of these fundamental types, if the sign of either member of the equation be changed, the curve is simply turned over, and if x and y be interchanged, the curve is revolved through 90 degrees.

Some more complicated cases will now be taken up. The following general principles will be of use.

I. When the equation of an algebraic curve is rationalized and cleared of fractions, if the constant term be absent the origin is on the curve; and the terms of the first degree, equated to zero, give the equation of the tangent at the origin.

II. If the constant term and terms of the first degree be absent, the origin is a double point; and the terms of the second degree, equated to zero, give the equation of the pair of nodal tangents.

III. If all the terms below the third degree be absent, the origin is a triple point; and the terms of the third degree, equated to zero, furnish the equation of the three tangents at the multiple point. Similarly, in general.

For, let the equation be of the form

$$f(x, y) \equiv ax + by + (cx^2 + dxy + cy^2) + \cdots = 0, \quad (1)$$

then the tangent at the origin will be represented by the equation

$$y = \left(\frac{dy}{dx}\right)x, \quad (2)$$

in which $\dfrac{dy}{dx}$ is to be obtained from the relation

$$\frac{\partial f}{\partial x} + \frac{\partial f}{\partial y} \cdot \frac{dy}{dx} = 0; \quad (3)$$

hence, eliminating $\dfrac{dy}{dx}$ between the last two equations, the equation of the tangent at the origin becomes

$$x\frac{\partial f}{\partial x} + y\frac{\partial f}{\partial y} = 0; \quad (4)$$

but at the point (0, 0),

$$\frac{\partial f}{\partial x} = a, \quad \frac{\partial f}{\partial y} = b, \quad (5)$$

hence the equation of the tangent at this point is

$$ax + by = 0. \quad (6)$$

Again, if the constants a, b be zero, the expression for $\dfrac{dy}{dx}$, given by (3), is indeterminate, and the slope at the origin is to be obtained from the quadratic (Art. 161),

$$\frac{\partial^2 f}{\partial y^2}\left(\frac{dy}{dx}\right)^2 + 2\frac{\partial^2 f}{\partial x\, \partial y}\left(\frac{dy}{dx}\right) + \frac{\partial^2 f}{\partial x^2} = 0; \quad (7)$$

hence eliminating $\dfrac{dy}{dx}$ between (2) and (7), there results

$$\dfrac{\partial^2 f}{\partial y^2} \cdot y^2 + 2\dfrac{\partial^2 f}{\partial x\,\partial y} \cdot xy + \dfrac{\partial^2 f}{\partial x^2} \cdot x^2 = 0, \qquad (8)$$

which is then the equation of the pair of tangents at the origin; but at the point (0, 0),

$$\dfrac{\partial^2 f}{\partial x^2} = 2\,c, \quad \dfrac{\partial^2 f}{\partial x\,\partial y} = d, \quad \dfrac{\partial^2 f}{\partial y^2} = 2\,e\,;$$

hence the equation of the pair of tangents at the origin is

$$cx^2 + d\,xy + cy^2 = 0 \qquad (9)$$

Similarly proceed in general.

168. Another proof. The equation that gives the abscissas of the intersections of the line $y = mx$ with the given curve is

$$(a + bm)x + (c + dm + em^2)x^2 + \cdots = 0.$$

There will be two intersections at the origin if $a + bm = 0$, that is, if $m = -\dfrac{a}{b}$. Hence the tangent at the origin is $y = -\dfrac{a}{b}x.$

Again, if $a = 0$, $b = 0$, every line through the origin will meet the curve in two coincident points; and in this case the origin will be a double point. If, moreover, m be so taken that $c + dm + em^2 = 0$, the line $y = mx$ will meet the curve in three coincident points at the origin; hence the equation of the pair of nodal tangents is to be found by eliminating m between $y = mx$ and $c + dm + em^2 = 0$, and is therefore $cx^2 + d\,xy + ey^2 = 0$; and so on.

169. Illustration. Oblique branch through origin. Expansion of y in ascending powers of x. Given the equation

$$y - 2x + 3x^2 + 4xy - 5y^2 + 6x^3 + y^3 = 0,$$

to expand y in ascending powers of x, and thence to trace the locus in the vicinity of the origin.

The first approximation to the value of y, obtained by omitting terms of order x^2, is $y = 2x$, which is the equation of the tangent, and gives the direction of the curve at the origin. In approaching the origin along the curve, the variables x and y are infinitesimals of the same order, and their ratio $\dfrac{y}{x} \doteq 2$.

To obtain the second approximation to the value of y, test for the order and value of the infinitesimal $y - 2x$, by comparing it with x^2, thus

$$\frac{y-2x}{x^2} = -3 - 4 \cdot \frac{y}{x} + 5\left(\frac{y}{x}\right)^2 - 6x - \left(\frac{y}{x}\right)^2 \cdot y,$$

$$\doteq 9 \text{ when } x \doteq 0,\ y \doteq 0,\ \frac{y}{x} \doteq 2,$$

hence $y = 2x + 9x^2$, with an error above the second order of smallness.

Since the second-order term $9x^2$ is positive, the curve is situated above the tangent $y = 2x$ on both sides of the origin. If desired, the third-order term can be obtained by substituting $2x + 9x^2$ for y in the second and third order terms of the given equation, and collecting the coefficient of x^3. The third approximation is then $y = 2x + 9x^2 + 130x^3$. This shows that the curve is above the parabola $y = 2x + 9x^2$ on the right of the origin, and below it on the left. These two curves have contact of the second order at the origin.

(2) Trace in the vicinity of the origin the curve

$$-2x^2 + xy + y^2 + x^3 - 2y^3 + x^4 - 2x^3y = 0.$$

Here the origin is a double point, at which the tangent, obtained by factoring $-2x^2 + xy + y^2 = 0$, are $y - x = 0$,

$y + 2x = 0$; hence on one branch $\dfrac{y}{x} \doteq 1$, and on the other $\dfrac{y}{x} \doteq -2$; thus the branches are oblique, and on each branch x and y are infinitesimals of the same order.

The second approximation to the equation of each branch is to be obtained by taking account of the third-order terms in the given equation, thus

$$(y - x)(y + 2x) = -x^3 + 2y^3;$$

then, on the first branch the comparison of $y - x$ with x^2 gives

$$\frac{y - x}{x^2} = \frac{-1 + 2\left(\dfrac{y}{x}\right)^3}{2 + \dfrac{y}{x}} \doteq \frac{1}{3}, \qquad \left[\dfrac{y}{x} \doteq 1\right.$$

hence the branch has the approximate equation $y = x + \tfrac{1}{3}x^2$, which shows that it lies above the tangent $y = x$ on both sides of the origin.

The third approximation, obtained by writing the given equation in the form

$$y - x = \frac{-x^3 + 2y^3 - x^4 + 2x^3y}{y + 2x},$$

substituting for y the second approximation and dividing as far as x^3, is $y = x + \tfrac{1}{3}x^2 + \tfrac{35}{27}x^3$; which shows that the first branch is above the parabola $y = x + \tfrac{1}{3}x^2$ on the right, and below it on the left of the origin.

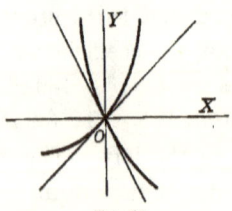

Fig 80.

On the second branch the comparison of $y + 2x$ with x^2 gives

$$\frac{y + 2x}{x^2} = \frac{-1 + 2\left(\dfrac{y}{x}\right)^3}{\dfrac{y}{x} - 1} \doteq \frac{17}{3}, \qquad \left[\dfrac{y}{x} \doteq -2\right.$$

hence its approximate equation is $y = -2x + \tfrac{17}{3}x^2$. The third approximation is $y = -2x + \tfrac{17}{3}x^2 - 38x^3$.

Both branches are shown in Fig. 80.

170. Branches touching either axis. Trace, near the origin, the curve
$$xy^2 + x^3y - y^4 - 2x^5 = 0.$$

The y-axis is a single tangent, and the x-axis is a double one; thus the origin is a triple point.

To determine the form of the curve near the origin, the method of Art. 169 will not apply, as x, y are not infinitesimals of the same order on either branch. Here a method of trial will be employed. Suppose the terms xy^2 and y^4 are of the same order on one branch, then x and y^2 are of the same order, $i.e.$, y is of the same order as $x^{\frac{1}{2}}$, hence the terms in the given equation are of the respective orders

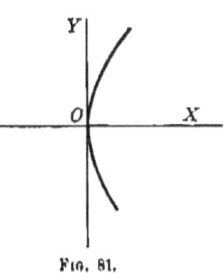

Fig. 81.

$$2,\ 3\tfrac{1}{2},\ 2,\ 5;$$

thus the terms selected are of the lowest order, and are therefore the controlling ones near the origin, showing that there is a branch having the approximate equation $xy^2 - y^4 = 0$. Removing the factor y^2, the equation of this branch is $x - y^2 = 0$; and the next term in its equation is given by

$$x - y^2 = \frac{2x^5 - x^3y}{y^2} = -y^5 + \cdots;$$

hence the branch is situated to the left of the parabola $x - y^2 = 0$ above the x-axis, and to the right below that axis.

Next suppose there is a branch for which y^4 and $2x^5$ are infinitesimals of the same order; then y has the same order as $x^{\frac{5}{4}}$, and the four terms have the orders

$$3\tfrac{1}{2}, \quad 4\tfrac{1}{4}, \quad 5, \quad 5;$$

hence there is no branch for which the two terms selected are the controlling ones.

Once more, suppose there is a branch on which xy^2 and x^3y are of the same order; then dividing by xy, it follows that y is of the same order as x^2, and the orders in x of the four terms are

$$5, \quad 5, \quad 8, \quad 5.$$

Therefore there is a branch on which the first, second, and fourth are the controlling terms, and its approximate equation is

$$xy^2 + x^3y - 2x^5 = 0,$$

which reduces to $\quad y^2 + x^2y - 2x^4 = 0,$

i.e., $\quad (y + 2x^2)(y - x^2) = 0.$

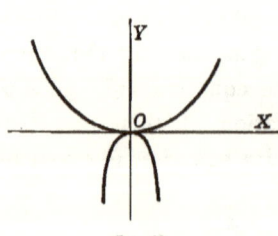

Fig. 82.

Hence the part of the curve in question consists of the two parabolas $y = x^2$, $y = -2x^2$.

Writing the given equation in the form

$$x(y - x^2)(y + 2x) = y^4,$$

the first of these branches has the equation

$$y - x^2 = \frac{y^4}{x(y + 2x^2)} = \text{(approx.)} \frac{x^8}{x(x^2 + 2x^2)} = \tfrac{1}{3}x^5;$$

hence the curve gets steeper than the approximate parabola on one side and flatter on the other. Similarly, the approximate equation of the other branch is

$$y + 2x^2 = \tfrac{16}{3} x^5.$$

Combining these two sets of results, the form of the curve in the vicinity of the origin is as given in Fig. 83.

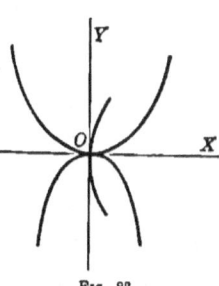

Fig. 83.

171. Two branches oblique; a third touching x-axis. Trace, in the vicinity of the origin, the curve

$$x^4 + x^2 y - y^3 = 0.$$

Since there are no terms below the third degree, the origin is a triple point, and the three tangents represented by the equation

$$x^2 y - y^3 = y(x - y)(x + y) = 0$$

have the separate equations

$$y = 0, \quad y = x, \quad y = -x.$$

To show roughly, without resorting to expansion, how the curve is related to these three lines, write its equation in the form

$$y(y^2 - x^2) = x^4.$$

First consider points near the origin on the branch that touches the line $y = 0$. Here $\lim\limits_{x \doteq 0} \dfrac{y}{x} = 0$, hence y is infinitesimal as to x, and the factor $(y^2 - x^2)$ is negative, but the term on the right is positive, hence the other factor on the left, y, is negative; thus the curve is below the line $y = 0$ on both sides of the y-axis.

298   *DIFFERENTIAL CALCULUS*   [Ch. XVIII.

Next consider points on the branch that touches the line $y = x$. When x is positive, y is positive, hence the factor $y^2 - x^2$ is also positive; thus y is greater than x, and the curve lies above the tangent $y = x$ in the first quarter. In the third quarter both x and y are negative, hence $y^2 - x^2$ is negative, and y is numerically less than x; thus the curve is above the tangent $y = x$ in the third quarter.

Lastly, consider points on the branch that touches the line $y = -x$. Here again $(y^2 - x^2)$ has the same sign as y, hence in the second quarter y is numerically greater than x, and the curve is above the tangent; but in the fourth quarter y is numerically less than x, and the curve is again above the tangent.

The position of the three branches can, however, be ascertained with greater accuracy from their approximate equations, obtained by the method of expansion: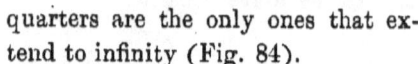

$$y = x + \tfrac{1}{2}x^2 - \tfrac{5}{8}x^3 + \cdots; \quad y = -x + \tfrac{1}{2}x^2 + \tfrac{5}{8}x^3 \cdots.$$

The form of the infinite branches will be considered later, and it will appear that the branches in the first and second quarters are the only ones that extend to infinity (Fig. 84).

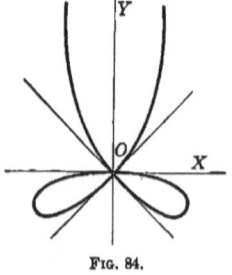

Fig. 84.

Form of curve in vicinity of point (a, b). The form in the vicinity of any point (a, b) can be found by first transforming to (a, b) as origin, and proceeding as in Arts. 169–171. This is equivalent to expanding the given function $f(x, y)$ in powers of $x - a$, $y - b$; and then expressing the small number $y - b$ in ascending powers of the small number $x - a$.

Remark on expansion of implicit functions. The methods

of Arts. 169–171 are often practically useful in purely algebraic operations. They may evidently be applied to any implicit relation between x and y, when the object is to express either variable explicitly in terms of the other, in the vicinity of two given corresponding values ($x = a$, $y = b$), to any required degree of approximation.

EXERCISES

Examine the following curves in the vicinity of the origin and find two or three terms of the expansion of y in ascending powers of x:

1. $y^2 = x^4 + x^5$;
2. $y^2 = 2x^2y + x^5$;
3. $y^2 = 2x^2 + x^3$;
4. $y^3 = x^3 - x^4$;
5. $y^2 - x^2 = x^4 - 2x^2y - x^8$;
6. $9xy = 2x^3 + 2y^3$.

7. In No. 6, for the vicinity of the point (1, 2), expand $y - 2$ in ascending powers of $x - 1$.

172. Approximation to form of infinite branches. It has been shown that in the vicinity of the origin, the approximate form of each branch of the curve could be obtained by examining the different suppositions regarding the relative orders of the infinitesimals x and y, in consequence of which two or more terms of the equation should become infinitesimals of like order, and compared with these all the other terms should be of higher order, and could therefore be neglected in writing down the first approximation to the value of y in ascending powers of x.

On a similar principle the approximate form of each branch of the curve at great distances from the origin can be obtained by examining the different suppositions, regarding the relative orders of the infinites x and y, in consequence of which two or more terms should become infinites of the same order, and in comparison with these all the other terms should be infinites of lower order, and could therefore be neglected in writing down the first approximation to the value of y in descending powers of x.

Ex. 1. Take the curve traced near the origin in Art. 171,

$$y(x^2 - y^2) + x^4 = 0.$$

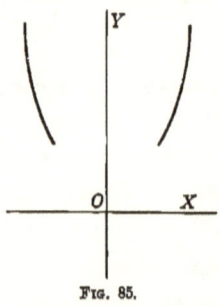

Fig. 85.

Here the supposition that y is an infinite of the same order as $x^{\frac{4}{3}}$ makes the terms y^3 and x^4 infinites of order 4, and the term yx^2 an infinite of order $3\frac{1}{3}$; thus there is an infinite branch which has the approximate equation $y^3 = x^4$, and hence passes out of the field in the manner shown in Fig. 85.

Again, the supposition that y is of the same order as x^2 makes yx^2 and x^4 infinites of order 4, and the term y^3 an infinite of order 6, which cannot be neglected in comparison. Hence there is no infinite branch on which y is approximately proportional to x^2.

Similarly the third supposition does not correspond to an infinite branch.

Ex. 2. Consider the curve that was traced near the origin in Art. 170,

$$xy^2 + x^3y - y^4 - 2x^5 = 0.$$

The supposition that y is of the same order as $x^{\frac{5}{4}}$ makes the four terms of the orders $3\frac{1}{2}$, $4\frac{1}{4}$, 5, 5; hence there is an infinite branch whose approximate equation is $y^4 + 2x^5 = 0$. The form of the curve is shown in Fig. 86.

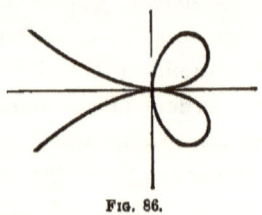

Fig. 86.

Hyperbolic and parabolic branches. Expansion in descending series. On an infinite branch the coördinates x and y may behave as follows:

1. One of the coördinates may approach a finite number, and the other become infinite. The branch has then a horizontal or vertical asymptote (Art. 130), and is thus a horizontal or vertical hyperbolic branch (Fig. 40).

2. The coördinates may become infinites of the same order. Then $\dfrac{y}{x} \doteq m$, a finite number; hence there is, in general, an oblique asymptote, that is, the infinite branch is, in general, an oblique hyperbolic branch. [In a special case it is an oblique parabolic branch. See Exs. 4, 5.]

3. The coördinates may become infinites of different orders. If y is an infinite of higher order than x, there is a parabolic branch on which the tangent tends to become vertical (Figs. 85, 86), — called a vertical parabolic branch. If y is of lower order than x, there is a horizontal parabolic branch (Fig. 81).

The test for Case (1) has been given in Art. 131. Case (2) comes under the head of oblique asymptotes; but it may be conveniently treated along with Case (3) by the method of this Article. The test for Case (2) is to observe whether there are two or more terms of the highest degree in x and y. If so, the supposition that y is of the same order as x makes these the controlling terms.

Ex. 3. Test for oblique infinite branches the fourth-degree curve

$$x^4 + x^3 y + 2 y^3 = x^3 + 3 x^2 - y^3.$$

Here there are two fourth-degree terms, and the supposition that y and x are infinites of the first order makes these the controlling terms; hence there is an oblique branch on which $\dfrac{y}{x} \doteq -1$. On putting the first approximation, $y = -x$, in the third-degree terms, and dividing by x^3, there results, for the second approximation, $y = -x + 3$; and this, when used in the same way, gives the third approximation, $y = -x + 3 - \dfrac{16}{x} + \cdots$. Thus the branch is hyperbolic, having the oblique asymptote $y = -x + 3$. There is also a pair of vertical parabolic branches, on which

$$\frac{y}{x^{\frac{3}{2}}} \doteq \pm \frac{1}{\sqrt{2}}.$$

Ex. 4. Test in the same way the cubic curve

$$(y - 2x)^2(y + x) = 5x^2 + xy + 5y^2 + 3x - 7y + 8,$$

in which the terms of the third degree have a square factor.

Corresponding to the single factor $y + x$ there is, as before, a hyperbolic branch whose equation is $y = -x + 1 - \dfrac{7}{9x} + \cdots$.

The equation of the branch corresponding to the square factor is given by

$$(y - 2x)^2 = \frac{5x^2 + xy + 5y^2 + 3x - 7y + 8}{y + x}$$

The first approximation, $y = 2x$, used on the right, gives $(y - 2x)^2 = 9x$; and the second approximation, $y = 2x \pm 3x^{\frac{1}{2}}$, used in the same way, gives $y = 2x \pm 3x^{\frac{1}{2}} + 2 + \cdots$ in descending powers of $x^{\frac{1}{2}}$. Hence the branch on which $\dfrac{y}{x} \doteq 2$ has no linear asymptote. The curvilinear asymptote of lowest degree is the second-degree curve $(y - 2x - 2)^2 = 9x$. There are thus two oblique parabolic branches.

Ex. 5. When the terms of highest degree have a factor repeated three times, show that the corresponding expansion of y descends in powers of $x^{\frac{1}{3}}$, and that the asymptote of lowest degree is a cubic curve.

The method of successive approximation in descending series can also be used in Case (3), when once the first approximation has been obtained by the method of comparison given above.

Ex. 6. In the curve of Fig. 85, the first approximation is $y = x^{\frac{4}{3}}$. Substituting this on the right of $y^3 = x^4 + x^2 y$, and taking cube root, the second approximation is $y = x^{\frac{4}{3}} + \frac{1}{3} x^{\frac{2}{3}} + \cdots$, in descending powers of $x^{\frac{2}{3}}$. For the third term it is easiest to let $y = x^{\frac{4}{3}} + \frac{1}{3} x^{\frac{2}{3}} + p$, substitute in $y^3 = x^4 + x^2 y$, and determine p so that the coefficients shall be equal as far down as x^2; then $p = -\frac{1}{27}$; and $y = x^{\frac{4}{3}} + \frac{1}{3} x^{\frac{2}{3}} - \frac{1}{27} + \cdots$.

Ex. 7. In Ex. 5 of Art. 171 show that on two branches the controlling terms are $y^2 + 2x^2 y - x^4$; that is, $[y - x^2(\sqrt{2} - 1)][y + x^2(\sqrt{2} + 1)]$, and that the equations of these branches are

$$y = (\pm \sqrt{2} - 1)x^2 \mp \frac{x}{2\sqrt{2}} \pm \frac{7}{16\sqrt{2}} + \cdots.$$

Remark on implicit functions. By this method, when any implicit algebraic relation between x and y is given, the value of either variable for large values of the other can be computed by descending series, with small relative error.

Transcendental Cartesian Curves. A number of figures of important transcendental curves are shown on pp. 237–238, and in A. G., p. 211 ff. They are traced by tabulating y, with assistance from $\dfrac{dy}{dx}$, $\dfrac{d^2y}{dx^2}$.

EXERCISES

Apply the methods of this article to the equations at end of Art. 171. In No. 6 compute the value of y when $x = 20$, by descending series.

173. Curve tracing : polar coördinates. In tracing curves defined by polar equations there is, as in the case of Cartesian equations, no fixed method of procedure.

If, as usually happens, the equation can be solved for ρ, successive values may be given to θ, and the corresponding values of ρ computed and tabulated. In constructing the table it is useful to record at what values of θ the radius vector ρ has turning values. The critical values of θ for this purpose are, as usual, determined from the equations $\dfrac{d\rho}{d\theta} = 0$, $\dfrac{d\rho}{d\theta} = \infty$; and are separately tested by observing whether the derivative changes its sign.

Next should be noted the asymptotic directions, which correspond to those values of θ, if any, at which ρ passes through an infinite value. The distance of the asymptote from the infinite radius vector is given in magnitude and sign by the corresponding value of the polar subtangent $\sigma = \rho^2 \dfrac{d\theta}{d\rho}$. Again, if ρ tends to a definite limit, as θ becomes infinite, there is a circular asymptote.

304   *DIFFERENTIAL CALCULUS*   [Ch. XVIII.

On sketching the path of the point (ρ, θ) from the tabulated record, greater accuracy in the direction of the curve at any point may be obtained by computing the slope of the tangent line to the radial direction, from the relation $\tan \psi = \rho \dfrac{d\theta}{d\rho}$. The same result can be achieved by tabulating the values of σ, the polar subtangent, not merely for asymptotic directions, but for other convenient values of θ. Assistance in tracing the curve may sometimes be obtained by noticing whether there are any axes of symmetry.

Ex. 1. Trace the locus of the equation

$$\rho = a(\sec 2\theta + \tan 2\theta) = a\frac{1 + \sin 2\theta}{\cos 2\theta}.$$

Here $\dfrac{d\rho}{d\theta} = 2\,a\sec 2\theta\,(\tan 2\theta + \sec 2\theta) = 2\,a\dfrac{1+\sin 2\theta}{\cos^2 2\theta}$,

$\tan \psi = \rho \dfrac{d\theta}{d\rho} = \tfrac{1}{2}\cos 2\theta$,

$\sigma = \rho \tan \psi = \tfrac{1}{2}a(1 + \sin 2\theta)$,

whence the following table may be constructed, and the locus traced:

θ	ρ	$\dfrac{d\rho}{d\theta}$	$\tan \psi$	σ	
0	a	$2\,a$	.5	$.5\,a$	
$\tfrac{1}{8}\pi$	$3.7\,a$	$14.8\,a$	.25	$.93\,a$	
$\tfrac{1}{4}\pi$	$\pm\infty$	$\pm\infty$	± 0	a	asymptote
$\tfrac{3}{8}\pi$	$-3.7\,a$	$14.8\,a$	$-.25$	$.93\,a$	
$\tfrac{1}{2}\pi$	$-a$	$2\,a$	$-.5$	$.5\,a$	
$\tfrac{5}{8}\pi$	∓ 0	a	∓ 0	± 0	
π	a	$2\,a$	.5	$.5\,a$	
$\tfrac{7}{8}\pi$	$3.7\,a$	$14.8\,a$	.25	$.93\,a$	
$\tfrac{5}{4}\pi$	$\pm\infty$	$\pm\infty$	± 0	a	asymptote
...	...	...	...	...	

As θ increases from 0 to $\frac{1}{4}\pi$, the tracing point P moves from A to B, and as θ increases by successive steps to $\frac{1}{2}\pi$, $\frac{3}{4}\pi$, π, $\frac{5}{4}\pi$, $\frac{3}{2}\pi$, $\frac{7}{4}\pi$, 2π, P

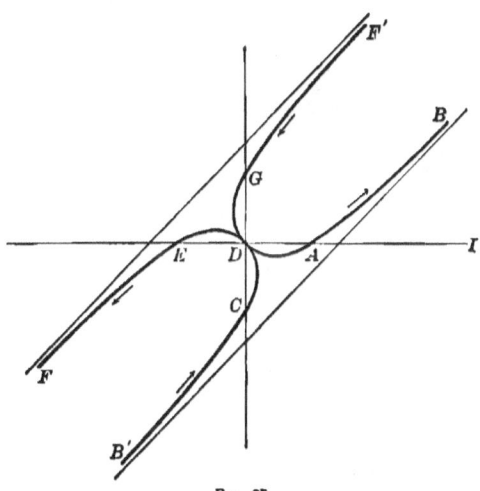

Fig. 87.

moves respectively from B' to C, C to D, D to E, E to F, F' to G, G to D, D to A. The lines $\theta = \frac{1}{4}\pi$, $\theta = \frac{3}{4}\pi$ are axes of symmetry.

Ex. 2. Transform to polar coördinates the equation

$$(x^2 + y^2)^2 - 2\,ay(x^2 + y^2) = a^2 x^2,$$

and then trace the curve.

On putting $x = \rho \cos \theta$, $y = \rho \sin \theta$, dividing by ρ^2, and solving the quadratic for ρ, there results

$$\rho = a(\sin \theta \pm 1).$$

First take the upper sign; then

$$\frac{d\rho}{d\theta} = a \cos \theta, \quad \tan \psi = \frac{\rho\, d\theta}{d\rho} = \frac{\sin \theta + 1}{\cos \theta},$$

and the following table is easily computed.

The figure is shown in Art. 108. If the lower sign be taken, the same curve will be traced in a different order. The line $\theta = \frac{1}{2}\pi$ is an axis of symmetry.

θ	ρ	$\dfrac{d\rho}{d\theta}$	$\tan \psi$	
0	a	a	1	
$\tfrac{1}{4}\pi$	$1.7\,a$	$.71\,a$	2.41	
$\tfrac{1}{3}\pi$	$1.87\,a$	$.5\,a$	3.73	
$\tfrac{1}{2}\pi$	$2\,a$	$^+_-0$	$^+_-\infty$	ρ a maximum, $\psi = \tfrac{1}{2}\pi$.
$\tfrac{3}{4}\pi$	$1.7\,a$	$-.71\,a$	-2.41	
π	a	$-a$	-1	
$\tfrac{5}{4}\pi$	$.29\,a$	$-.71\,a$	$-.41$	
$\tfrac{3}{2}\pi$	$^+_-0$	$^-_+0$	$^-_+0$	ρ a minimum, $\psi = 0$, origin a cusp.
$\tfrac{7}{4}\pi$	$.29\,a$	$.71\,a$	$.41$	
$\tfrac{11}{6}\pi$	$.5\,a$	$.87\,a$	$.58$	
2π	a	a	1	

EXERCISES

Trace the following curves:

1. $y = \dfrac{x}{1 + x^2}$.
2. $y^4 = 2\,x^2 + x^3$.
3. $y^2 = x^4 + x^5$.
4. $y^2(x - a) = (x + a)\,x^2$.
5. $x^2y^2 = a^2(x^2 - y^2)$.

6. Show that the curve $y^2 = x^3 - x^4$ has two branches which are both tangent to the axis of x at the origin.

7. Determine the direction of the curve $y^3 = x^2(x - a)$ at each point where it crosses the axis of x.

8. Trace the curve $y^3 - axy - b^2x = 0$ in the neighborhood of the origin.

9. Show that the curve $\rho = 1 + \sin 5\theta$ consists of 5 equal loops.

10. Trace the curve $\rho \cos 2\theta = a$.

Find its asymptotes and lines of symmetry.

11. Trace the curve $\rho = a(\tan \theta - 1)$.

12. Trace the curves $y = e^x$, $y = e^{\frac{1}{x}}$, $\dfrac{y-1}{y} = e^{\frac{1}{x}}$, $\dfrac{y-1}{y} = e^{\frac{1}{x-3}}$.

13. Find the points of inflexion of the curve $y = e^{-x^2}$.

This curve is known as the *probability* curve (Fig. 49).

14. $\rho = a + \sin \tfrac{3}{2}\theta$.   15. $\rho = a(1 - \tan \theta)$.

CHAPTER XIX

ENVELOPES

174. Family of curves. The equation of a curve,

$$f(x, y) = 0,$$

usually involves, besides the variables x and y, certain coefficients that serve to fix the size, shape, and position of the curve. The coefficients are called constants with reference to the variables x and y, but it has been seen in previous chapters that they may take different values in different problems, while the form of the equation is preserved. Let a be one of these "constants"; then if a be given a series of numerical values, and if the locus of the equation be traced, corresponding to each special value of a, a series of curves is obtained, all having the same general character, but differing somewhat from each other in size, shape, or position. A system of curves so obtained by letting one of the constant letters assume different numerical values in the fixed form of equation $f(x, y) = 0$ is called a *family* of curves.

Thus if h, k be fixed, and p be arbitrary, the equation $(y-k)^2 = 2p(x-h)$ represents a family of parabolas, having the same vertex (h, k), and the same axis $y = k$, but having an arbitrary latus rectum. Again, if k be the arbitrary constant, this equation represents a family of parabolas having parallel axes, the same latus rectum, and having their vertices on the same line $x = h$.

The presence of an arbitrary constant a in the equation of a curve is indicated in functional notation by writing the equation in the form $f(x, y, a) = 0$. The quantity a, which is constant for the same curve but different for different curves, is called the *parameter* of the family. The equations of two neighboring members are then written

$$f(x, y, a) = 0, \quad f(x, y, a + h) = 0,$$

in which h is a small increment of a; and these consecutive curves can be brought as near to coincidence as desired by diminishing h.

175. Envelope of a family of curves. The locus of the points of ultimate intersection of consecutive curves of a family, when these curves approach nearer and nearer to coincidence, is called the *envelope* of the family.

Let $\quad f(x, y, a) = 0, \quad f(x, y, a + h) = 0 \quad$ (1)

be two curves of the family. By the theorem of mean value (Art. 66)

$$f(x, y, a+h) = f(x, y, a) + h \frac{\partial f}{\partial a}(x, y, a+\theta h), \quad (2) \ [0 < \theta < 1$$

but the points common to the two curves satisfy equations (1), and therefore also satisfy $\dfrac{\partial f}{\partial a}(x, y, a + \theta h) = 0$. Hence, in the limit, when $h \doteq 0$, it follows that $\dfrac{\partial f}{\partial a}(x, y, a) = 0$ is the equation of a curve passing through the ultimate intersection of the curve $f(x, y, a) = 0$ with its consecutive curve. This determines for any assigned value of a a definite point of ultimate intersection on the corresponding member of the family. The locus of all such points is then to be obtained by eliminating the parameter a between the equations

$$f(x, y, a) = 0, \ \frac{\partial f}{\partial a}(x, y, a) = 0.$$

The resulting equation is of the form $F(x, y) = 0$, and represents the fixed envelope of the family.

176. The envelope touches every curve of the family.

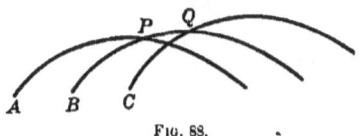

Fig. 88.

I. *Geometrical proof.* Let A, B, C be three consecutive curves of the family.; let A, B intersect in P; B, C intersect in Q. When P, Q approach coincidence, PQ will be the direction of the tangent to the envelope at P; but since P, Q are two points on B that approach coincidence, hence PQ is also the direction of the tangent to B at P; thus B and the envelope have a common tangent at P; similarly for every curve of the family.

II. *More rigorous analytical proof.* Let $\frac{\partial}{\partial \alpha} f(x, y, \alpha) = 0$ be solved for α, in the form $\alpha = \phi(x, y)$; then the equation of the envelope is

$$f(x, y, \phi(x, y)) = 0.$$

Equating the total x-derivative to zero,

$$\frac{\partial f}{\partial x} + \frac{\partial f}{\partial y}\frac{dy}{dx} + \frac{\partial f}{\partial \phi}\left(\frac{\partial \phi}{\partial x} + \frac{\partial \phi}{\partial y}\frac{dy}{dx}\right) = 0;$$

but $\frac{\partial f}{\partial \phi} = \frac{\partial f}{\partial \alpha} = 0$, hence the slope of the tangent to the envelope at the point (x, y) is given by

$$\frac{\partial f}{\partial x} + \frac{\partial f}{\partial y}\frac{dy}{dx} = 0;$$

but the same equation defines the direction of the tangent to the curve $f(x, y, \alpha) = 0$ at the same point. Therefore a point of ultimate intersection on any member of the family is a point of contact of this curve with the envelope.

Ex. Find the envelope of the family of lines

$$y = mx + \frac{p}{m}, \tag{1}$$

obtained by varying m.

Differentiate (1) as to m,

$$0 = x - \frac{p}{m^2}. \tag{2}$$

Hence the line (1) meets its consecutive line where it meets (2). To eliminate m, solve (2) for m, substitute in (1), and square; then the locus of the ultimate intersections is the fixed parabola

$$y^2 = 4\,px.$$

177. Envelope of normals of a given curve. The evolute (Art. 158) was defined as the locus of the center of curvature. The center of curvature was shown to be the point of intersection of consecutive normals (Art. 151), hence by Art. 174, the envelope of the normals is the evolute.

Ex. Find the envelope of the normals to the parabola $y^2 = 4\,px$.

The equation of the normal at (x_1, y_1) is

$$y - y_1 = -\frac{y_1}{2\,p}(x - x_1),$$

or, eliminating x_1 by means of the equation $y_1^2 = 4\,px_1$,

$$y - y_1 = \frac{y_1^3}{8\,p^2} - \frac{xy_1}{2\,p}. \tag{1}$$

The envelope of this line, when y_1 takes all values, is required.

Differentiating as to y_1,

$$-1 = \frac{3\,y_1^2}{8\,p^2} - \frac{x}{2\,p},$$

$$y_1^2 = \frac{4\,p}{3}(x - 2\,p).$$

Substituting this value for y_1 in (1), the result,

$$27\,py^2 = 4\,(x - 2\,p)^3,$$

is the equation of the required evolute.

178. Two parameters, one equation of condition. In many cases a family of curves may have two parameters which are connected by an equation. For instance, the equation of the normal to a given curve contains two parameters, x_1, y_1, which are connected by the equation of the curve. In such cases one parameter may be eliminated by means of the given relation, and the other treated as before.

When the elimination is difficult to perform, both equations may be differentiated as to one parameter α, regarding the other parameter β as a function of α, giving four equations from which α, β, and $\dfrac{d\beta}{d\alpha}$ may be eliminated, and the resulting equation will be that of the desired envelope.

Ex. 1. Find the envelope of the line
$$\frac{x}{a} + \frac{y}{b} = 1,$$
the sum of its intercepts remaining constant.

The two equations are
$$\frac{x}{a} + \frac{y}{b} = 1,$$
$$a + b = c.$$

Differentiate as to a,
$$\frac{-x}{a^2} - \frac{y}{b^2}\frac{db}{da} = 0,$$
$$1 + \frac{db}{da} = 0;$$

eliminate $\dfrac{db}{da}$, then $\dfrac{x}{a^2} = \dfrac{y}{b^2}$, therefore

$$\frac{\frac{x}{a}}{a} = \frac{\frac{y}{b}}{b} = \frac{\frac{x}{a} + \frac{y}{b}}{a+b} = \frac{1}{c}, \quad \text{hence} \quad a = \sqrt{cx}, \quad b = \sqrt{cy};$$

therefore
$$\sqrt{x} + \sqrt{y} = \sqrt{c}$$
is the equation of the desired envelope.

Ex. 2. Find the envelope of the family of coaxial ellipses having a constant area.

Here
$$\frac{x^2}{a^2} + \frac{y^2}{b^2} = 1;$$

$$ab = k^2.$$

For symmetry, regard a and b as functions of a single parameter t, then

$$\frac{x^2}{a^3} da + \frac{y^2}{b^3} db = 0,$$

$$b\,da + a\,db = 0;$$

hence
$$\frac{x^2}{a^2} = \frac{y^2}{b^2} = \frac{1}{2},$$

$$a = \pm x\sqrt{2}, \quad b = \pm y\sqrt{2},$$

and the envelope is the pair of rectangular hyperbolas $xy = \pm \tfrac{1}{2} k^2$.

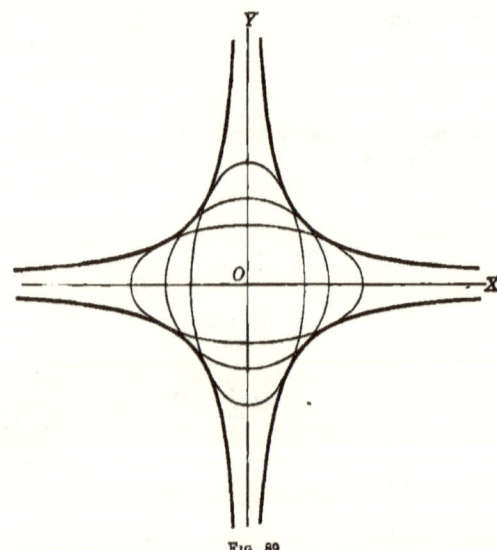

Fig. 89

Note. A family of curves with a single parameter may have no envelope; i.e., consecutive curves may not intersect; e.g., the family of concentric circles $x^2 + y^2 = r^2$, obtained by giving r all possible values.

EXERCISES

1. Find the envelope of the parabolas $y^2 = \dfrac{a^2}{c}(x - a)$, a being a parameter.

2. A straight line of fixed length a moves with its extremities in two rectangular axes; find its envelope.

3. Ellipses are described with common centers and axes, and having the sum of the semi-axes equal to c. Find their envelope.

4. Find the envelope of the straight lines having the product of their intercepts on the coördinate axes equal to k^2.

5. Find the envelope of the lines $y - \beta = m(x - a) + r\sqrt{1 + m^2}$, m being a variable parameter.

6. What is the evolute of the envelope of Ex. 5?

7. Circles are described on successive double ordinates of a parabola as diameters; show that their envelope is an equal parabola. Find what part of this system of circles does not admit of an envelope.

8. Show that the envelope of
$$f(x, y)a^2 + \phi(x, y)a + \psi(x, y) = 0$$
is
$$4f(x, y) \cdot \psi(x, y) - [\phi(x, y)]^2 = 0.$$

9. Find the curve whose tangents have the general equation
$$y = mx \pm \sqrt{am^2 + bm + c}.$$

10. Prove that the circles which pass through the origin and have their centers on the equilateral hyperbola
$$x^2 - y^2 = a^2$$
envelop the lemniscata $(x^2 + y^2)^2 = 4a^2(x^2 - y^2)$.

11. If in Ex. 10 the locus of the centers of circles passing through the origin be the parabola $y^2 = 4ax$, the envelope will be the cissoid
$$y^2(x + 2a) - x^3 = 0.$$

12. Show that a family of curves having two independent parameters has no envelope.

APPENDIX

NOTE A (P. 29)

Let $y = f(x)$ be a function which is continuous and increasing from $x = a$ to $x = b$; and let $f(a) = A, f(b) = B$.

Let the inverse function be written $x = \phi(y)$; then it is proposed to show that $\phi(y)$ is a continuous function of y from $y = A$ to $y = B$.

Let h be any assigned positive number numerically less than $b - a$; then, since $f(x)$ is an increasing function,

$$f(x + h) - f(x)$$

preserves its sign unchanged when x and $x + h$ both lie anywhere in the interval from a to b. Let the smallest value that this difference can take for the assigned value of h be

$$f(x + h) - f(x) = k; \qquad (1)$$

then when $\qquad x' > x + h,$

$$f(x') - f(x) > k. \qquad (2)$$

Consequently, if

$$f(x') - f(x) < k, \qquad (3)$$

then x' must be less than $x + h$,

i.e., $\qquad x' - x < h; \qquad (4)$

or, putting $\quad f(x) = y, f(x') = y', x = \phi(y), x' = \phi(y'),$

(3) and (4) may be written thus:

if $\qquad y' - y < k, \qquad (5)$

then $\qquad \phi(y') - \phi(y) < h,$ the assigned number. $\qquad (6)$

Hence, $\phi(y)$ is a continuous function of y throughout the stated interval.

A similar proof applies to intervals in which $f(x)$ is a decreasing function. Hence:

For every interval in which a function is continuous there exists an interval in which the inverse function is continuous.

NOTE B (P. 60)

To prove $\lim\limits_{m \doteq \infty}\left(1 + \dfrac{1}{m}\right)^m = e$, a finite determinate number when m is not restricted to be a positive integer.

The case in which m is a positive integer was considered on p. 59; and it will be seen by applying the tests of Art. 54 that the series there given is unconditionally convergent.

When m is not a positive integer, let it be supposed to lie between the two positive integers p and $p+1$, that is,

$$p < m < p+1;$$

then
$$\frac{1}{p} > \frac{1}{m},$$

$$1 + \frac{1}{p} > 1 + \frac{1}{m},$$

$$\left(1 + \frac{1}{p}\right)^p > \left(1 + \frac{1}{m}\right)^p,$$

$$\left(1 + \frac{1}{p}\right)^p \cdot \left(1 + \frac{1}{m}\right)^{m-p} > \left(1 + \frac{1}{m}\right)^m. \qquad (1)$$

Again,
$$\frac{1}{p+1} < \frac{1}{m},$$

$$1 + \frac{1}{p+1} < 1 + \frac{1}{m},$$

$$\left(1 + \frac{1}{p+1}\right)^{p+1} < \left(1 + \frac{1}{m}\right)^{p+1},$$

$$\left(1 + \frac{1}{p+1}\right)^{p+1} \cdot \left(1 + \frac{1}{m}\right)^{m-p-1} < \left(1 + \frac{1}{m}\right)^m. \qquad (2)$$

Hence, from (1) and (2),

$$\left(1+\frac{1}{p+1}\right)^{p+1}\left(1+\frac{1}{m}\right)^{m-p-1} < \left(1+\frac{1}{m}\right)^{m} < \left(1+\frac{1}{p}\right)^{p}\left(1+\frac{1}{m}\right)^{m-p}. \quad (3)$$

It will now be shown that when p, m, $p+1$ all $\doteq \infty$, the first and third members of these inequalities have the common limit e. For, since the exponents $m-p$, $m-p-1$ are finite,

$$\left(1+\frac{1}{m}\right)^{m-p} \doteq 1, \quad \left(1+\frac{1}{m}\right)^{m-p-1} \doteq 1;$$

but since p, $p+1$ are infinite positive integers,

$$\left(1+\frac{1}{p}\right)^{p} \doteq e, \quad \left(1+\frac{1}{p+1}\right)^{p+1} \doteq e; \qquad \text{[p. 59.}$$

hence e is the common limit of the first and last members of (3), and is therefore also the limit of the intermediate member,

i.e., $$\lim_{m \doteq \infty}\left(1+\frac{1}{m}\right)^{m} = e.$$

Finally, let m be any negative number, say $-p$, then

$$\left(1+\frac{1}{m}\right)^{m} = \left(1+\frac{1}{p}\right)^{-p} = \left(\frac{p-1}{p}\right)^{-p}$$

$$= \left(\frac{p}{p-1}\right)^{p} = \left(1+\frac{1}{p-1}\right)^{p}.$$

Writing k for $p-1$,

$$\left(1+\frac{1}{m}\right)^{m} = \left(1+\frac{1}{k}\right)^{k+1} = \left(1+\frac{1}{k}\right)^{k}\left(1+\frac{1}{k}\right); \quad (4)$$

but when $m \doteq -\infty$ and $k \doteq +\infty$,

$$1+\frac{1}{k} \doteq 1, \quad \left(1+\frac{1}{k}\right)^{k} \doteq e;$$

therefore, by (4),

$$\lim_{m \doteq \infty}\left(1+\frac{1}{m}\right)^{m} = e.$$

NOTE ON HYPERBOLIC FUNCTIONS

Definitions and direct inferences. For the present purpose the hyperbolic cosine and sine may be defined analytically in terms of the exponential function, as follows:

$$\cosh x = \tfrac{1}{2}(e^x + e^{-x}), \quad \sinh x = \tfrac{1}{2}(e^x - e^{-x}), \qquad (1)$$

and the hyperbolic tangent, cotangent, secant, and cosecant are then defined by the equations

$$\left. \begin{aligned} \tanh x &= \frac{\sinh x}{\cosh x}, \quad \coth x = \frac{\cosh x}{\sinh x}, \\ \operatorname{sech} x &= \frac{1}{\cosh x}, \quad \operatorname{csch} x = \frac{1}{\sinh x}. \end{aligned} \right\} \qquad (2)$$

Among the six functions there are five independent relations, so that when the numerical value of one of the functions is given, the values of the other five can be found. Four of these relations consist of the four defining equations (2). The fifth is derived from (1) by squaring and subtracting, giving

$$\cosh^2 x - \sinh^2 x = 1. \qquad (3)$$

By a combination of some of these equations other subsidiary relations may be obtained; thus, on dividing (3) successively by $\cosh^2 x$, $\sinh^2 x$, and applying (2), it follows that

$$\left. \begin{aligned} 1 - \tanh^2 x &= \operatorname{sech}^2 x, \\ \coth^2 x - 1 &= \operatorname{csch}^2 x. \end{aligned} \right\} \qquad (4)$$

Equations (2), (3), (4) will readily serve to express the value of any function in terms of any other. For example, when $\tanh x$ is given,

$$\coth x = \frac{1}{\tanh x}, \quad \operatorname{sech} x = \sqrt{1 - \tanh^2 x},$$

$$\cosh x = \frac{1}{\sqrt{1 - \tanh^2 x}}, \quad \sinh x = \frac{\tanh x}{\sqrt{1 - \tanh^2 x}}.$$

Ex. 1. From equations (1) prove

$\cosh(-x) = \cosh x, \; \sinh(-x) = -\sinh x,$

$\cosh 0 = 1, \; \sinh 0 = 0, \; \cosh \infty = \infty, \; \sinh \infty = \infty.$

Ex. 2. From equations (3), (4) show that
$$\cosh x > \sinh x, \quad \cosh x > 1, \quad \tanh x < 1.$$

Ex. 3. Prove that $\tanh x \doteq \pm 1$, when $x \doteq \pm \infty$.

Ex. 4. By direct substitution from (1) verify the addition formulas
$$\sinh(x \pm y) = \sinh x \cosh y \pm \cosh x \sinh y,$$
$$\cosh(x \pm y) = \cosh x \cosh y \pm \sinh x \sinh y;$$
and hence derive the conversion formulas
$$\cosh x + \cosh y = 2 \cosh \tfrac{1}{2}(x+y) \cosh \tfrac{1}{2}(x-y),$$
$$\sinh x + \sinh y = 2 \sinh \tfrac{1}{2}(x+y) \cosh \tfrac{1}{2}(x-y); \text{ etc.}$$

Show that the corresponding formulas for the circular functions could be verified by their exponential expressions (p. 101).

Ex. 5. Prove the identities: $\sinh 2x = 2 \sinh x \cosh x$,
$$\cosh 2x = \cosh^2 x + \sinh^2 x = 1 + 2\sinh^2 x = 2\cosh^2 x - 1.$$

Ex. 6. Prove $\cosh nx + \sinh nx = e^{nx} = (\cosh x + \sinh x)^n$.

Derivatives of hyperbolic functions. By differentiating (1),

$$\frac{d}{dx} \cosh x = \tfrac{1}{2}(e^x - e^{-x}) = \sinh x, \tag{5}$$

$$\frac{d}{dx} \sinh x = \tfrac{1}{2}(e^x + e^{-x}) = \cosh x; \tag{6}$$

hence $\quad \dfrac{d}{dx} \tanh x = \dfrac{d}{dx} \dfrac{\sinh x}{\cosh x} = \dfrac{\cosh^2 x - \sinh^2 x}{\cosh^2 x} = \operatorname{sech}^2 x. \tag{7}$

Also, $\quad \dfrac{d^2}{dx^2} \cosh x = \cosh x, \quad \dfrac{d^2}{dx^2} \sinh x = \sinh x. \tag{8}$

$$\frac{d^2}{dx^2} \cosh mx = m^2 \cosh mx, \quad \frac{d^2}{dx^2} \sinh mx = m^2 \sinh mx. \tag{9}$$

It thus appears that the functions $\sinh x$, $\cosh x$ reproduce themselves in two differentiations, just as the functions $\sin x$, $\cos x$ produce their opposites in two differentiations. In this connection it may be noted that the frequent appearance of the hyperbolic (and circular) functions in the solution of physical problems is due to the fact that they answer the question: What function has its second derivative equal to a positive (or negative) constant multiple of the function itself?

Ex. 7. Eliminate the constants by differentiation from the equation $y = A \cosh mx + B \sinh mx$, and prove $\dfrac{d^2y}{dx^2} = m^2 y$.

Ex. 8. Prove $\dfrac{d}{dx} \coth x = -\operatorname{csch}^2 x$, $\dfrac{d}{dx}\operatorname{sech} x = \operatorname{sech} x \tanh x$.

Expansions. By applying Maclaurin's theorem, using (5), (6), (8), or else by substituting the developments of e^x, e^{-x}, in (1), the following series are obtained:

$$\left.\begin{aligned}\cosh x &= 1 + \frac{x^2}{2!} + \frac{x^4}{4!} + \frac{x^6}{6!} + \cdots \\ \sinh x &= x + \frac{x^3}{3!} + \frac{x^5}{5!} + \frac{x^7}{7!} + \cdots\end{aligned}\right\} \quad (10)$$

By means of these series, which are available for all finite values of x, the numerical values of $\cosh x$, $\sinh x$ can be computed and tabulated for successive values of x.*

Derivatives of inverse hyperbolic functions.

Let $\quad y = \sinh^{-1} x$, then $x = \sinh y$;

$$dx = \cosh y \, dy = \sqrt{1 + x^2}\, dy;$$

hence $\quad \dfrac{d}{dx} \sinh^{-1} x = \dfrac{1}{\sqrt{1 + x^2}}.$ \hfill (11)

Similarly, $\quad \dfrac{d}{dx} \cosh^{-1} x = \dfrac{1}{\sqrt{x^2 - 1}}.$ \hfill (12)

Again, let $\quad y = \tanh^{-1} x$, then $x = \tanh y$,

$$dx = \operatorname{sech}^2 y \, dy = (1 - \tanh^2 y)\, dy = (1 - x^2)\, dy;$$

therefore $\quad \dfrac{d}{dx} \tanh^{-1} x = \dfrac{1}{1 - x^2}\Big]_{x<1}.$ \hfill (13)

Similarly, $\quad \dfrac{d}{dx} \coth^{-1} x = \dfrac{1}{x^2 - 1}\Big]_{x>1}.$ \hfill (14)

Ex. 9. Prove $\dfrac{d}{dx}\operatorname{sech}^{-1} x = \dfrac{d}{dx}\cosh^{-1}\dfrac{1}{x} = \dfrac{-1}{x\sqrt{1 - x^2}};$

$\dfrac{d}{dx}\operatorname{csch}^{-1} x = \dfrac{d}{dx}\sinh^{-1}\dfrac{1}{x} = \dfrac{-1}{x\sqrt{1 + x^2}}.$

---

* See Tables, p. 162, Merriman and Woodward's "Higher Mathematics."

APPENDIX

Ex. 10. Prove* $d \sinh^{-1}\dfrac{x}{a} = \dfrac{dx}{\sqrt{x^2+a^2}}$, $d \cosh^{-1}\dfrac{x}{a} = \dfrac{dx}{\sqrt{x^2-a^2}}$;

$d \tanh^{-1}\dfrac{x}{a} = \dfrac{a\,dx}{a^2-x^2}\Big]_{x<a}$, $d \coth^{-1}\dfrac{x}{a} = -\dfrac{a\,dx}{x^2-a^2}\Big]_{x<a}$.

Relation of hyperbolic functions to hyperbolic sectors.

In the circle $x^2 + y^2 = a^2$, let A be the area of the sector included between the radii drawn to the points $(a, 0)$, (x, y); and let θ be the included angle; then, by geometry,

$$2A = a^2\theta = a^2 \sin^{-1}\dfrac{y}{a} = a^2 \cos^{-1}\dfrac{x}{a}.$$

Again, it is shown in Integral Calculus by means of the derivatives in Ex. 10, that in the hyperbola $x^2 - y^2 = a^2$, if $2A'$ be the area of the sector included between the radii drawn to the points $(a, 0)$, (x, y), then $2A' = a^2 \sinh^{-1}\dfrac{y}{a} = a^2 \cosh^{-1}\dfrac{x}{a}$.

Thus the hyperbolic functions are related to hyperbolic sectors as the circular functions are related to circular (and elliptic) sectors.†

Expansions of inverse hyperbolic functions.

By the method of Art. 67,

$$\sinh^{-1}x = x - \dfrac{1}{2}\dfrac{x^3}{3} + \dfrac{1}{2}\dfrac{3}{4}\dfrac{x^5}{5} - \cdots. \qquad [-1 < x < 1] \qquad (15)$$

Another series, convergent when $x > 1$, is obtained by writing the derivative in the form

$$\dfrac{d}{dx}\sinh^{-1}x = (x^2+1)^{-\frac{1}{2}} = \dfrac{1}{x}\left(1+\dfrac{1}{x^2}\right)^{-\frac{1}{2}}$$

$$= \dfrac{1}{x}\left(1 - \dfrac{1}{2}\dfrac{1}{x^2} + \dfrac{1}{2}\dfrac{3}{4}\dfrac{1}{x^4} - \cdots\right),$$

hence, $\quad \sinh^{-1}x = A + \log x + \dfrac{1}{2}\dfrac{1}{2}\dfrac{1}{x^2} - \dfrac{1}{2}\dfrac{3}{4}\dfrac{1}{4}\dfrac{1}{x^4} + \cdots, \qquad (16)$

---

* These derivatives will be found useful in the "Integral Calculus."

† For a treatment of the hyperbolic functions from this point of view, see Merriman and Woodward.

where A is a constant, which is shown later to be equal to $\log 2$;

similarly, $\quad \cosh^{-1}x = A + \log x - \dfrac{1}{2}\dfrac{1}{2x^2} - \dfrac{1}{2}\dfrac{3}{4}\dfrac{1}{4x^4} - \cdots,\quad$ (17)

which is always available for computation, since $\cosh^{-1}x$ is a real number only when $x > 1$.

Ex. 11. Prove that $\tanh^{-1}x = x + \dfrac{1}{3}x^3 + \dfrac{1}{5}x^5 + \cdots$, and that this series is always available when $\tanh^{-1}x$ is real, *i.e.*, when $-1 < x < 1$.

Logarithmic expressions for inverse hyperbolic functions.

Let $\qquad x = \cosh y$, then $\sqrt{x^2-1} = \sinh y$,

$\qquad\qquad x + \sqrt{x^2-1} = \cosh y + \sinh y = e^y,$

hence $\qquad y = \cosh^{-1}x, = \log(x + \sqrt{x^2-1}).\qquad$ (18)

Similarly, $\qquad \sinh^{-1}x = \log(x + \sqrt{x^2+1}).\qquad$ (19)

Also $\qquad \mathrm{sech}^{-1}x = \cosh^{-1}\dfrac{1}{x} = \log\dfrac{1+\sqrt{1-x^2}}{x},\qquad$ (20)

$\qquad\qquad \mathrm{csch}^{-1}x = \sinh^{-1}\dfrac{1}{x} = \log\dfrac{1+\sqrt{1+x^2}}{x}.\qquad$ (21)

Again, let $\qquad x = \tanh y = \dfrac{e^y - e^{-y}}{e^y + e^{-y}},$

therefore $\qquad \dfrac{1+x}{1-x} = \dfrac{e^y}{e^{-y}} = e^{2y},$

i.e., $\qquad 2y = \log\dfrac{1+x}{1-x};$

hence $\qquad \tanh^{-1}x = \dfrac{1}{2}\log\dfrac{1+x}{1-x},\qquad$ (22)

and $\qquad \coth^{-1}x = \tanh^{-1}\dfrac{1}{x} = \dfrac{1}{2}\log\dfrac{x+1}{x-1}.\qquad$ (23)

Ex. 12. Show from (18), (19) that, when $x \doteq \infty$,

$\qquad \sinh^{-1}x - \log x \doteq \log 2, \quad \cosh^{-1}x - \log x \doteq \log 2,$

and hence that the constant A in (16), (17) is equal to $\log 2$.

APPENDIX

Graphs of hyperbolic functions. The student is advised to sketch the graphs of these functions from their definitions and fundamental properties. Aid is also obtained from the values of their first and second derivatives.

Ex. 13. The curve $y = \sinh x$ has an inflexion at the origin, the slope of the tangent being unity; the bending is upwards to the right and downwards to the left of the origin.

Ex. 14. The curve $y = \cosh x$ is symmetrical as to the y-axis, and has a minimum ordinate at $x = 0$.

Ex. 15. Show that the curve $y = \tanh x$ has two asymptotes $y = \pm 1$.

Ex. 16. Using the graphs, give approximate solutions of the transcendental equations, $\tanh x = x - 1$, $\cosh x = x + 2$, $\sinh x = \frac{3}{2}x$, $\cos x \cosh x = 1$.

Ex. 17. The equation of the catenary is $\frac{y}{c} = \cosh \frac{x}{c}$; show that the derivative of the arc is $\frac{ds}{dx} = \cosh \frac{x}{c}$, and hence that $\frac{s}{c} = \sinh \frac{x}{c}$.

Gudermanian function. When two variables x, y are so related that $\sec y = \cosh x$, then y is called the *Gudermanian function* of x, and is denoted by $gd\, x$. The angle whose radian measure is equal to $gd\, x$ is called the *Gudermanian angle* of x.

Ex. 18. Show that the six hyperbolic functions of x can be expressed as circular functions of $gd\, x$: *e.g.*, $\cosh x = \sec gd\, x$, $\sinh x = \tan gd\, x$, etc.

Ex. 19. The curve $y = gd\, x$ has asymptotes $y = \pm \frac{1}{2}\pi$.

Ex. 20. Prove $\frac{d}{dx} gd\, x = \operatorname{sech} x$, $\frac{d}{dx} gd^{-1} x = \sec x$.

NOTE ON INTERPOLATION BY TAYLOR'S THEOREM

Two ordinates given; to compute an intermediate ordinate. In the curve $y = f(x)$, let the ordinate at $x = a$ be y_1, and let the ordinate at $x = a + h$ be y_2. If y_1, y_2 be given numerically, it is required to compute the ordinate y at the intermediate point $x = a + \epsilon h$, where $\epsilon < 1$.

Consider the three equations,

$$y_1 = f(a), \tag{1}$$
$$y_2 = f(a+h) = f(a) + hf'(a), \quad [\text{neglecting } h^2 f''(a)] \tag{2}$$
$$y = f(a+\epsilon h) = f(a) + \epsilon h f'(a); \tag{3}$$

then from (1), (2), $hf'(a) = y_2 - y_1$; hence, by (3),

$$y = y_1 + \epsilon(y_2 - y_1). \tag{A}$$

The neglect of the term $h^2 f''(a)$ in (2) is justified either when h^2 is very small, or when $f''(a)$ is zero. The former is the case when the given ordinates are very close together. The latter is the case when $f(x)$ is of the first degree, i.e., when the locus $y = f(x)$ is a straight line; hence (A) gives accurately the ordinate of the straight line joining two given points on a curve. Thus the neglect of h^2 comes to the same thing as considering the curve straight in the vicinity of $x = a$.

Three equidistant ordinates given; to compute an intermediate ordinate. Let the ordinate at $a - h$ be y_1, at a be y_2, at $a + h$ be y_3; it is required to find the ordinate at $a + \epsilon h$, where $-1 < \epsilon < 1$. In this case, neglecting $h^3 f'''(a)$,

$$y_1 = f(a-h) = f(a) - hf'(a) + \tfrac{1}{2} h^2 f''(a), \tag{4}$$
$$y_2 = f(a), \tag{5}$$
$$y_3 = f(a+h) = f(a) + hf'(a) + \tfrac{1}{2} h^2 f''(a), \tag{6}$$
$$y = f(a+\epsilon h) = f(a) + \epsilon h f'(a) + \tfrac{1}{2} \epsilon^2 h^2 f''(a). \tag{7}$$

From (4), (5), (6), $\quad y_1 - 2y_2 + y_3 = h^2 f''(a); \tag{8}$

and from (4), (6), $\quad y_3 - y_1 = 2hf'(a); \tag{9}$

hence, substituting for $hf'(a)$, $h^2 f''(a)$, in (7),

$$y = y_2 + \tfrac{1}{2}\epsilon(y_3 - y_1) + \tfrac{1}{2}\epsilon^2(y_1 - 2y_2 + y_3). \tag{B}$$

This interpolation formula gives accurately the ordinate of a parabola whose equation is of the form $y = A + Bx + Cx^2$; for then $f'''(x) = 0$. Hence the neglect of h^3 is the same in its effect as considering the curve $y = f(x)$ parabolic in the vicinity of $x = a$.

Four equidistant ordinates given; to compute an intermediate ordinate. In this case let $2h$ be the common distance, then the problem may be stated thus: given the numerical values of

$$y_1 = f(x - 3h), \quad y_2 = f(x - h),$$
$$y_3 = f(x + h), \quad y_4 = f(x + 3h),$$

to compute the value of $y = f(x + \epsilon h)$, where $-3 < \epsilon < 3$. [It is better if the given ordinates can be arranged such that $-1 < \epsilon < 1$.]

By a slight extension of the former method, neglecting $h^4 f^{iv}(a)$, and denoting $y_1 - 2y_2 + y_3$ by q, $y_1 - 3y_2 + 3y_3 - y_4$ by r, the result is found to be

$$y = \tfrac{1}{2}(y_1 + y_4) - \tfrac{9}{8}(2q - r) + \tfrac{1}{2}\epsilon(y_3 - y_2 - \tfrac{1}{24}r)$$
$$+ \tfrac{1}{8}\epsilon^2(2q - r) - \tfrac{1}{48}\epsilon^3 r. \quad (C)$$

This gives accurately the ordinate of a cubic parabola whose equation is of the form $y = A + Bx + Cx^2 + Dx^3$, which may be regarded as a third approximation to the equation of the given curve in the vicinity of $x = a$.

ANSWERS

Art. 16

1. $2x - 2$. 2. $6x - 4$. 3. $-\dfrac{1}{4x^2}$. 4. $4x^3 - \dfrac{6}{x^3}$.

Art. 18

1. $15y^2 - 2$. 2. $14t - 4 - 33t^2$. 3. $12u^2 - 2$. 4. $4x - 5$.

Art. 20

2. Inc. from $-\infty$ to $\tfrac{1}{3}$; dec. from $\tfrac{1}{3}$ to 1; inc. from 1 to $+\infty$; $\tfrac{1}{3}$ and 1.
3. Two. $+1$ at $x = \tfrac{1}{2} \pm \sqrt{\tfrac{5}{12}}$; -1 at $x = \tfrac{1}{2} \pm \sqrt{\tfrac{1}{12}}$.
4. $\pm \tan^{-1} \tfrac{4}{27}$.

Art. 21

2. $72 x^5 - 20\tfrac{1}{4} x^2$.

Art. 28

1. $\dfrac{c}{2\sqrt{x}}$.

2. $\dfrac{a^2}{(a^2 - x^2)^{\frac{3}{2}}}$.

3. $\dfrac{m(b+x) + n(a+x)}{(a+x)^{m+1}(b+x)^{n+1}}$.

4. $\dfrac{\sqrt{a}\,(\sqrt{x} - \sqrt{a})}{2\sqrt{x}\,\sqrt{x+a}\,(\sqrt{a} + \sqrt{x})}$.

5. $\dfrac{1}{\sqrt{1-x^2}\,(1-x)}$.

6. $\dfrac{ny}{x\sqrt{1-x^2}}$.

7. $\dfrac{4a^{\frac{1}{2}} + 3x^{\frac{1}{2}}}{4\sqrt{x}\,\sqrt{a^{\frac{1}{2}} + x^{\frac{1}{2}}}}$.

8. $(x-b)(x-c)^2 + (x-a)(x-c)^2 + 2(x-a)(x-b)(x-c)$.

9. $\dfrac{2x^3 - 4x}{(1-x^2)^{\frac{3}{2}}(1+x^2)^{\frac{1}{2}}}$.

10. $\dfrac{-2nx^{n-1}}{(x^n - 1)^2}$.

13. $\dfrac{-b^2 x}{a^2 y} = -\dfrac{bx}{a\sqrt{a^2 - x^2}}$.

Art. 33

1. $\dfrac{8x - 7}{4x^2 - 7x + 2}$.

2. $4 e^{4x+5}$.

3. $-\dfrac{e^{\frac{1}{1+x}}}{(1+x)^2}$.

4. $x^{n-1} + nx^{n-1} \log x$.

5. $\dfrac{1}{2\sqrt{x}} - \dfrac{1}{2\sqrt{x}(\sqrt{x}+1)}$

6. $1 - y^2$.

7. $\dfrac{e^x}{(1+e^x)^2}$.

8. $y - 3x^2 e^x$.

9. $\dfrac{1}{x \log x}$.

10. $x^x e^{x^x}(1 + \log x)$.

11. $2 x a^{x^2} \log a$.

12. $\dfrac{1}{\log a} \cdot \dfrac{12x\sqrt{2+x} - 1}{2(3x^2 - \sqrt{2+x})\sqrt{2+x}}$.

13. $x^x(1 + \log x)$.

14. $\dfrac{-(x-1)^{\frac{2}{3}}(7x^2 + 30x - 97)}{12(x-2)^{\frac{7}{4}}(x-3)^{\frac{10}{3}}}$.

Art. 40

1. $10 x \cos 5 x^2$.
2. $14 \sin 7x \cos 7x$.
3. $\tan^4 x - 1$.
4. $2 \cos 2x$.
5. $-\dfrac{1}{x^2} \log a \cdot a^{\frac{1}{x}} \cdot \sec^2(a^{\frac{1}{x}})$.
6. $\sec x$.

7. $-2 \csc 2x$.
8. $n \sin^{n-1} x \sin (n+1)x$.
9. $\tan^2 x$.
10. $\cos 2u \dfrac{du}{dx}$.
11. $y\left(\dfrac{\sin x}{x} + \cos x \log x\right)$.
12. $\cos (\sin u) \cos u \dfrac{du}{dx}$.

Art. 47

1. $\sin^{-1} x + \dfrac{x}{\sqrt{1-x^2}}$.
2. $\sec^2 x \tan^{-1} x + \dfrac{\tan x}{1+x^2}$.
3. $\dfrac{1}{\sqrt{1-2x-x^2}}$.
4. $\dfrac{2(1-x^2)}{1+6x^2+x^4}$.
5. $\dfrac{e^x}{1+e^{2x}}$.
6. $-\dfrac{1}{x\sqrt{1-\log^2 x}}$.
7. $\dfrac{-1}{\cos^{-1} x \cdot \sqrt{1-x^2}}$.
8. $\dfrac{4x}{\sqrt{1-4x^4}}$.
9. $\dfrac{1}{\sqrt{2ax-x^2}}$.

10. $\dfrac{-2x}{1+(x^2-5)^2}$.
11. $\dfrac{1}{\sqrt{1-x^2}}$.
12. $\dfrac{-1}{2x\sqrt{9x-1}}$.
13. $\dfrac{\cos \log x}{x}$.
14. $\cot x$.
15. $\dfrac{x \cos x^2}{\sqrt{\sin x^2}}$.
16. $\dfrac{e^{\cos \frac{1}{x}}}{x^2} \sin \dfrac{1}{x}$.
17. $\dfrac{e^{\tan^{-1} x}}{1+x^2}$.
18. $-\cos(\cos x) \sin x$.

ANSWERS 329

Page 72. Miscellaneous Exercises

1. $\dfrac{e^x - e^{-x}}{e^x + e^{-x}}$

2. $n\left(\dfrac{x}{n}\right)^{nx}\left(1 + \log \dfrac{x}{n}\right)$.

3. $-2 \csc 2x$.

4. $(2x - 5)e^{2x} + 4(x+1)e^x + 1$.

5. $\dfrac{e^x(1 - x) - 1}{(e^x - 1)^2}$.

6. $-2xe^{-x^2}\cos x - e^{-x^2}\sin x$.

7. $x^{\sin^{-1} 2x}\left(\dfrac{\sin^{-1} 2x}{x} + \dfrac{2\log x}{\sqrt{1 - 4x^2}}\right)$.

8. $\dfrac{-2}{\sqrt{1 - x^4}}$

9. $\dfrac{4 \cos(2\log x^2 - 7)}{x}$.

10. 1.

11. $2\tan t + e^{\sec t}\sec t \tan t$.

12. For all values.

15. x, y are determined from $a^2 y = \pm b^2 x$ and equation of curve.

16. $\dfrac{dy}{dx} = -\dfrac{2xy^3 - 5x^4y^2 + 12x}{3x^2y^2 - 2y - 5}$.

Art. 51

1. $12(x^2 - 2x + 1)$.

2. $4[(x - 2)e^{2x} + (x + 2)e^x]$.

3. 0.

4. $\dfrac{6}{x}$.

5. $\dfrac{-8(e^x - e^{-x})}{(e^x + e^{-x})^3}$.

6. $2x \log x + \dfrac{2x}{3} - \dfrac{10}{x^3}$.

7. $6 \tan^4 x$.

10. $\dfrac{\cos x}{(1 - \sin x)^2}$.

11. $\dfrac{4a^3}{(a^2 + x^2)^2}$.

13. $\dfrac{3a^2}{4\sqrt{x}(x - a)^{\frac{5}{2}}}$

14. $\dfrac{(n - 1)!}{x}$

15. $(-1)^{n-1}(n - 1)!\left[\dfrac{1}{(1 - x)^n} - \dfrac{1}{(1 + x)^n}\right]$.

16. $\dfrac{4!}{(1 - x)^5}$.

17. $\dfrac{d^2y}{dx^2} = y^2(\tan 2x + 1)$.

18. $\dfrac{3 - y}{(2 - y)^3} e^{2y}$.

19. $\dfrac{-2(5 + 8y^2 + 3y^4)}{y^9}$.

20. $\dfrac{-24x}{(1 + 2y)^5}$.

21. $\dfrac{(e^{x+y} - 1)(x - y)}{(e^y + 1)^3}$.

22. $y\dfrac{(2 - y)e^y + 2x}{(e^y + x)^3}$.

23. $\dfrac{-2a^3xy}{(y^2 - ax)^3}$.

24. $(-1)^n 2^{n-1} n!\left\{\dfrac{1}{(2x - 1)^{n+1}} - \dfrac{1}{(2x + 1)^{n+1}}\right\}$.

25. $\dfrac{(-1)^{n-1}(n - 1)!}{2}\left\{\dfrac{1}{(x + a)^n} - \dfrac{1}{(x - a)^n}\right\}$.

26. $e^x(x+n)$.

27. $a^{n-2}e^{ax}(a^2x^2 + 2\,anx + n(n-1))$.

28. $\dfrac{2(-1)^{n-1}(n-3)!}{x^{n-2}}$.

29. $\dfrac{n!}{(x+1)^{n+1}}$.

30. $-8\,e^x\cos x$.

Art. 57

3. $e^2 + e^2(x-2) + \dfrac{e^2}{2!}(x-2)^2 + \cdots$

4. $3 + 4(x-1) + (x-1)^2 + (x-1)^3$.

5. $-6 + 4(y-3) + 3(y-3)^2$.

Art. 64

3. $R_3\left(\dfrac{\pi}{6}\right) = \dfrac{\pi^3}{3!\cdot 6^3}\sin\left(\dfrac{\theta\pi}{6} + \dfrac{3}{2}\pi\right)$.

4. $ax - \dfrac{a^3x^3}{3!} + \dfrac{a^5x^5}{7} - \dfrac{a^7x^7}{7!}\sin\left(a\theta x + \dfrac{7\pi}{2}\right)$.

Art. 69

2. $\dfrac{1}{a(\sqrt{3}+\sqrt{2})}$.

Art. 70

1. 1.   2. 0. ∞

Art. 72

3. $\tfrac{2}{3}$.   4. 4.   5. $\tfrac{2}{3}$.   7. $\tfrac{1}{4}$.   8. -4.

Art. 75

1. 1.
2. 2.
3. 4.
4. $-\tfrac{1}{5}$.
5. $\tfrac{1}{10}$.
6. 2.
7. $\dfrac{1}{n}$.
8. $-\tfrac{1}{5}$.
9. $\log\dfrac{a}{b}$.
10. 2.

11. -1.
12. $\dfrac{m}{3}$.
13. $\dfrac{2}{m^2}$.
14. $\dfrac{\sqrt{2}}{\sqrt{3}+1}$.
15. $-\tfrac{3}{2}$.
16. 2.
17. $-\tfrac{2}{3}$.
18. 1.
19. $\dfrac{-1}{2\sqrt{2}}$.

20. $-\tfrac{1}{2}$.
21. -3.
22. m.
23. 0.
24. ∞.
25. 5.
26. $\dfrac{-4}{\pi}$.
27. 0.
28. 0.
29. 0.

30. $\dfrac{2}{\pi}$.
31. $\tfrac{1}{3}$.
32. 0.
33. $\tfrac{1}{2}$.
34. $\tfrac{7}{8}$.
35. $\dfrac{\pi^2}{8}$.
37. a.
38. $\dfrac{4a}{\pi}$.
39. $\tfrac{128}{81}$.

Art. 77

1. $e^{-\tfrac{1}{2}}$.

2. $e^{-\tfrac{a^2}{2\beta^2}}$.

3. 1.

4. $\dfrac{1}{e}$.

5. $\dfrac{1}{e}$.

6. 1.

7. 1.

8. $a_1 a_2 \cdots a_n$.

ANSWERS

Art. 86

1. -5, min.; -7, max. 2. 2, min.; $\frac{8}{5}$, max. 3. $1, -\frac{1}{3}$, min.; $\frac{1}{2}$, max.
7. $\frac{1}{e}$, min. 8. e, max. 9. $\frac{a}{4}$, min.
10. $(n + \frac{1}{4})\pi$, max.; $(n - \frac{1}{4})\pi$, min.; n any integer.
12. $2n\pi$, min., and also $\tan^{-1} \pm \sqrt{\frac{2}{3}}$ for angles in 2 and 3 quarter. $(2n+1)\pi$, $\tan^{-1} \pm \sqrt{\frac{2}{3}}$, 1 and 4 quarter, max.
13. $(2n + 1)\frac{\pi}{2}$, min.; $\sin^{-1}\frac{1}{4}$, max. 14. No max. nor min.
15. Min. for value of x which satisfies the equation $(x - 1)e^{3x} - 2x^2 = 0$. It is between $1\frac{1}{11}$ and $1\frac{1}{12}$.

Art. 87

6. $\frac{4}{27}\pi a^2 h = \frac{4}{9}$ vol. of cone. 9. Isosceles.
7. $\frac{4}{3}\sqrt{\frac{2a^3P}{3}}$. 10. Isosceles.
8. Half that of paraboloid. 11. Radius of circle is equal to height of rectangle.
12. Breadth $= \frac{a}{2\sqrt{3}}$, thickness $= \frac{a}{\sqrt{6}}$.
15. Height is equal to diameter of base.
16. $\sqrt{b(c+b)}$. 17. $(a^{\frac{2}{3}} + b^{\frac{2}{3}})^{\frac{3}{2}}$.
18. Sine of semi-vertical angle is $\frac{1}{3}$. 19. $\sqrt{2}$. 20. $\frac{3a}{4}$.
22. One mile from destination. 23. $30°$.
24. Side parallel to wall is double the other. 25. $\frac{3a}{2}$.

Art. 90

3. $.00145$. 5. $32\sqrt{3}$. 6. $2ab$. 7. ± 2.
8. $\frac{\pi a^2}{4}$, a being diameter. 9. 2. 10. 1 and 5.

Art. 96

1. $dz = \frac{\partial f(x, y)}{\partial x} \cdot dx + \frac{\partial f(x, y)}{\partial y} \cdot \phi'(x)dx$; substitute $\phi(x)$ for y.
2. $\frac{dz}{dx} = -2\sqrt{1 - 4x^2}\left(\frac{1 - 4x^2 + 16x^4}{1 - 4x^2}\right)$.

Art. 97

3. $\frac{dy}{dx} = -\frac{ax + hy + g}{hx + by + f}$. 4. $\frac{dy}{dx} = \frac{x^3}{y^3}$.

Art. 99

1. $\dfrac{du}{dx} = \dfrac{u}{a} + e^{\frac{x}{u}}\left[\dfrac{a}{x} - \dfrac{a^2}{x^2}\cos\left(\dfrac{a\log\frac{x}{a}}{x}\right)\left(\dfrac{1 - \log\frac{x}{a}}{1}\right)\right].$

Art. 105

1. $x + y + x^2 + xy + \tfrac{1}{3}x^3 - \tfrac{1}{2}xy^2 - \tfrac{1}{3}y^3 \cdots$.
2. $x^4 + 4x^3y + 6x^2y^2 + 4xy^3 + y^4$.
3. $\sqrt{x}\tan y + \dfrac{h\tan y}{2\sqrt{x}} + k\sqrt{x}\sec^2 y + \dfrac{1}{2}\left(\dfrac{-h^2}{4(x+\ell h)^{\frac{3}{2}}}\tan(y + \theta k)\right.$
$\left. + \dfrac{hk\sec^2(y+\theta k)}{\sqrt{x+\ell h}} + k^2 2\sqrt{x+\theta h}\sec^2(y+\theta k)\tan(y+\theta k)\right).$
4. $-25 + (x-2)^2 + (y-3)^2 + (z-1)^2$.

Art. 107

4. $x = 0,\ y = 0$.   7. The three parts are equal.
8. $6a^2$; the parallelopiped is a cube.
9. $\dfrac{x}{a} = \dfrac{y}{b} = \dfrac{1 \pm \sqrt{1 + a^2 + b^2}}{a^2 + b^2}$; with the upper sign there is a maximum; with the lower, a minimum.
10. $x = y = \dfrac{3\pi}{2}$, min.; $x = y = \dfrac{\pi}{6}$, max.
11. $\dfrac{a^3}{27}$.   12. $ax = by = cz = \sqrt[3]{\dfrac{abcg}{n}}$.

Art. 108

3. $\dfrac{a^2b^2}{a^2+b^2}$.   5. Min., $x = \pm a$.
6. Max., $x = a\sqrt[3]{2}$; min., $x = 0$.   7. Max. for $x = \dfrac{3a}{2}$.

Art. 117

2. $y = \pm\dfrac{3\sqrt{3}x}{8} - \dfrac{a}{8},\ y = \mp\dfrac{8x}{3\sqrt{3}} + \dfrac{41}{36}a$.
3. (a) $xx_1 + yy_1 = c^2$;   (b) $xy_1 + x_1y = 2k^2$;
(c) $(2x_1y_1 + y_1^2)x + (x_1^2 + 2x_1y_1)y = 3a^3$;   (d) $y - y_1 = \cot x_1(x - x_1)$.
5. $x = 2 \pm \sqrt{\tfrac{1}{17}}$.   8. $P = \sqrt[3]{ax_1y_1},\ a$.
15. At $(0, 0)$, $90°$. At the other points, $45°$.   16. $\dfrac{2ax - x^2}{3a - x}$.   17. $\dfrac{y^2}{a}$.

Art. 120

3. Subtangent $= \rho\tan\alpha$; subnormal $= \rho\cot\alpha$.   4. $90°$.
5. $\psi = \theta$; $\phi = 2\theta$.

Art. 126

1. $\sqrt{\dfrac{a^2 - e^2x^2}{a^2 - x^2}}$, $\dfrac{b}{a}\sqrt{a^2 - x^2}$, $\dfrac{2\pi b}{a}\sqrt{a^2 - e^2x^2}$, $\dfrac{\pi b^2}{a^2}(a^2 - x^2)$.

2. $\sqrt{\dfrac{a+x}{x}}$, $2\sqrt{ax}$, $4\pi\sqrt{a^2 + ax}$, $4\pi ax$.   5. a, $\dfrac{a^2 \cos^2\theta}{2}$

Art. 130
4. $x = 0, y = 0$.

Art. 131
6. $y = a$; $x = 0$; and the oblique asymptote $x + y = a$.

Art. 135
2. $y = 0$, $y + x = -1$, $y - x = -1$.
3. $x - y = -1$, $x + y = 1$, $x + 2y = 0$.   4. $y = x$.

Art. 136
8. $xy + a^2 = 0$, $x^2 y - a^3 = 0$.   9, 10, 11 are given in text.

Art. 137
1. $x = -a$, $y = -b$, $y = x + b - a$.   9. $x + y = \frac{2}{3}a$.
2. $x = -2a$, $x = a$.   10. $y = 0$.
3. $x = \pm 1$, $y = \pm 1$.   11. $x = 0$, $y = 0$, $x + y = 0$.
4. $x = y \pm 1$, $x + y = \pm 1$.   12. $x = \pm a$.
5. $x = \pm a$, $x = y \pm a\sqrt{2}$.   13. $x^2 = 0$; two parabolic branches.
6. $y = x$.   7. $x = 2a$.   14. $y = 0$.
8. $x = 2a$, $x + a = \pm y$.   15. $y = 0$, $x = y$, $x = y \pm 1$.

Art. 139
1. Parallel to initial line; a units above it.
2. One, perpendicular to initial line, at distance a left of pole.
3. Their equations are $\rho \sin\left\{(2k+1)\dfrac{\pi}{2a} - \theta\right\} = \dfrac{b}{a}\csc\left\{(2k+1)\dfrac{\pi}{2}\right\}$.
4. $\dfrac{a}{2\rho} = \pm\cos\theta - \sin\theta$.   5. $\rho \sin\theta = 2a$.

Art. 143
3. $x = (\frac{4}{9})^6 \cdot a$.   5. $(4, \frac{4}{3}\sqrt{3})$.

Art. 156

1. Second.
2. First.
3. Second.
4. Second.
6. $a = -1$.
7. Third.

10. $\dfrac{2(x+a)^{\frac{3}{2}}}{\sqrt{a}}$.

11. $\dfrac{(a^4y^2 + b^4x^2)^{\frac{3}{2}}}{a^4b^4}$.

12. $\dfrac{y^2}{c}$.

13. $\dfrac{(c^2+y^2)^{\frac{3}{2}}}{cy}$.

14. $\dfrac{(x+y)^{\frac{3}{2}}}{\sqrt{a}}$.

15. $3\sqrt[3]{axy}$.

16. $\sec x$.

Art. 157.

2. $\dfrac{a^m}{(m+1)\rho^{m-1}}$.

3. $\dfrac{a}{2}$.

4. $\dfrac{a^2}{3\rho}$.

5. $\rho\sqrt{1+a^2}$.

6. $\dfrac{a(5 - 4\cos\theta)^{\frac{3}{2}}}{9 - 6\cos\theta}$.

7. $-\dfrac{\rho^3}{a^2}$.

Art. 159

1. $a = 2a + 3x$, $\beta = -\dfrac{2x}{\sqrt{a}}$.

2. $a = -x - \dfrac{9x^2}{2a}$, $\beta = 4\left(x + \dfrac{a}{3}\right)\sqrt{\dfrac{x}{a}}$.

3. $a = x + 3(xy^2)^{\frac{1}{3}}$, $\beta = y + 3(x^2y)^{\frac{1}{3}}$.

4. $a = x - \dfrac{y}{a}\sqrt{y^2 - a^2}$, $\beta = 2y$.

5. $(a+\beta)^{\frac{2}{3}} - (a-\beta)^{\frac{2}{3}} = (4a)^{\frac{2}{3}}$.

Art. 164

1. $(0, 0)$, $x \pm y = 0$.
2. $(a, -a)$, $2(x - a) = \pm 3(y + a)$.
3. $(0, 0)$, $x = 0$, $y = 0$.
4. $(0, 0)$, $x \pm y = 0$.
9. When it is made numerically smaller.

Art. 178

1. $y^2 = \dfrac{4}{27} x^3$.
2. $x^{\frac{2}{3}} + y^{\frac{2}{3}} = a^{\frac{2}{3}}$.
3. $x^{\frac{2}{3}} + y^{\frac{2}{3}} = c^{\frac{2}{3}}$.
4. $(x - y)^2 + 4ky = 0$.
5. $(x - a)^2 + (y - \beta)^2 = r^2$.
6. The point (a, β).
9. $4(ay^2 + bxy + cx^2) = 4ac - b^2$.

INDEX

(The numbers refer to pages)

Absolute value, 84.
Absolutely convergent, 84.
Acceleration, 157.
Actual velocity, 151.
Algebraic expression, 3.
 operation, 1.
Argument, 4.
Asymptote, 221.
Asymptotic circle, 240.
Average curvature, 84.
 velocity, 151.

Beman, 113.
Bending, 243.

Catenary, 211.
Cauchy, 94.
Center of curvature, 255.
Change of variable, 198.
Circle, asymptotic, 240.
 of curvature, 255.
Cissoid, 211.
Commutative, 2.
Comparison of infinitesimals, 21.
Computation of π, 111.
Concave, downwards, 241.
 upwards, 241.
Conditionally convergent, 85.
Conjugate point, 281.
Constant, 7.
Contact, 252.
Continuity of an algebraic function, 30.
 of a^x, 31.
 of $\log x$, 31.
 of $\sin x$, $\cos x$, 32.
Continuous function, 7.
 variable, 7.
Criteria for continuous function, 29.
Critical value, 134.
Curvature, 27.
Cusp, 279.

Decreasing function, 43.
Definition of continuity, 8.
 of curvature, 261.
 of nth derivative, 73.
De Moivre, 101.
Dependent variable, 7.
Derivative, 37.
 of arc, 216.
 of area, 40.
 of surface, 218.
 of volume, 218.
 partial, 160.
 total, 160.
Determinate value, 117.
Differentiable, 43.
Differential, 156.
Differentiation, 41.
 of inverse function, 48.
Discontinuity, 8.
Distributive, 3.

Elementary forms of curves, 289.
Entire, 4.
Envelopes, 308.
Equiangular spiral, 125.
Euler, 172.
 theorem, 171.
Even contact, 171.
Evolute, 267.
Explicit function, 4.
Exponential curve, 211.
 function, 58.
Expression, 3.

Family of curves, 307.
Form of remainder, 95.
Function, 4.
Functional differentiation, 47.
Fundamental problem, 37.
 theorem, 20.

335

336

INDEX

General exp. func., 82.
Generating function, 82.
Geometric applications, 212.
   illus. of a der., 39.

Hyperbolic branches, 221.
   functions, 318.

Implicit function, 4.
Incommensurable power, 58.
Increasing function, 43.
Increment, 8.
Independent variable, 7.
Indeterminate form, 115.
Infinite, 10.
Infinitesimal, 10.
Inflexion, 243.
Integral expression, 3.
Interval of convergence, 82, 90.
   of equivalence, 82.
Inverse function, 5, 58.
   operation, 1.
Involute, 267, 272.
Irrational, 4.

Klein, 113.

Leibnitz, 75.
   theorem, 75.
Limit, 9.
Logarithmic function, 58.

Maclaurin, 87.
   theorem, 87.
Maximum, 132, 185.
Mean value, 107.
Measure of curvature, 261.
Minimum, 132, 185.
Modulus, 60.
Montferier, 113.
Multiple point, 277.

Naperian base, 60.
Natural base, 60.
Newton, 75.
Node, 277.
Non-unique derivative, 43.
Normal, 208.
   length, 209.
Number, 1.

Oblique asymptotes, 227.
Odd contact, 254.
Operation, 1.

Order of contact, 252.
   infinite, 19.
   infinitesimal, 19.
Osculating circle, 255.
Osgood, 82, 85.

Parabolic branches, 221, 300.
Parallel curves, 272.
Parameter, 308.
Partial derivative, 160.
Perry, 190.
Polar coördinates, 212.
   normal length, 213.
   subnormal, 213.
   subtangent, 213.
   tangent length, 213.
Polynomials, 141.
Principal infinitesimal, 19.
Probability curve, 238, 306.
Process of differentiation, 42.

Radius of curvature, 255.
Rate, 152.
Rational expression, 4.
Rectilinear asymptote, 221.
Relative error, 101.
Remainder, 86.
Rolle, 85.

Semicubical parabola, 290.
Series, 81.
Shanks, 113.
Simple exponential functions, 58.
Simultaneous increments, 33.
Singular points, 275.
   values, 117, 275.
Slope of a line, 40.
Smith, 113.
Stationary tangent, 244.
Stirling, 85.
Subnormal, 209.
Subtangent, 209.
Successive differentiation, 73.
   operations, 2.
Sum of a series, 83.
Surd expression, 4.
Symbol of approach, 11.
   of an increment, 8.
   for inverse functions, 5.
   of a function, 5.
Symmetric expression, 4.

Table of derivatives, 71.
Tangent, 208.

Tangent length, 209.
Taylor, 87.
Test for convergence, 84.
Test for increasing function, 45.
Theorems on infinitesimals, 12, 16.
Total curvature, 261.
Total differential, 164.
Tractrix, 211.
Transcendental, expression, 3.
 operation, 1.

Transformed expression, 4.
Trigonometric functions, 58.
Turning value, 132.

Unconditionally convergent, 89.
Uniform velocity, 151.
Unique derivative, 43.

Variable, 7.
Vectorial angle, 213.

www.ingramcontent.com/pod-product-compliance
Lightning Source LLC
Chambersburg PA
CBHW030745250426
43672CB00028B/715